Transdermal Applications of Minimally Invasive Drug Delivery Systems

Malakapogu Ravindra Babu
Monica Gulati • Vandana B. Patravale
Hélder Santos • Kamal Dua
Sachin Kumar Singh
Editors

Transdermal Applications of Minimally Invasive Drug Delivery Systems

Current Trends and Future Perspectives

Springer

Editors
Malakapogu Ravindra Babu
Department of Pharmacy, School
of Medical and Allied Sciences
Galgotias University
Greater Noida, Uttar Pradesh, India

Vandana B. Patravale
Department of Pharmaceutical Sciences
and Technology
Institute of Chemical Technology
Mumbai, Maharashtra, India

Kamal Dua
Discipline of Pharmacy
University of Technology Sydney
Sydney, NSW, Australia

Monica Gulati
School of Pharmaceutical Sciences
Lovely Professional University
Jalandhar, Punjab, India

Hélder Santos
Department of Biomedical Engineering
University of Groningen
Groningen, The Netherlands

Sachin Kumar Singh
School of Pharmaceutical Sciences
Lovely Professional University
Jalandhar, Punjab, India

ISBN 978-981-96-6274-6 ISBN 978-981-96-6275-3 (eBook)
https://doi.org/10.1007/978-981-96-6275-3

This Springer imprint is published by the registered company Springer Nature Singapore Pte Ltd.
The registered company address is: 152 Beach Road, #21-01/04 Gateway East, Singapore 189721, Singapore

Preface

The field of drug delivery has witnessed significant advancements, with transdermal applications emerging as a significant focus of research and innovation. As the healthcare landscape evolves, the demand for minimally invasive, patient-friendly, and efficient drug delivery systems continues to grow. Transdermal systems have carved a niche by offering solutions that bypass traditional oral and injectable routes, minimizing first-pass metabolism and improving patient compliance. This book, *Transdermal Applications of Minimally Invasive Drug Delivery Systems: Current Trends and Future Perspectives*, delves into the transformative potential of these technologies.

With contributions from leading experts across the globe, the chapters explore diverse facets of minimally invasive drug delivery systems like thermal ablation, acoustic cavitation, and magnetic hyperthermia techniques. This book also takes the readers to the delivery of hydrogels and macromolecules by minimally invasive drug delivery systems. Emerging technologies such as microneedles, nanocarriers, and iontophoresis are discussed, highlighting their ability to enhance drug permeation, targeting, and therapeutic efficacy. Furthermore, this book examines the challenges faced in regulatory approval, scalability, and patient adoption, offering insights into overcoming these hurdles.

This comprehensive work also casts a forward-looking lens on the future of transdermal systems, envisioning how personalized medicine, wearable technologies, and bioresponsive platforms could redefine the field. By bridging the gap between research and clinical application, it serves as a valuable resource for researchers, clinicians, and industry professionals.

We hope this book inspires innovation and collaboration, advancing minimally invasive drug delivery systems toward improving global healthcare outcomes.

Greater Noida, Uttar Pradesh, India — Malakapogu Ravindra Babu
Jalandhar, Punjab, India — Monica Gulati
Mumbai, Maharashtra, India — Vandana B. Patravale
Groningen, The Netherlands — Hélder Santos
Sydney, NSW, Australia — Kamal Dua
Jalandhar, Punjab, India — Sachin Kumar Singh

Contents

About the Editors

Malakapogu Ravindra Babu is currently serving as a Professor, Department of Pharmacy, School of Medical and Allied Sciences, Galgotias University, Greater Noida, Uttar Pradesh, India. Previously, he served as a Professor and Research Coordinator, Department of Pharmaceutics, School of Pharmaceutical Sciences, Lovely Professional University, Punjab, India. He completed his PhD in Pharmaceutical Sciences from Jawaharlal Nehru Technological University Anantapur (JNTUA), Andhra Pradesh, India. Dr. Babu has a vast experience of more than 14 years in teaching and research. His key research area includes the development of minimally invasive drug delivery systems, such as microneedles, iontophoresis, and transdermal patches. He has a rich experience in formulating nano drug delivery systems. He has guided 12 M.Pharm projects and 35 B.Pharm projects. He is currently working on microneedle technology and various lipid-based nanoparticulate delivery systems. Dr. Babu serves as a reviewer for a number of reputed journals, including the *International Journal of Pharmaceutics*, *Drug Delivery and Translational Research*, *Journal of Drug Delivery Science and Technology*, *Indian Journal of Pharmaceutical Education and Research*, *European Journal of Pharmacology* as well as for a couple of journals of Bentham Publishers. Dr. Babu has published several book chapters with reputed publishers and is currently acting as lead editor for several books with international publishers.

Monica Gulati is the Registrar and Executive Dean at the School of Pharmaceutical Sciences, Lovely Professional University, India, with more than 25 years of experience in teaching, industry, research, and administration. Having completed her B.Pharm, M.Pharm, and PhD from UIPS, Panjab University, Chandigarh, Dr. Gulati served for 2 years in the R&D department of Panacea Biotec. Later, she shifted to academia, serving for more than 19 years at LPU in various academic and administrative roles. Her research expertise lies in enhancing the bio-performance of drug molecules, phytochemicals, probiotics, metabiotics, and other biotherapeutics using conventional and novel dosage forms, especially in the area of colon targeting. She holds a keen interest in the quality control parameters of delivery systems. She has more than 150 high-impact publications and 15 invited book chapters to her credit, with an h-index of 28 and an RG score of 39.05. She has been granted four Indian patents and three Australian patents. She also serves on the editorial and review board of more than 25 journals.

Vandana B. Patravale is currently a Professor of Pharmaceutics at the Institute of Chemical Technology, Mumbai, India. She has more than 30 years of teaching and research experience. She has over 200 refereed publications with over 10,000 citations (h-index 52, i10 130), 30 granted patents, 20 patents in pipeline, and 2 trademark registries. She has published 2 books and 35 book chapters with international publishers. Dr. Patravale has been active in teaching, research, and service throughout her career. She is an independent director of a device manufacturing company and is listed among the top 2% of the world's scientists in pharmacy. Her areas of research include the development of nanocarriers with a major emphasis on infectious diseases, cancer, and neurodegenerative disorders, as well as medical device development, nanodiagnostics, and nanovaccines. She was awarded the Topmost Healthcare Leader Global award (Education) 2022, the Abdul Kalam Technology Innovation National Fellowship 2021, a Fellowship of the Indian Chemical Society 2020, the Kukreja Oration Award 2020, APTI's Dr. Manjushree Pal Best Pharmaceutical Scientist Award 2019, the Shri Amrut Mody Distinguished Researcher Award 2018, the OPPI Women Scientist Award 2015, the Bill and Melinda Gates Grant Award 2015, the Best Pharmaceutical Scientist Award 2014, the VASVIK Award 2013, the Veneto Nanotech Award 2013, APTI's Best Teacher Award 2012, a Fellowship of the Maharashtra Academy of Sciences 2011, and the K.H. Garda Distinguished Researcher Award 2009. She serves as Convener, APTI Women forum; Vice President, CRS Indian Chapter; and Editor, CRS IC and APTI Women Forum newsletters and is a member of the editorial board of peer-reviewed journals. She is actively collaborating with researchers and industries both within country and abroad and has completed Indo-Swiss, Indo-Japan, and Indo-UK projects. She is executing all major grants from the Indian Government focusing on product development. She has done several technology transfers to industry, including drug eluting stents being marketed in more than 65 countries.

Hélder Santos, is a Full Professor in the Department of Biomedical Engineering at the University of Groningen, The Netherlands. He is also Research Director at the University of Helsinki's Faculty of Pharmacy. Prof. Santos is also the co-founder of the startups Capsamedix Oy and Medixmicro Oy and Coordinator of MSCA-ITN P4 FIT network in tendon repair (~4.1 m€). He holds Visiting Professorships at the Shanghai Jiao Tong University School of Medicine and the University of Tartu. Prof. Santos' research interests include the development of nanoparticles, nanomedicines, and biomaterials for biomedical applications, particularly for cancer and heart disease treatment. His lab makes a unique bridge between medical engineering, pharmaceutical nanotechnology, and tissue engineering by combining unique techniques to develop novel therapeutic formulations for translation into the clinic. He is a co-author of over 450 publications (with over 20,400 citations; h-index = 86) and holds 6 patents. He has delivered more than 220 invited talks around the world. He is also a member of the editorial boards of *Advanced Healthcare Materials*, *Advanced Therapeutics*, *Chemical Society Reviews*, *Journal of Controlled Release*, *VIEW*, *Journal of Functional Materials*, *Drug Delivery Letters*, *Current Nanoscience*, *Pharmaceutical Nanotechnology*, *AIMS Materials Science*, *Frontiers* *in Bioengineering* and Biotechnology, *Materials, Frontiers in Biomaterials, Advances in Materials Science and Engineering, Precision Nanomedicine, PLOS ONE, Journal of Healthcare Engineering*, and *BioMed Research International*. He is an Associate Editor of *Nano Select, Smart Materials in Medicine*, and *Drug Delivery and Translational Research*, among others. Prof. Santos has received prestigious awards and grants, including the "Talent Prize in Science" in 2010, awarded by the Portuguese Government, the ERC Starting Grant in 2013, the ERC Proof-of-Concept Grant in 2018, the Young Researcher Award in 2013 by the Faculty of Pharmacy, the Academy of Finland Award for Social Impact in 2016, and the CRS Young Investigator Award in 2021.

Kamal Dua is an Associate Professor, Discipline of Pharmacy, Graduate School of Health, University of Technology Sydney (UTS) and has a research experience of over 12 years working in the field of drug delivery targeting inflammatory diseases. Dr. Dua is also a Node Leader of Drug Delivery Research at the Centre for Inflammation at Centenary Institute/UTS, where the targets identified from the research projects are pursued to develop novel formulations as the first step toward translation into clinics. Dr. Dua conducts research in two complementary areas: drug delivery and immunology, specifically addressing how these disciplines can advance each other helping the community to live longer and healthier. This is evidenced by his extensive publication record in reputable journals. Dr. Dua's research interests focus on harnessing the pharmaceutical potential of modulating critical regulators, such as interleukins and microRNAs, and developing new and effective drug delivery formulations for the management of inflammation in chronic airway diseases and cancer.

Sachin Kumar Singh is a Professor and chief of domain for Pharmaceutical Analysis and Pharmaceutical Chemistry at the School of Pharmaceutical Sciences, Lovely Professional University, India. He holds a Master's degree in Pharmacy with a specialization in Pharmaceutical Analysis from J.S.S. College of Pharmacy, Ooty, Tamil Nadu, India, and a PhD degree in Pharmaceutical Sciences from Karpagam University, Coimbatore, India. He has completed one DST-SERB funded project on the development of SNEDDS of Polypeptide-k for the treatment of diabetes mellitus. Dr. Singh has published more than 350 research papers with a cumulative impact factor of 1200, authored 3 books, and co-authored 16 book chapters. He has been granted eight Indian patents. Dr. Singh has reviewed more than 100 research papers upon invitation from international publishers. He has supervised 15 PhD students and 35 M.Pharm students and is currently guiding 3 master's and 2 PhD students. Currently, his h-index is 37 with more than 6000 citations. In January 2020, he received Dr. B.C. Deb Memorial Award for the popularization of science by the Indian Science Congress Association at the 107th Indian Science Congress held at UAS Bengaluru. He is the first recipient from the field of pharmaceutical sciences to receive this award. He is also a recipient of the 2014 Bharat Shiksha Ratan Award, presented by the Global Society for Health and Educational Growth, Delhi, India. His core area lies in improving the oral bioavailability of poorly water-soluble drugs using various formulation approaches, including nanotechnology, complexation, solid dispersion, liquisolid technology, as well as the development of non-invasive formulations for topical applications.

Introduction to Minimally Invasive Drug Delivery System (MIDDS) 1

Tejaswini Kolipaka, Etikala Amulya, Ganesh Vambhurkar, Shalini Shukla, and Saurabh Srivastava

Abstract

This chapter delves into transdermal approaches, including invasive (penetrating through stratum corneum) and non-invasive (drug diffusion) techniques, highlighting their underlying mechanism. This work provides a comparative understanding of minimally invasive drug delivery systems (MIDDS) with invasive as well as non-invasive techniques. The MIDDS technique enhances the penetration ability of macromolecules and hydrophilic substances, a challenge with non-invasive techniques. Further, a brief discussion on the potential challenges associated with invasive techniques is included. We have emphasised how the MIDDS can overcome the challenges by improving drug delivery efficiency and patient compliance with limited risk. Finally, this chapter provides insight into different MIDDS, such as microneedles, velocity-based devices, and laser- and radiofrequency-induced microporation techniques. In addition, it highlights newer aspects, such as a combination of microneedles with sonophoresis, iontophoresis, and vibratory actuation.

Keywords

Skin · Transdermal drug delivery · Minimally invasive techniques · Microneedles · Sonophoresis

T. Kolipaka · E. Amulya · G. Vambhurkar · S. Shukla · S. Srivastava (✉)
Pharmaceutical Innovation and Translational Research Lab (PITRL), Department of Pharmaceutics, National Institute of Pharmaceutical Education and Research (NIPER), Hyderabad, Telangana, India
e-mail: saurabh@niperhyd.ac.in

M. R. Babu et al. (eds.), *Transdermal Applications of Minimally Invasive Drug Delivery Systems*, https://doi.org/10.1007/978-981-96-6275-3_1

1.1 Transdermal Drug Delivery System

The transdermal drug delivery system (TDDS) delivers the drugs into the systemic circulation via the skin. Few drugs administered through the oral route undergo high first-pass metabolism, whereas drugs administered through parenteral is an invasive technique (Lambert and Laurent 2008). As a result, TDDS overcomes the pitfalls of oral and systemic routes.

TDDS are broadly classified into three generations. The first generation includes formulations, namely, spray, ointments, gel, and creams, that are fabricated for lipophilic drugs having a low molecular weight (<500 Da) with a therapeutic activity at minimal doses (in mg) (Prausnitz et al. 2004). For example, the FDA-approved oestradiol transdermal spray (Evamist) in 2007 for vasomotor symptoms has a molecular weight and dose strength of 272.4 Da and 1.53 mg/spray, respectively (US-FDA 1975). In the first generation, permeation enhancers were often used to enhance permeation, but their use is limited based on safety. For example, FDA permitted only 0.5% w/w use of polyethylene glycol 1000 (CAS Number: 25322683) in gel (D'souza and Shegokar 2016). Moving forward, the second-generation TDDS aims to enhance the drug's penetrating ability by reversibly damaging stratum corneum (SC) with the help of driving forces, thereby protecting deeper tissue damage. However, this generation does not have an impact in delivering macromolecules and hydrophilic compounds, which can be overcome by the third-generation TDDS. The minimally invasive drug delivery systems (MIDDS) are part of third-generation TDDS (Joshi et al. 2023; Prausnitz and Langer 2008; Lee et al. 2018). The MIDDS overcomes the challenge of the non-invasive route by enhancing penetration to the target site.

1.2 Skin: The Most Vital Organ

The skin controls the first defence mechanism to protect against the external environment. It covers 15% of total body mass and stretches an area of around 1.5–2 m^2 (Phatale et al. 2022). It comprises three different layers, with the outermost being the epidermis, as shown in Fig. 1.1. The surface layer of the epidermis consists of dead keratinocytes and lipids, known as SC. Underneath SC lies the stratum lucidum, composed of keratinocytes (two to three layers). As these cells die, they migrate outward and form the SC layer. Further progressing in deeper layers, it consists of stratum granulosum (keratohyalin granules) and stratum spinosum (eight to ten layers of keratinocytes, Langerhans cells, and desmosomes). The Langerhans cells are antigen-presenting cells which engulf exogenous substances via phagocytosis. Finally, the deepest layer of the epidermis comprises keratinocyte primary cells, melanocytes, and Merkel cells. The Merkel cells are sensory units that function as touch receptors. Beneath the epidermis, the second layer consists of the dermis having papillary and reticular layers. The thin papillary layer contains loosely arranged connective tissue above reticular layer. The reticular layer consists

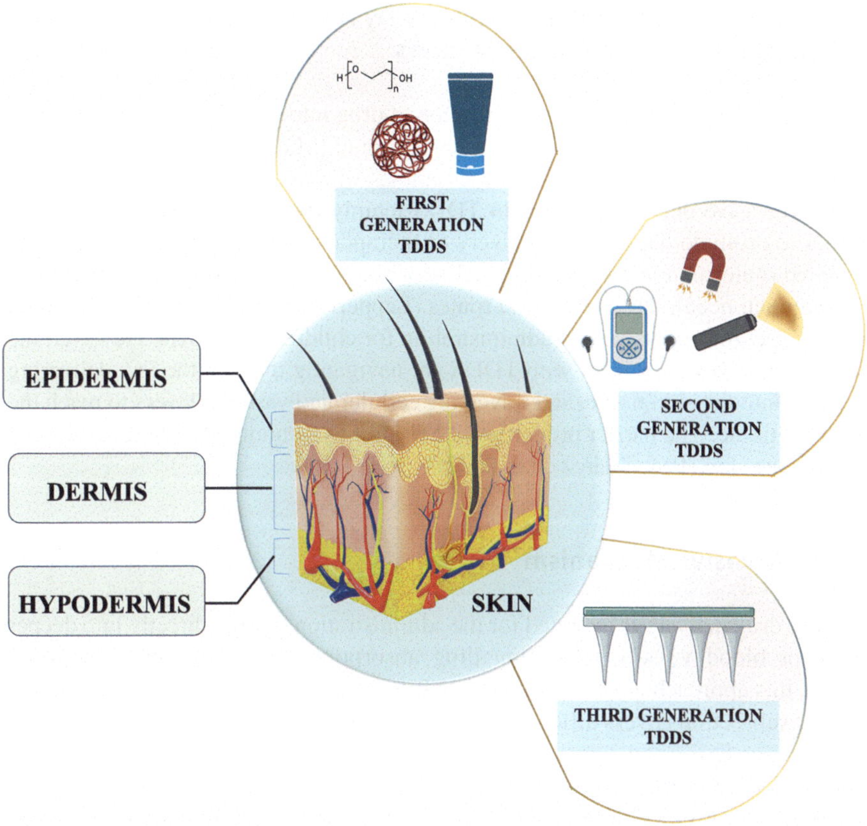

Fig. 1.1 The three generations of TDDS

collagen fibres and connective tissue with a dense lymphatic and blood vessels. The dermis layer plays a vital role in immune protection and consists of leucocytes, phagocytes, and mast cells. Underneath the dermis lies the fathomless skin layer, the hypodermis or subcutaneous layer. This layer consists of a higher content of proteoglycans and glycosaminoglycans (Ramadon et al. 2022). The skin is a potential target site in drug delivery, yet the covalent bond between the components of the SC layer, i.e. between dead keratinocytes and lipids, acts as a barrier for exogenous substances, including drugs. According to the present understanding, there are two ways: altering the skin microenvironment and providing an external force to enhance the penetration. However, acting on molecule may not be sufficient; rather, bringing up techniques that reversibly damage the SC with minimal pain, irritation, and risk of infection could provide an effective solution (Prausnitz et al. 2004). This chapter delves with various types of MIDDS and their advantages over the invasive techniques.

1.3 Mechanism of Drug Delivery by Invasive and Non-invasive techniques

Invasive route involves direct administration of drug into subcutaneous, intramuscular, intradermal, intraarterial, and intravenous route (Mathias and Sridharan 2017). Section 1.3.1 gives a brief idea on these routes.

Non-invasive delivery systems in TDDS mainly focus on bypassing the skin's barrier to promote the efficient delivery of therapeutic agents. TDDS is one of the explored routes in non-invasive drug delivery via the skin, offering an alternative to conventional needle-based injection routes (Leppert et al. 2018). It causes minimal patient discomfort and ease of administration for children and elders. However, the major challenges associated with TDDS are navigating through the SC, delivering the drug into the skin tissue, and traversing cellular and vascular layers to reach the intended tissue. The barrier properties of SC limit the amount of drug that can pass through the skin (Peña-Juárez et al. 2022).

1.3.1 Invasive Mechanism

Invasive drug delivery is intended for the administration of drug directly into deeper tissue or blood vessels, enhancing drug absorption through localised delivery. Hence, this approach ensures localised delivery with enhanced therapeutic activity. Further subsection covers different invasive routes and its site of action.

1.3.1.1 Intravenous Route (IV)

The drug administered through the IV route delivers the drugs directly into the systemic circulation. This route is preferred when a rapid response is required, when serum drug concentrations must be precisely controlled, or when the drugs are less absorbed via the gastrointestinal tract (Vargason et al. 2021). It is also recommended for patients suffering from impaired mental status or severe nausea and vomiting. IV administration is widely used to deliver small molecules, antibodies, proteins, and peptides (Sultatos 2007).

1.3.1.2 Intramuscular Injection (IM)

IM injection administers drugs into the deeper layers of specific muscles at an angle of 75–90°. The bulky muscles have an excellent blood supply, which allows the drug to reach the target site via systemic circulation, avoiding first-pass metabolism (Saxen 2016). An IM injection is less invasive than an IV injection and usually takes shorter time as the muscles offer a larger, more accessible site than veins (Jeremy and Wright 2012). Drugs injected into the muscle can also be given as depot injections, ensuring a prolonged and gradual drug release over time.

1.3.1.3 Subcutaneous Route

The subcutaneous route involves administering the drug into the fatty tissues located beneath the epidermis and dermis (Usach et al. 2019). Drug delivery by this route enhances bioavailability and bypasses pitfalls via oral route. Therefore, this method is mainly preferred for drugs unsuitable for oral administration (Scioli Montoto and Ruiz 2021). The subcutaneous injections result in slower drug absorption than IV or IM injections, leading to side effects, such as pain and infection at the subcutaneous injection site (Kim et al. 2017).

1.3.1.4 Intraarterial

The purpose of the intraarterial injection is to concentrate the drug at a targeted site, such as a tumour or infection. It tends to minimise the systemic side effects by providing localised drug delivery (Maddison et al. 2008). Intraarterial administration of drugs is not a common practice. This approach is mainly utilised to treat gliomas and deliver localised chemotherapy treatments.

1.3.1.5 Intrathecal

Drugs can be administered intrathecally, which involves injecting the drug into the spinal canal or subarachnoid space to enter the cerebrospinal fluid (CSF). This route can deliver drugs for pain relief, chemotherapy, and spinal anaesthesia (Fowler et al. 2020). Moreover, drugs that combat specific infections, particularly post-neurosurgical, are delivered through this route. Intrathecal injections bypass the blood-brain barrier and, hence, are preferred over the oral route (Gartziandia et al. 2016).

1.3.2 Non-Invasive Mechanism

1.3.2.1 Hypodermis

In the earlier Sect. 1.2, we had a brief outline on the skin composition. Hence, a drug that permeates the skin layers passively needs to cross the SC framework, i.e. dead keratinocytes and lipids organised in brick-and-mortar alignment. Hence, there are two possibilities for the permeation of the drug, either via transepidermal (through SC) or transappendageal (through appendages such as hair follicles and eccrine ducts), as represented in Fig. 1.2.

1.3.2.2 Transepidermal Route

In transepidermal diffusion, fluid lipids play a crucial role by occupying spaces that allow lipid and amphiphilic molecules to migrate and be absorbed into deeper layers (Barry 2001). The transepidermal route is further classified into transcellular and intercellular. The techniques that enhance permeation using both mechanisms involve electroporation and photomechanical waves. The electroporation technique implies the high-voltage electric pulses that range from 5 and 500 V for short durations, forming minute pores in the SC that enhance permeation and ease drug absorption (Charoo et al. 2010). The pulse duration governs the transport route, either transcellular or intercellular. High voltage with a short duration of pulse promotes transcellular, while low voltage with an extended pulse duration favours intercellular transport (Denet et al. 2004). Supporting the earlier statement, Rita

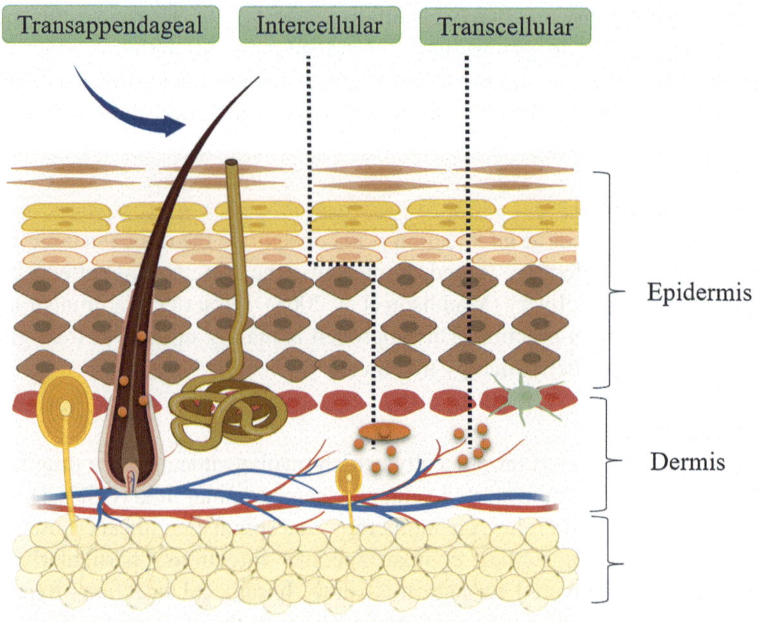

Fig. 1.2 Non-invasive mechanism: transappendageal, intercellular, and transcellular route

et al. executed a comparative research between high voltage (100 V) with a short pulse duration (1 ms) and medium voltage (30 V) with a long pulse duration (100 ms). The earlier parameters enhance the lateral molecular transport of energy from the dissipated area to the local transport area. Henceforth, the permeation occurs *via* the transcellular route, whereas later parameters moves through the intercellular route(Pliquett et al. 1998). Similarly, photomechanical waves induce low radiation (5–7 J/cm^2) and enhance the penetration range to 50–400 μm to ensure effective transmission (Jeong et al. 2021).

Transcellular

The transcellular route involves drug diffusion across SC cells, where they must penetrate membranes consisting of lipid bilayers. In this mechanism, the permeant travels directly through the skin, sequentially traversing protein-rich corneocytes and intercellular lipids (Valenta et al. 2000). Corneocytes contain moisture and microfibrillar keratin, enclosed by a cornified layer consisting of tightly cross-linked protein structures. It is generally believed that molecule partitioning into these areas is restricted, leading to this route being mostly discounted as a significant pathway (Kushner et al. 2007). However, it is important to mention that various ex vivo experimental studies indicate that transcellular pathways may significantly contribute to the absorption of certain active compounds (Iliopoulos et al. 2022). Hydrophobic drugs primarily utilise this pathway as the lipid complex in the SC cell membranes possesses hydrophobic properties (Barbero and Frasch 2009).

Intercellular Pathway

In the intercellular pathway, drugs must navigate across the lipid matrix in the intercellular spaces surrounding keratinocytes within the SC. Small molecules and hydrophilic substances travel through this route to reach the vascular capillaries within the dermis (Zhang et al. 2017). This diffusion pathway follows a complex path, approximately 500 μm, which is over 20 times longer compared to the thickness of SC. Moreover, it is a predominant pathway for drug absorption, which largely relies on the drug molecule having the solubility in lipidic and aqueous phase (Hadgraft 2001).

The hydrophilic molecules lacking lipophilicity can incorporate chemical enhancers that enhance the drug permeability through this pathway. These are inactive molecules that partition into, diffuse through, and interact with the SC (Haque and Talukder 2018). Numerous chemicals have been found to enhance the permeability of SC, such as alcohols, poly-alcohols, pyrrolidones, amines, amides, sulphoxide, alkanes, fatty acids, esters, terpenes, surfactants, and phospholipids (Karande and Mitragotri 2009).

The second-generation TDDS, such as sonophoresis and iontophoresis, involves ultrasonic waves (20 kHz to 16 MHz) and current (below 0.5 mA/cm^2), respectively. Sonophoresis technique is utilised in the delivery of various drugs across different categories, including excipients (mannitol) and higher molecular mass entities like insulin. This technique is independent of drug parameters such as dissociation constant, solubility, and ionisation constants or electrical properties like hydrophilicity (Merino et al. 2003). The effectiveness of iontophoresis is influenced by the polarity, charge, and molecule mobility, along with the properties of the electrical cycle and the formulation (Moarefian et al. 2020).

1.3.2.3 Transappendageal Route

The transappendageal route is the second mechanism for drug absorption across skin, which refers to movement of drug transport via sweat glands or hair follicles. This route is crucial for transporting polar or ionisable substances and aids in delivering macromolecules that encounter challenges passing via epidermal cells owing to their molecular size with differing partition properties (Verma et al. 2016). However, it is important to highlight that sebum, an oily substance secreted by the sebaceous glands in hair follicles, can obstruct permeation (Zoabi et al. 2021). In addition, the utilisation of this pathway is restricted, as it covers only about 0.1% of the skin area, much smaller than the area available for the transepidermal pathway. Nanoparticles have been shown to effectively transport drugs via hair follicles (Yu et al. 2021). Therefore, apart from diffusion, the transappendageal pathway is recognised as a crucial mechanism in skin penetration (Hadgraft 2001). The small average size of nanocarriers primarily determines follicular penetration. Among the nanocarriers, liposomes outstand itself for this route. However, challenges such as fragility, drug leakage, and poor stability pave the way for modification in the lipidic system. To address these issues, ethosomes and transferosomes have been developed which are distinguished from liposomes by the inclusion of ethanol and edge activator, respectively. Ethanol provides increased stability and flexibility to ethosomes, making it easier to penetrate the skin and fit between lipid layers and cells (Emanet and Ciofani 2023). At the same time, the edge activator helps in the deformability of vesicles, which helps in improved transdermal drug delivery (Matharoo et al. 2024).

1.4 Limitations of Invasive Delivery of Drugs

The invasive systems often have pitfalls that need to be taken into consideration:

- *Trained personnel*: The invasive systems need controlled administration by trained staff that increases cost and limits ease of access.
- *Patient in compliance*: The invasive systems have predictable pain, so patients might be less willing to comply with the dosage regimen.
- *Aseptic conditions*: To avoid the infection, the procedure requires aseptic maintenance, which decreases the complication and risk of infection (Manchanda et al. 2020).

The various pitfalls studied in different routes are listed below:

1.4.1 Subcutaneous Route

- The complications at the injection site, such as pain and irritation.
- Due to the low absorption rate at the subcutaneous site, frequent changes of the injection site are necessary to prevent the accumulation of unabsorbed drugs, which can damage the tissue.

1.4.2 IM Route

- The most possible complications observed include hematoma formation and abscess at the site of injection.
- The drug quantity to be injected depends on the muscle mass.

1.4.3 IV Route

- Air embolism due to the possibility of air drawn via an intravenous line.
- Thrombophlebitis due to trauma in the vein.
- At the site of catheter introduction, there might be an infection (Jain 2020).

1.5 The Importance and Merits of Minimal Invasion in Drug Delivery

Conventional strategies in transdermal delivery possess certain drawbacks like conventional injectable systems that cause pain and discomfort to patients. This results in reduced patient compliance. Upon injecting frequently, it may lead to oedema. For conventional invasive, there is a need for specialised personnel. MIDDS remodels the landscape of drug delivery, making it more effective. They minimise the risks associated with the traditional techniques.

- Minimally invasive systems like microneedles minimise the pain as well as anxiety associated with the needles, thereby improving patient compliance in comparison to the other traditional techniques (Ganeson et al. 2023).
- Its needle-free nature and lack of massive punctures lower the complications (Ramadon et al. 2022).
- These systems improve the permeability across the skin and enable the absorption of hydrophilic as well as lipophilic drugs.
- They help in sustained drug delivery and also minimise the dosing frequency.
- They enable precise drug delivery, thereby enhancing the therapeutic efficacy.
- By reducing the need for hospitalisation and surgery, minimally invasive systems lower healthcare costs (Bastawrous et al. 2023).
- Faster recovery process.
- Self-administration is feasible.
- They help in maintaining the controlled drug release pattern.

1.6 Various Types of Minimally Invasive Drug Delivery Systems for Transdermal Applications

MIDDS for transdermal delivery have achieved tremendous interest due to their ability to improve drug absorption, alleviate discomfort, and improve patient compliance.

1.6.1 Microneedles

Microneedles are MIDDS consisting of hundreds of micro-size needles organised on a small patch. Upon application to the skin, these microneedles penetrate the different layers of the epidermis without reaching the dermis and its pain-sensitive nerves (Fig. 1.3a). The skin gets temporarily disrupted, thereby facilitating effective drug delivery. These microneedles bypass the barrier layer by piercing SC. The drug gets deposited in the epidermis or the upper dermis layer and then enters the systemic circulation. Subsequently, the drug reaches functional site to elucidate the desired therapeutic response (Larrañeta et al. 2016).

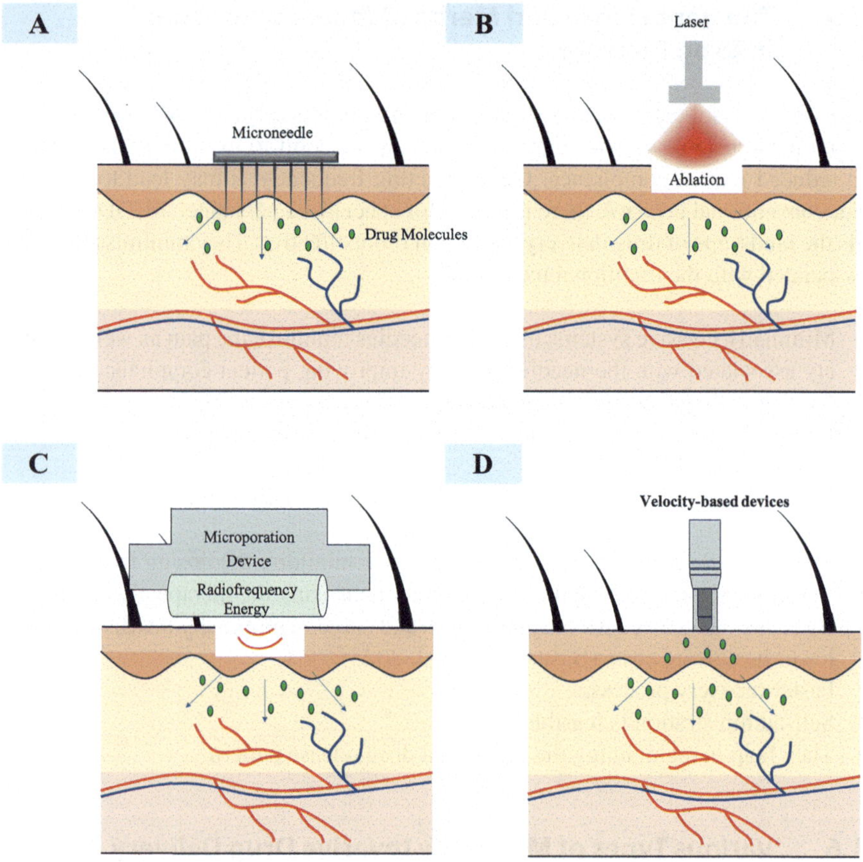

Fig. 1.3 Types of MIDDS: (**a**) microneedle, (**b**) laser ablation, (**c**) radiofrequency-induced microporation, and (**d**) velocity-based devices

Based on the material and type of needle employed, the microneedles come in various sizes. Typically, the microneedle dimensions vary depending on the requirement, including a 150–1500 μm in length, a width of about 50–250 μm, and a 1–25 μm tip thickness. However, longer and thicker needles penetrate deep into the dermis layer, which potentiates the nerve damage, leading to pain. Hence, needles of up to 1500 μm length are generally used to deliver the therapeutic moieties, as the thickness of the epidermis is up to 1500 μm. Microneedle tips are available in different shapes, including triangular, cylindrical, pointed, pentagonal, and octagonal (Waghule et al. 2019).

As per the FDA guidelines, the safety and effectiveness of the medical devices/products need to be taken into consideration. To fulfil the FDA clinical requirements, the materials used for microneedle fabrication must be safe, biocompatible, and stable. In line with these requirements, materials commonly explored for the fabrication of microneedles include inorganic non-metallic materials, metals, and

Table 1.1 List of various formulating techniques for different kinds of microneedles

Types of microneedles	Fabrication techniques
Metal microneedles	Metal electroplating methods, laser cutting, and wet etching
Silicon microneedle	Anisotropic wet etching, isotropic etching, dicing a silicon substrate, acid etching, and silicon dry-etching process
Ceramic microneedles	Sintering lithography and ceramic micro-moulding
Polymer microneedles	Photolithography
Dissolving microneedles	Micro-moulding
Hollow microneedles	Micro-electromechanical systems techniques: laser micromachining, an integrated lithographic moulding technique, deep reactive ion etching of silicon, deep X-ray photolithography, wet chemical etching, and micro-fabrication

polymers (Oliveira et al. 2024). Silicon and ceramics are generally used inorganic nonmetals. The metals used in microneedle preparation include aluminium, titanium, and stainless steel. In addition, metals like nickel and palladium and alloys, such as palladium-cobalt alloys, are also utilised. Microneedles made of metals possess excellent physical and mechanical properties. However, the nondegradability and inflexibility represent their significant drawbacks. Polycarbonate, polylactide, acrylate, and polyvinylpyrrolidone are the various polymers used in microneedle fabrication. The polymer's short processing cycle, cost-effectiveness, and ease of formability make it ideal for large-scale production. Additionally, the biocompatibility and biodegradability make it highly suitable for application in medical field (Luo et al. 2023).

Different categories of microneedles are fabricated and explored in drug delivery, including solid microneedles, dissolving microneedles, coated microneedles, and hydrogel microneedles. Each microneedle has its unique method of drug delivery into the epidermis. Few microneedles are designed to create pores in SC, while others are pre-coated with drug solution on the microneedle surface. Certain microneedles are dissolvable, whereas others are pre-filled with the drug solution (Waghule et al. 2019).

The choice of fabrication technique for microneedles depends on the type, material, and geometry of the microneedle, as illustrated in Table 1.1.

1.6.2 Laser Thermal Ablation

Laser techniques are generally employed to treat various skin-related conditions. The laser thermal ablation involves the selective removal of the SC while preserving the deeper tissues. The hydrophilic and lipophilic drugs showed enhanced penetration through different layers of skin. Various phases of laser ablation of the skin include thermal application on the skin (37–60 °C), protein coagulation like collagen and elastin (60–65 °C), moisture drying (90–100 °C), and its vaporisation (>100 °C). Various factors, including controlled wavelength, tissue thickness, tissue absorption coefficient, pulse length, pulse repetition rate, and pulse energy, play a role in the degree of subcutaneous barrier disruption (Fig. 1.3b).

The major components of the medical laser employed for thermal ablation include a pump, lasing material, and optical cavity. On the basis of the laser beam structure, the laser conformations are of two types: full and fractional beam. The major pitfall associated with the full beam is that it causes damage to the whole targeted area. In contrast, the damage associated with the fractional beam is less, i.e. damage to smaller areas of targeted area. However, the healing rate with the former is faster rather than the latter.

The crucial mechanisms involved in laser ablation include photothermolysis, direct ablation, and mechanical waves. The principle of photothermolysis involves vaporisation and ablation. In this, the skin's moisture or any other related components absorb the laser radiation, which causes the elevation of skin temperature leading to ablation. This technique utilises carbon dioxide (CO_2) laser (Huang et al. 2017). In the following type, i.e. direct ablation, the laser radiation directly interacts with the skin, expulsing skin portions. In this technique, erbium-doped yttrium aluminium garnet (Er:YAG) is used as a lasing material, absorbed by the skin's water content, and forms micropores. Unlike the earlier technique, Er:YAG causes a minimal skin damage, making it a desired approach for a cutaneous delivery. The last technique involves the mechanical wave generation by irradiating the material in contact with the skin. Transdermal drug delivery is achieved by creating the subcutaneous layer's transient pores. On the whole, the healing rate is faster in comparison to the aforementioned mechanisms (Parhi and Mandru 2021).

1.6.3 Radiofrequency-Induced Microporation

The radiofrequency-assisted device comprises a microelectrode array that consists of 100 microelectrodes/cm^2 and a patch containing the drug. In this technique, an array of microelectrodes is mounted directly onto the skin, generating an electric current within the 100–500 kHz frequency range. The radiofrequency-exposed skin region generates skin, leading to ablation of the epidermal cells (Fig. 1.3c). The drug delivery rate can be controlled by altering the number and depth of microchannels, which are confined to the superficial skin layers, to minimise the pain.

This device possesses a multifaceted application in various treatments, such as malignant tissue removal, electrosurgery, and in the field of cosmetic procedures like skin tightening and body contouring. Moreover, it helps in hydrophilic and macromolecule delivery by creating a micron-sized aqueous channel in the subcutaneous layer of skin (Parhi and Mandru 2021).

1.6.4 Velocity-Based Devices

Velocity-based devices are systems that utilise a power source, namely, spring or compressed gas, to deliver the drugs. These systems utilise a high-velocity jet having a velocity range from 100 to 200 m/s to deliver the drugs via the transdermal route. In the early 1930s, Arnold Sutermeister explored the concept of jet injectors. A jet injector avoids the use of needles (Fig. 1.3d) and delivers the doses in a controllable manner, resulting in enhanced delivery consistency and improved patient compliance (Schlich et al. 2022). These are devices are utilised in the administration of small molecules and vaccines. Their ability to deliver precise doses and minimise pain made them an appealing alternative to traditional needle-based injections. These velocity-based devices are of various types, such as liquid jet, powder jet, and solid jet injectors.

Liquid jet injectors are devices in which the liquid propels via a nozzle having an orifice radius in the range of 25–180 μm, which is significantly lesser than the outer diameter of standard hypodermic. These liquid jet injectors come in two types: single-dose disposable cartridge and multiuse nozzle. These systems deliver the drug into different skin layers including subcutaneous, intradermal, and intramuscular. This is facilitated mainly by varying the orifice diameter or changing the jet velocity. These needle-free devices eliminate the concerns regarding safe needle disposal and accidental needle-stuck injuries. Nevertheless, there are higher chances of cross-contamination, which occurs due to the splashback of interstitial liquid from the skin to the nozzle. Henceforth, taking this into account, the multiuse nozzle jet injectors have been discontinued, and these devices are now used only when multidose drug delivery is needed to the same individual, e.g. Tjet® device delivers somatropin (human growth hormone, hGH) marketed as TEV-TROPIN® (Mitragotri 2006).

Powder jet injectors offer considerable advantages over the liquid jet injectors. They enable the delivery of solid drugs or vaccines to the skin. They offer enhanced stability, require simplified transport, and eliminate the need for cold storage. The improved stability reduces the associated costs, making the powder jet injectors more economical. These are generally formulated using nanoparticles or microparticles that have active moiety or lyophilised drugs and antigens. However, their acceptance is limited due to intermittent pain and bruising experienced by the patients (Arora 2017).

The solid jet injectors include the following basic components: a power source, a drug-loaded compartment, and a nozzle. The compressed gas acts as a power source, the drug-loaded compartment contains solid-drug formulation, and the nozzle directs the flow of particles towards the skin. When the actuation mechanism gets triggered, the expansion of compressed gas occurs and expels the drug powder through the nozzle. Upon exiting the nozzle, the particles achieve higher velocity, enabling them to penetrate the skin effectively. Upon contact with the skin, the particles create micro-holes, facilitating drug delivery into the SC or the epidermis. The delivery of particles across the SC is affected by specific parameters, including particle properties and the impact velocity. The size and density of particles play a crucial role in skin penetration. For DNA vaccination, the particle size is recommended to be 0.5–3 μm. Higher velocities typically yield better penetration and drug deposition (Alkilani et al. 2015).

1.6.5 Microdermabrasion

The microdermabrasion technique is an abrasive mechanism that involves the removal of SC to enhance drug permeability. Precisely, this technique is used in skin-resurfacing to treat conditions like scarring, photoaging, and hyperpigmentation. In this process, solid inert particles like aluminium oxide crystals combined with negative pressure (typically 15 mmHg) are applied to remove the SC (Lee et al. 2003; Karve and Banga 2024).

1.7 Avant-Garde Combination Therapy

1.7.1 Microneedle Combined with Iontophoresis

Similar to the earlier combination, microneedles and iontophoresis also enhance drug release. Iontophoresis involves oppositely charged electrodes to ease drug delivery; for instance, a positively charged drug is delivered by placing the anode on the skin, and vice versa for the negatively charged drug. The presence of current and its intensity drive the drug through the skin, enabling rapid and efficient drug delivery. However, this technique faces a drawback in delivering the macromolecule with a molecular weight greater than 13 KDa; therefore, combining other techniques with iontophoresis showed a promising outcome. In this section, we exemplify the combination of iontophoresis and microneedle, as shown in Fig. 1.4a. Microneedle creates a way for hydrophilic and macromolecules, and iontophoresis drives these

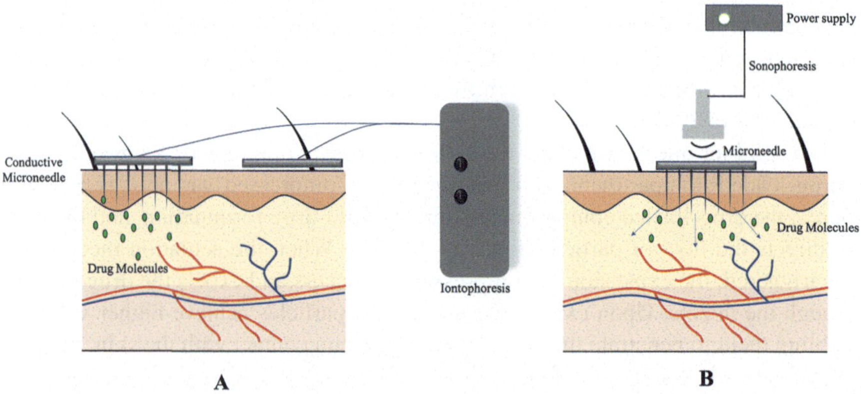

Fig. 1.4 Combination technique: (**a**) microneedle and iontophoresis and (**b**) microneedle and sonophoresis

entities and minimises their lag time for permeation across the skin (Zhou et al. 2017). A study by Vijay Kumar and his co-workers evaluated the penetration ability of human growth hormones using individual techniques and combinations. The results depicted higher penetration of combined techniques with a current density of 0.5 mA/cm^2, achieved ~2.5-fold higher than microneedle (Kumar and Banga 2012). Another study by Ryan F. Donnelly et al. evaluated the combination therapy in delivering FITC-BSA at an in vivo scale. The result concluded ~4.8-fold higher release than microneedle by the end of 2 h ($p < 0.001$) (Donnelly et al. 2014).

1.7.2 Microneedle Combined with Sonophoresis

The microneedle pierces SC and delivers in the epidermis, while low-frequency sonophoresis (20–100 KHz) induces cavitation, as shown in Fig. 1.4b. The combinatorial activity of two mechanisms improves the delivery of larger molecular weight and hydrophilic molecules. For example, Tao Han et al. have explored the individual and combination effects of microneedle and sonophoresis. The results summarised ~2.09- and ~2.25-fold enhancement compared to solid microneedles of 1.5 mm depth and 15 W ultrasound, respectively (Han and Das 2013). In another study, Boonnada Pamornpathomkul and co-workers analysed the permeation of fluorescein isothiocyanate-labelled bovine serum albumin (FITC-BSA) via a combination technique. The results have concluded continuous flux of 4 W/cm^2 for 2 m in combination with microneedle showed approximately fivefold enhancement than only current flux (Pamornpathomkul et al. 2015).

1.7.3 Microneedle Combined with Vibratory Actuation

Vibratory actuation is another technology that minimises microneedle deformation. By combining these techniques, it is possible to control the force and stress exerted on the microneedle. This combination is specifically beneficial when fabricating microneedles from polymers with lower Young's modulus, such as polymeric microneedles. Ming et al. performed ex vivo studies to measure the decrease in penetration force of microneedle in the presence and lack of vibratory actuation. The outcomes summarised a 70% decrease in insertion force in the existence of vibratory actuation (Yang and Zahn 2004). Another assessment by Tingting et al. reported the microneedle of 550 μm ~2.25-fold higher penetration in the presence of 200 Hz than in the absence of vibration (Kang et al. 2022). Vibration can be induced using either a linear resonant actuator (LRA) or an eccentric rotating mass (ERM). A study was executed by Vahid et al. to analyse the influence of the actuator on the penetration depth of the polymeric dissolving microneedles. The results concluded the penetration depth of microneedle was enhanced by ~1.37-fold for LRA than ERM at 200 Hz (Ebrahiminejad et al. 2024).

1.8 Recent Advances in Minimally Invasive Delivery of Drugs

1.8.1 Nanoneedles

In recent times, nanoneedles have been investigated in drug delivery via transcutaneous route. These are smaller in size, with a higher surface area. Like microneedles, the nanoneedles penetrate the skin with minimal damage to cells, owing to the release of the drug in the intracellular space (Rabiei et al. 2020). Lately, Liu et al. explored the combination of electroporation with nanoneedle to minimise tissue damage. They combined triboelectric nanogenerators (applied voltage = 20 V) with silicon nanoneedles (5 × 5 mm) to evaluate the penetration of FITC-labelled dextran. Their results concluded that dextran penetrated up to 23 μm, which is ~2.09- and ~3.83-fold higher than using only triboelectric nanogenerators and nanoneedle, respectively (Liu et al. 2019).

1.8.2 Thermoporation

Thermal ablation includes a temperature rise in the particular region of skin, resulting in the tissue vaporisation of SC that creates a pathway for drug permeation (Parhi and Mandru 2021). In the earlier section, we had a detailed discussion on radiofrequency ablation, chemical ablation, and laser microporation. Similarly, the thermoporation technique involves metallic filaments that are heated up by passing electric current, when they are in contact with skin layer. This leads to vaporisation SC layer and formation of channel. Naoto et al. performed a study on insulin and fluorescein isothiocyanate-dextran delivery using a thermoporation device, PassPort™. The results concluded 960-fold higher permeation of isothiocyanate-dextran in the presence of PassPort™ (4 ms and 200 micropores/cm^2). Similarly, insulin delivery has been shown prolonged release up to 8 h, whereas the subcutaneous injection reached its Cmax at 30 m and the minimal dose after 6 h. Hence, the results summarise utility of PassPort™ in the drug permeation enhancement (Ono et al. 2022).

1.9 Commercialisation

The process of commercialising of any product involves stringent guidelines. Once a product has proven safety and efficacy in the preclinical assessment, it enters to clinical studies, where its safety and efficacy are evaluated in human populations. After the successful completion of these randomised trials, the product is marketed, and its adverse effects are monitored in a larger population. In post-market surveillance, suboptimal adherence becomes a prominent factor—a challenge that is less existing in the phase II and phase III, where the population size is smaller and adherence is more closely managed. Completion of all these paves the way for the product to enter the market (Le Flohic et al. 2024). Figure 1.5 lists the FDA-approved commercialised TDDS to date with their market status.

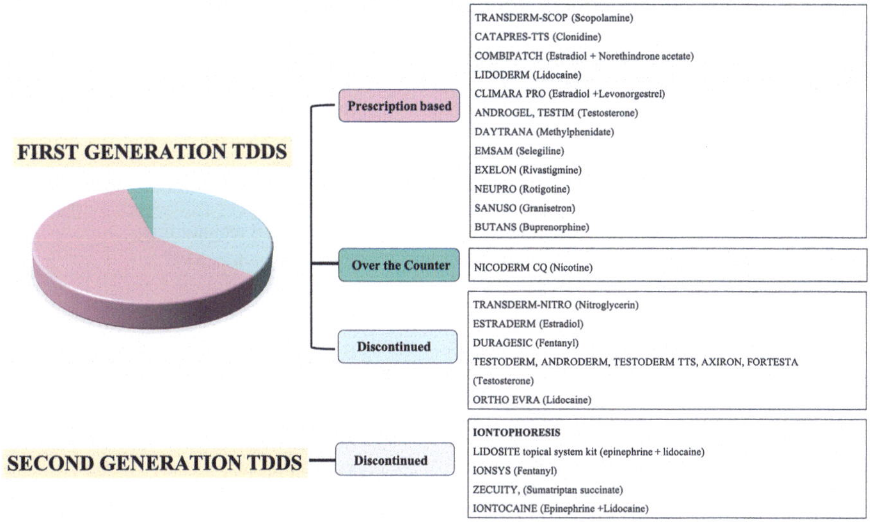

Fig. 1.5 The market potential of FDA-approved drugs

Specifically, any device to enter the market needs to fulfil the FDA prerequisites. Initially, a device is classified into one of the three classes based on its potential risk. For devices classified under I (lowest risk) and II (moderate risk), the sponsor must provide a substantial equivalent data to an existing device and submit a 510 (k). For class III (high risk) devices, sponsors must show reasonable assurance in the preclinical safety and efficacy data and proceed with a premarket approval (PMA). Both 510 (k) and PMA-based devices undergo clinical evaluation. Later, FDA reviews the 510 (k) and PMA submissions and has an interactive review session with the applicant. Finally, device is marketed after ensuring its registration, adheres to good manufacturing practice, and follows labelling rules and no adulteration or misbranded (Lauer et al. 2017).

Here, we have tabulated information on the commercialised products of minimally invasive TDDS in Table 1.2.

1.10 Conclusion

MIDDS is a painless technique in which the SC is reversibly damaged without affecting deeper tissues or pain-sensitive nerves. This needle-free approach reduces the need for surgical procedures, such as implants, while allowing regulated drug delivery. Hence, it reduces the risk associated with invasive techniques. The present chapter delves into the introduction of MIDDS. It begins with a brief overview of TDDS, highlighting their advantages over oral and systemic routes and their classification. The chapter then covers the structure of the skin, invasive and non-invasive mechanisms, and the benefits of MIDDS over other conventional techniques.

Table 1.2 Listed commercialised products of different MIDDS

S. no	Techniques	Commercialised product
1	Chemical heating	Synera®
		Rapydan®
2	Thermoporation	PassPort™ system
		Tixel system
3	Radiofrequency	ViaDor®
4	Laser microporation	P.L.E.A.S.E technology
		Epiture Easytouch™
5	Velocity-based devices	Sumavel®
		Dose Pro®
		Biojector®
		Zetajet
		Injex30
		Stratis®
		Shireen Porejet
		PenJect®
		Tjet®
6	Microneedle	Darmaroller®
		MicroHyala®
		Dermapen®
		Microcore®
		VaxMat®
		Micro-Trans®
		Macroflux®
		MicronJet®
		Drugmat®
		Nanoject®
		IDflu®
		Soluvia®

Further, it will provide a detailed description of various types and possible combinations for enhancing drug delivery. This chapter also covers recent advances in MIDDS, such as nanoneedles and thermoporation techniques. Finally, the chapter ends with an overview of the commercialised product available in the market.

Acknowledgement The authors would like to acknowledge the support extended by the Department of Pharmaceuticals (DoP), Ministry of Chemicals and Fertilizers, Govt. of India, to Pharmaceutical Innovation and Translational Research Lab (PITRL), Department of Pharmaceutics, National Institute of Pharmaceutical Education and Research (NIPER) Hyderabad, India.

Declaration of Competing Interest Reported none.

References

Alkilani AZ, McCrudden MT, Donnelly RF (2015) Transdermal drug delivery: innovative pharmaceutical developments based on disruption of the barrier properties of the stratum corneum. Pharmaceutics 7(4):438–470

Arora A (2017) Liquid and powder jet injectors in drug delivery: mechanisms, designs, and applications. In: Percutaneous penetration enhancers physical methods in penetration enhancement, pp 221–230

Barbero AM, Frasch HF (2009) Pig and Guinea pig skin as surrogates for human in vitro penetration studies: a quantitative review. Toxicol In Vitro 23(1):1–13

Barry BW (2001) Novel mechanisms and devices to enable successful transdermal drug delivery. Eur J Pharm Sci 14(2):101–114

Bastawrous AL, Shih IF, Li Y, Khalil M, Almaz B, Cleary RK (2023) Health-care expenditures are less for minimally invasive than open colectomy for colon cancer: a US commercial claims database analysis. Surg Endosc 37(8):6278–6287

Charoo NA, Rahman Z, Repka MA, Murthy SN (2010) Electroporation: an avenue for transdermal drug delivery. Curr Drug Deliv 7(2):125–136

D'souza AA, Shegokar R (2016) Polyethylene glycol (PEG): a versatile polymer for pharmaceutical applications. Expert Opin Drug Deliv 13(9):1257–1275

Denet AR, Vanbever R, Préat V (2004) Skin electroporation for transdermal and topical delivery. Adv Drug Deliv Rev 56(5):659–674

Donnelly RF, Garland MJ, Alkilani AZ (2014) Microneedle-iontophoresis combinations for enhanced transdermal drug delivery. Methods Mol Biol 1141:121–132

Ebrahiminejad V, Malek-khatabi A, Faraji RZ (2024) Influence of low-frequency vibration and skin strain on insertion mechanics and drug diffusion of PVA/PVP dissolving microneedles. Adv Mater Technol 9(4):2301272

Emanet M, Ciofani G (2023) Ethosomes as promising transdermal delivery systems of natural-derived active compounds. Adv NanoBiomed Res 3(10):2300020

Fowler MJ, Cotter JD, Knight BE, Sevick-Muraca EM, Sandberg DI, Sirianni RW (2020) Intrathecal drug delivery in the era of nanomedicine. Adv Drug Deliv Rev 165:77–95. https://pubmed.ncbi.nlm.nih.gov/32142739/

Ganeson K, Alias AH, Murugaiyah V, Amirul AAA, Ramakrishna S, Vigneswari S (2023) Microneedles for efficient and precise drug delivery in cancer therapy. Pharmaceutics 15(3):744

Gartziandia O, Herran E, Pedraz JL, Igartua M, Hernandez RM (2016) Nanotechnology-based drug-delivery systems releasing growth factors to the CNS: focusing on neurodegenerative disorders. In: Nanobiomaterials in drug delivery: applications of Nanobiomaterials, pp 371–402

Hadgraft J (2001) Skin, the final frontier. Int J Pharm 224(1–2):1–18

Han T, Das DB (2013) Permeability enhancement for transdermal delivery of large molecule using low-frequency sonophoresis combined with microneedles. J Pharm Sci 102(10):3614–3622

Haque T, Talukder MMU (2018) Chemical enhancer: a simplistic way to modulate barrier function of the stratum corneum. Adv Pharm Bull 8(2):169–179

Huang Y, Lui H, Zhao J, Wu Z, Zeng H (2017) Precise spatially selective photothermolysis using modulated femtosecond lasers and real-time multimodal microscopy monitoring. Theranostics 7(3):513–522

Iliopoulos F, Sil BC, Evans CL (2022) The role of excipients in promoting topical and transdermal delivery: current limitations and future perspectives. Front Drug Deliv 2:1049848

Jain KK (2020) An overview of drug delivery systems. Methods Mol Biol 2059:1–54

Jeong WY, Kwon M, Choi HE, Kim KS (2021) Recent advances in transdermal drug delivery systems: a review. Biomater Res 25(1):24

Jeremy C, Wright DJB (2012) Long acting injections and implants. In: Long acting injections and implants. Springer, New York

Joshi N, Azizi Machekposhti S, Narayan RJ (2023) Evolution of transdermal drug delivery devices and novel microneedle technologies: a historical perspective and review. JID Innov 3(6):100225

Kang T, Zhao J, Lin L, Zhu L, Zhao Z, Huang Y et al (2022) The synergistic effect of mechanical vibration for skin puncturing using polymeric microneedles. J Drug Deliv Sci Technol 71:103334
Karande P, Mitragotri S (2009) Enhancement of transdermal drug delivery via synergistic action of chemicals. Biochim Biophys Acta Biomembr 1788(11):2362–2373
Karve T, Banga AK (2024) Comparative evaluation of physical and chemical enhancement techniques for transdermal delivery of linagliptin. Int J Pharm 654:123992
Kim H, Park H, Lee SJ (2017) Effective method for drug injection into subcutaneous tissue. Sci Rep 7(1):9613
Kumar V, Banga AK (2012) Modulated iontophoretic delivery of small and large molecules through microchannels. Int J Pharm 434(1–2):106–114
Kushner J IV, Kim D, So PTC, Blankschtein D, Langer RS (2007) Dual-channel two-photon microscopy study of transdermal transport in skin treated with low-frequency ultrasound and a chemical enhancer. J Invest Dermatol 127(12):2832–2846
Lambert PH, Laurent PE (2008) Intradermal vaccine delivery: will new delivery systems transform vaccine administration? Vaccine 26(26):3197–3208
Larrañeta E, Lutton REM, Woolfson AD, Donnelly RF (2016) Microneedle arrays as transdermal and intradermal drug delivery systems: materials science, manufacture and commercial development. Mater Sci Eng R Rep 104:1–32
Lauer M, Barker JP, Solano M, Dubin J (2017) FDA device regulation. Mo Med 114(4):283–288
Le Flohic E, Vrijens B, Hiligsmann M (2024) The impacts of undetected nonadherence in phase II, III and post-marketing clinical trials: an overview. Br J Clin Pharmacol 90(8):1984–2003
Lee WR, Shen SC, Wang KH, Hu CH, Fang JY (2003) Lasers and microdermabrasion enhance and control topical delivery of vitamin C. J Invest Dermatol 121(5):1118–1125
Lee H, Song C, Baik S, Kim D, Hyeon T, Kim DH (2018) Device-assisted transdermal drug delivery. Adv Drug Deliv Rev 127:35–45
Leppert W, Malec-Milewska M, Zajaczkowska R, Wordliczek J (2018) Transdermal and topical drug administration in the treatment of pain. Molecules 23(3):681
Liu Z, Nie J, Miao B, Li J, Cui Y, Wang S et al (2019) Self-powered intracellular drug delivery by a biomechanical energy-driven triboelectric nanogenerator. Adv Mater 31(12):e1807795
Luo X, Yang L, Cui Y (2023) Microneedles: materials, fabrication, and biomedical applications. Biomed Microdevices 25(3):20
Maddison JE, Page SW, Dyke TM (2008) Clinical pharmacokinetics. In: Small animal clinical pharmacology. WB Saunders, Philadelphia, pp 27–40
Manchanda S, Das N, Chandra A, Shantanu Bandyopadhyay SC (2020) Chapter 2. Fabrication of advanced parenteral drug-delivery systems. In: Advances in pharmaceutical product development and research, drug delivery systems. Academic Press, Elsevier, New York, pp 47–84
Matharoo N, Mohd H, Michniak-Kohn B (2024) Transferosomes as a transdermal drug delivery system: dermal kinetics and recent developments. Wiley Interdiscip Rev Nanomed Nanobiotechnol 16(1):e1918
Mathias N, Sridharan S (2017) Alternate routes of administration. In: Translating molecules into medicines. AAPS advances in the pharmaceutical sciences series, vol 25. Springer, Cham, pp 391–419
Merino G, Kalia YN, Delgado-Charro MB, Potts RO, Guy RH (2003) Frequency and thermal effects on the enhancement of transdermal transport by sonophoresis. J Control Release 88(1):85–94
Mitragotri S (2006) Current status and future prospects of needle-free liquid jet injectors. Nat Rev Drug Discov 5(7):543–548
Moarefian M, Davalos RV, Tafti DK, Achenie LE, Jones CN (2020) Modeling iontophoretic drug delivery in a microfluidic device. Lab Chip 20(18):3310–3321
Oliveira C, Teixeira JA, Oliveira N, Ferreira S, Botelho CM (2024) Microneedles' device: design, fabrication, and applications. Macromolecules 4(2):320–355
Ono N, Iibuchi T, Todo H, Itakura S, Adachi H, Sugibayashi K (2022) Enhancement of skin permeation of fluorescein isothiocyanate-dextran 4 kDa (FD4) and insulin by thermalporation. Eur J Pharm Sci 170:106096

Pamornpathomkul B, Duangjit S, Laohapatarapant S, Rojanarata T, Opanasopit P, Ngawhirunpat T (2015) Transdermal delivery of fluorescein isothiocyanate-dextrans using the combination of microneedles and low-frequency sonophoresis. Asian J Pharm Sci 10(5):415–424

Parhi R, Mandru A (2021) Enhancement of skin permeability with thermal ablation techniques: concept to commercial products. Drug Deliv Transl Res 11(3):817–841

Peña-Juárez MC, Guadarrama-Escobar OR, Escobar-Chávez JJ (2022) Transdermal delivery Systems for Biomolecules. J Pharm Innov 17(2):319–332

Phatale V, Vaiphei KK, Jha S, Patil D, Agrawal M, Alexander A (2022) Overcoming skin barriers through advanced transdermal drug delivery approaches. J Control Release 351:361–380

Pliquett UF, Vanbever R, Preat V, Weaver JC (1998) Local transport regions (LTRs) in human stratum corneum due to long and short "high voltage" pulses. Bioelectrochem Bioenerg 47(1):151–161

Prausnitz MR, Langer R (2008) Transdermal drug delivery. Nat Biotechnol 26(11):1261–1268

Prausnitz MR, Mitragotri S, Langer R (2004) Current status and future potential of transdermal drug delivery. Nat Rev Drug Discov 3(2):115–124

Rabiei M, Kashanian S, Samavati SS, Jamasb S, McInnes SJP (2020) Nanomaterial and advanced technologies in transdermal drug delivery. J Drug Target 28(4):356–367

Ramadon D, McCrudden MTC, Courtenay AJ, Donnelly RF (2022) Enhancement strategies for transdermal drug delivery systems: current trends and applications. Drug Deliv Transl Res 12(4):758–791

Saxen MA (2016) Pharmacologic management of patient behavior. In: McDonald and Avery's dentistry for the child and adolescent, 10th edn, pp 303–327

Schlich M, Lai F, Sinico C, Pitzanti G, Pireddu R, Valenti D et al (2022) Needle-free jet injectors and nanosuspensions: exploring the potential of an unexpected pair. Pharmaceutics 14:1085

Scioli Montoto S, Ruiz ME (2021) Subcutaneous drug delivery. In: The ADME encyclopedia. Springer International, Cham, pp 1–8

Sultatos L (2007) Routes of drug administration. In: xPharm: The comprehensive pharmacology reference. Elsevier, Amsterdam, pp 1–3

Usach I, Martinez R, Festini T, Peris JE (2019) Subcutaneous injection of drugs: literature review of factors influencing pain sensation at the injection site. Adv Ther 36(11):2986–2996

US-FDA (1975) Evamist (estradiol transdermal spray) [Internet]. https://www.accessdata.fda.gov/drugsatfda_docs/label/2007/022014lbl.pdf

Valenta C, Siman U, Kratzel M, Hadgraft J (2000) The dermal delivery of lignocaine: influence of ion pairing. Int J Pharm 197(1–2):77–85

Vargason AM, Anselmo AC, Mitragotri S (2021) The evolution of commercial drug delivery technologies. Nat Biomed Eng 5(9):951–967

Verma A, Jain A, Hurkat P, Jain SK (2016) Transfollicular drug delivery: current perspectives. Res Rep Transdermal Drug Deliv 5:1–17

Waghule T, Singhvi G, Dubey SK, Pandey MM, Gupta G, Singh M et al (2019) Microneedles: a smart approach and increasing potential for transdermal drug delivery system. Biomed Pharmacother 109:1249–1258

Yang M, Zahn JD (2004) Microneedle insertion force reduction using vibratory actuation. Biomed Microdevices 6(3):177–182

Yu YQ, Yang X, Wu XF, Bin FY (2021) Enhancing permeation of drug molecules across the skin via delivery in nanocarriers: novel strategies for effective transdermal applications. Front Bioeng Biotechnol 9:646554

Zhang A, Jung EC, Zhu H, Zou Y, Hui X, Maibach H (2017) Vehicle effects on human stratum corneum absorption and skin penetration. Toxicol Ind Health 33(5):416–425

Zhou Y, Kumar V, Herwadkar A, Banga AK (2017) Transdermal delivery of peptides and proteins by physical methods. In: Percutaneous penetration enhancers physical methods in penetration enhancement, pp 423–437

Zoabi A, Touitou E, Margulis K (2021) Recent advances in nanomaterials for dermal and transdermal applications. Colloids Interfaces 5(1):18

Minimally Invasive Injectable Hydrogels

2

Suraj Wagh, Pooja Khairnar, Giriraj Pandey, Vivek Phatale, and Saurabh Srivastava

Abstract

Hydrogels have intrigued significant interest in medicine owing to their porous structure and tissue adhesiveness, rendering them desirable candidates for drug delivery, tissue regeneration, and cancer management. Before injection, hydrogels should be liquid, but after injection, they should jellify quickly and form a soft, solid material. Injectable hydrogels enable a minimally invasive strategy compared to traditional surgeries with reduced cost and speedy recovery. Injectable hydrogels can be introduced into the target location using a catheter or syringe in a minimally invasive way. Injectable hydrogels hold great potential as biomaterials due to their superior biocompatibility, biodegradability, permeability, stimuli-responsive properties, tissue extracellular matrix (ECM) mimicry, and tissue regeneration ability. Supramolecular hydrogels can undergo reversible sol-gel transition concerning environmental stimuli, and noncovalent cross-linking has shown promising scaffolds for therapy. This book chapter highlights a detailed overview of minimally invasive injectable hydrogel, including materials for their fabrication and the process of gelation with various cross-linking mechanisms. Additionally, special emphasis has been given to the supramolecular hydrogel with minimal invasion and the therapeutic application of minimal invasive hydrogel in drug delivery systems and tissue regeneration.

S. Wagh · P. Khairnar · G. Pandey · V. Phatale · S. Srivastava (✉)
Pharmaceutical Innovation and Translational Research Lab (PITRL), Department of Pharmaceutics, National Institute of Pharmaceutical Education and Research (NIPER), Hyderabad, Telangana, India
e-mail: saurabh@niperhyd.ac.in

M. R. Babu et al. (eds.), *Transdermal Applications of Minimally Invasive Drug Delivery Systems*, https://doi.org/10.1007/978-981-96-6275-3_2

Keywords

Injectable hydrogels · Minimally invasive · Stimuli-responsive · Cross-linking · Supramolecular hydrogels

2.1 Introduction

Hydrogels are three-dimensional polymeric networks with highly porous structures containing large water content (Li and Mooney 2016; Peppas et al. 2006). The porous nature of hydrogels holds water-soluble drugs for various purposes (Li et al. 2022). The sol–gel transition relies on covalent bonding or supramolecular interactions, resulting in a polymeric network. Hydrogels with diverse features can be achieved by varying the ratio of cross-linking agents, changing the polymer component, or introducing functional groups (Piantanida et al. 2019).

In situ forming hydrogels (ISFH) can be injected with a syringe or catheter before undergoing a sol–gel phase transition with respect to endogenous stimuli. Their injectability enables injectable hydrogels to be a minimally invasive approach since they offer the mode of implantation of hydrogels without surgery (Li et al. 2022). The effect of stimuli including pH, temperature, or ion concentration alteration can initiate the gelation. Water-insoluble polymers can be introduced as a liquid dissolved in a solvent, which can precipitate and form a semisolid depot. Several biomaterials have been used, including natural and synthetic biomaterials studied for preparing ISFH, including chitosan (Gao et al. 2022), alginate (Shen et al. 2022), polypeptides (Wan et al. 2021), and hyaluronic acid (Yang et al. 2022) and Pluronic, poly(ethylene glycol), poly(lactic-co-glycolic acid), and poly(vinyl alcohol). Alginate can undergo ionic cross-linking with calcium ions (Ca^{2+}) and interaction with carboxylate moieties of polymer chains forming an insoluble gel (Lee and Mooney 2012). Chemical or physical cross-linking leads to the development of injectable hydrogels. After gel formation, the hydrogel should release cargo to the surrounding environment or attract active components within their networks (Lee et al. 2021; Lo et al. 2019). Injectable hydrogel systems are ideal because of controlled release, prolonged action, and localization in tumor tissues, resulting in improved therapeutic efficiency and minimizing side effects of therapeutics to normal tissues (Qiu and Park 2001; Hoffman 2002). Supramolecular hydrogels are popular and held together by noncovalent interactions owing to their supramolecular chemistry approach such as host–guest interactions and self-assembly. Minimally invasive supramolecular hydrogels can be broadly classified based on their composition such as polymers and peptides (Lim et al. 2019; Dong et al. 2015).

This book chapter highlights the use of minimally invasive injectable hydrogels, including the gelation process with various cross-linking mechanisms, and the biomaterials required for the fabrication of hydrogels have been discussed. Furthermore, supramolecular hydrogels and their different types of hydrogels as minimally invasive approaches have been explained. Finally, the therapeutic applications of minimally invasive injectable hydrogels have been stated.

2.2 Materials for Preparation of Gelation

2.2.1 Natural Polymers

Natural polymers that include polysaccharides and proteins have excellent biocompatibility and biodegradability. Polysaccharides that include chitosan, alginate, cellulose, dextran, and hyaluronic acid and proteins such as gelatin, collagen, and fibrin can be used for hydrogel fabrication (Fig. 2.1).

2.2.1.1 Chitosan

Chitosan is a cationic polysaccharide derived from the deacetylation of chitin and consists of subunits of β-(1,4)-linked glucosamine and N-acetyl glucosamine (Tao et al. 2020). It has been extensively used owing to their biocompatibility, biodegradability, mucoadhesive, antimicrobial, and wound healing properties (Mellati et al. 2016a). Chitosan can be combined with ionic cross-linker such as β-glycerophosphate (GP) to fabricate an in situ thermosensitive gel (Chenite et al. 2000).

2.2.1.2 Collagen or Gelatin

Collagen is comprised of three subunits of polypeptide α-chains. Collagen offers mechanical strength to hydrogels and modulates to mimic native tissues for cellular

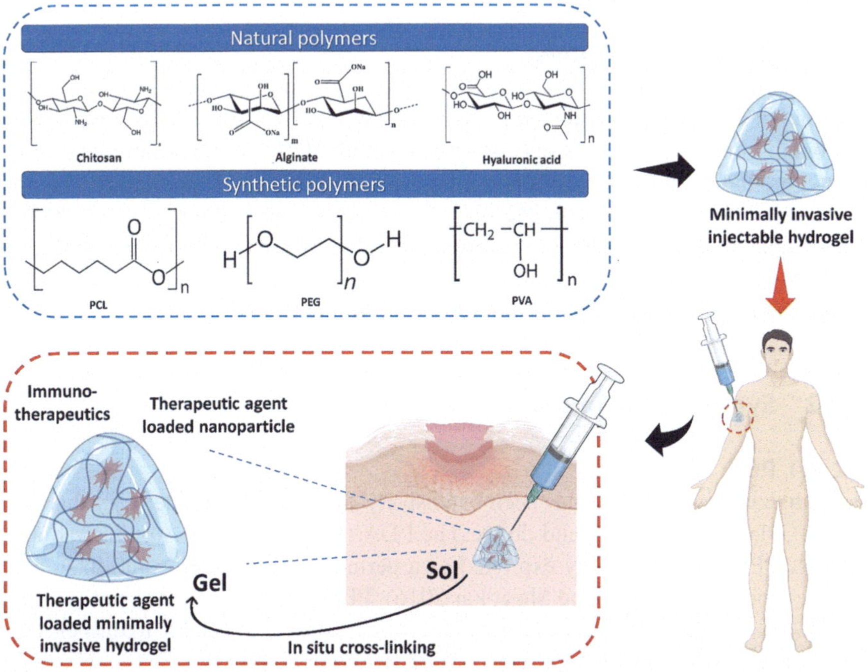

Fig. 2.1 Fabrication of minimally invasive injectable hydrogel using natural and synthetic polymers

migration (Montalbano et al. 2018). Collagen produces the triple-helical fibrous structure by self-assembly and provides tensile strength to collagen (Gajbhiye and Wairkar 2022). The degradation of collagen due to the presence of collagenase inside the body aids collagen-based hydrogels to have a wide demand in biomedical fields such as tissue regeneration. Gelatin is a water-soluble polypeptide derived from collagen. Collagen, upon hydrolysis, dissociates the triple-helical structure and decomposes into gelatin, leading to a biodegradable nature (Guillén-Carvajal et al. 2023).

2.2.1.3 Alginate

Alginate is obtained from brown seawoods consisting of β-(1,4)-mannuronate and α-l-guluronate subunits. Ionic cross-linking is a commonly employed method for gelation of gelatin in which divalent cations bound to guluronate residues, forming egg-box structures and leading to gel formation (Grant et al. 1973). Alginate can absorb significant amounts of biological fluids and escape immunogenicity, making alginate a suitable material for hydrogel fabrication (Li et al. 2023a).

2.2.1.4 Hyaluronic Acid

A linear polysaccharide hyaluronic acid (HA) is made up of β-1,4-D-glucuronic acid and β-1,3-N-acetyl-D-glucosamine subunits. HA is a non-sulfated glycosaminoglycan and constituent of ECM. HA is used as an injectable dermal filler owing to its biocompatibility and viscoelasticity (Wang et al. 2007; Burdick and Prestwich 2011).

2.2.1.5 Cellulose Derivatives

The naturally occurring cellulose is obtained from the source of plants and microorganisms. It is comprised of repeating structural units of 1,4-β-hydro glucose linked through β-ether (Oprea and Voicu 2020; Yang et al. 2021). Water-soluble hydroxypropyl methylcellulose (HPMC) is used broadly in biomedical applications owing to its stability over a varied range of pH and temperature (Bourges et al. 2002). Nanocellulose is a nanoscale biomaterial generated by the chemo-mechanical process and ultrasonication of cellulose. Nanocellulose was used as the matrix component to fabricate injectable nanocomposite gel (Mellati et al. 2021).

2.2.2 Synthetic Polymers

2.2.2.1 PEG

Polyethylene glycol (PEG) has been extensively explored in drug delivery through the PEGylation of proteins and drugs. The FDA (Food and Drug Administration)-approved PEG is extensively exploited as a permeation enhancer, solubilizer, and release modifier (D'souza and Shegokar 2016). PEG can be altered through several approaches for injectability. PEG-based thermoresponsive injectable hydrogels can be prepared through block copolymerization (Gong et al. 2009).

2.2.2.2 PNIPAAm

Poly(N-isopropyl acrylamide) (PNIPAAm) is the highest demanding thermosensitive polymer. Its lower critical solution temperature (LCST) is near proximity to the body temperature, especially used for preparing minimally invasive hydrogels (Kiamahalleh et al. 2017). Above LCST, molecules undergo coil-to-globule transformation because of the inclusion of hydrophilic and hydrophobic components in the molecular structure (Mellati et al. 2016b; Ortiz de Solorzano et al. 2020).

2.2.2.3 Pluronic

Pluronic® also known as poloxamers is another thermoresponsive biomaterial. Pluronic is made up of polypropylene oxide (PPO), a hydrophobic block connected to two hydrophilic polyethylene oxide (PEO) blocks. Above the critical micelle concentration, it self-assembles into micelles, and its biodegradable nature and biocompatibility enables a suitable candidate for tissue engineering and drug delivery (Akash and Rehman 2015; Zarrintaj et al. 2020).

2.2.2.4 PLGA, PGA, and PLA

Poly(lactide-co-glycolide) (PLGA), poly(glycolic acid) (PGA), and poly(lactic acid) (PLA) have been widely explored polymers for biomedical purposes. PLGA is produced from the monomers of glycolic and lactic acid by ring-opening phenomena. The biodegradability and biocompatibility of these polymers enable them to be approved by the US FDA for drug delivery. The biodegradability of PLGA can be monitored by regulating the proportion of glycolic and lactic acid (Anderson and Shive 1997; Sharma et al. 2016).

2.2.2.5 PVA

A linear polymer polyvinyl alcohol is synthesized by hydrolyzing polyvinyl acetate. High water solubility makes PVA a better candidate for biomedical purposes. Physicochemical attributes of PVA can be tuned by the degree of hydroxylation.

2.3 Injectable Covalent Precursors

In the area of regenerative medicine, the covalently linked hydrogels have extensive applications. Covalent cross-links are not reversible by nature, in contrast to the bonds that make up physically annealed hydrogels, and as a result they often produce more robust and mechanically stable materials. Because of their excellent stability, covalent hydrogels can complement larger types of tissues physically and function as scaffolds with greater mechanical strength in comparison to physical hydrogels. Previously, highly invasive surgical procedures were needed for the delivery of such materials, but nowadays mechanisms like delayed bonding or photo-induced radical polymerization are used to deliver the material to the targeted site via noninvasive injection.

Furthermore, it is not always necessary to employ radically propagated photoinitiation reaction methods in order to develop covalently linked hydrogels; instead, one can use injectable covalent precursors. Instead, some biocompatible chemicals are having slow kinetics, which allows ex vivo mixing afterward, injecting the precursor liquid at the targeted site, resulting in the solidification and annealing surrounding the tissue environment. Among these systems are the most well-known Schiff base polymerization (Wu et al. 2016), oxime linkage (Grover et al. 2012), Michael type addition (Jin et al. 2010), and acrylate homopolymerization (Shenoy et al. 2013). Redox techniques using compounds like tetramethylethylenediamine (TEMED) and ammonium persulfate (APS) can be used to replace the photosensitive reaction-based systems as that requires the utilization of the free radicals. Over the last decade, materials science has shown much interest in certain types of covalent precursor reactions known as "click" reactions. These click reactions are designed in such a way that one can get the maximum yield with the minimum step reaction and need only a few back-end purification steps (Xi et al. 2014).

Along with the previously mentioned oxime linkage (Grover et al. 2012) and Michael-type thiol-ene addition reactions (Jin et al. 2010), the most often employed click reactions include nitrile oxide-norbornene cycloadditions (Truong et al. 2015), diels-alder cycloaddition (Yu et al. 2013), and strain-promoted azide-alkyne cycloaddition (Hodgson et al. 2016). At the start various click reactions used copper catalysts. Unfortunately, it soon became evident that they were rather cytotoxic, which rendered them unsuitable for use in the tissue or cell environment. This has led to an entire field of study to develop the click reactions free of copper which enables biocompatibility and biorthogonality. The library of covalently linked precursors offers the adaptability and accessibility required for various drug delivery applications (Desai et al. 2015).

Covalent gels can be employed for the delivery of small molecules and are a simple method that is achieved just by combining precursors and active components before injection. This leads to the uniform distribution of the therapeutics within the polymer delivered, which can only be restricted by the solubility of the API. The field of injectable hydrogels is discovering more innovative injectable covalent precursors that can respond to various stimuli and degrade in a controlled pattern. Thus, this approach is considered as more promising owing to its ability to develop biocompatible, customizable in situ structures.

2.4 Process of Gelation

Injectable hydrogels can be introduced in the appropriate area or tissue using minimally invasive approaches. The material should form gel rapidly after injection to circumvent off-target spreading in the tissues. During injection, the injectable hydrogel must be fluid, i.e., its elastic modulus (G') < storage modulus (G''), and after injection it should form a solid ($G' > G''$). In situ gelation can occur by chemical or physical sol–gel transformation. The physical transitions rely on supramolecular chemistry including hydrophobic interactions and hydrogen bonding which are accountable for thermosensitive gelation, complex formation includes host–guest chemistry and gelation triggered by ionic interaction, and electrostatic

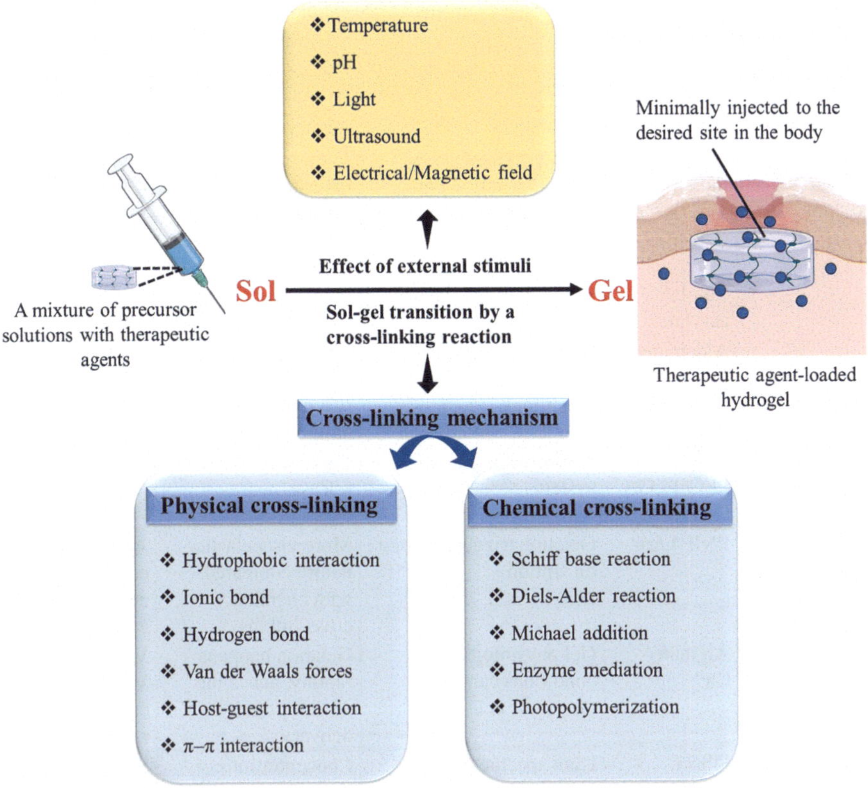

Fig. 2.2 Minimally invasive injectable hydrogels to deliver therapeutics in tissue engineering and disease management

interactions are responsive in pH-sensitive materials (Yesilyurt et al. 2016). Shear-thinning hydrogels act as elastic gels at resting conditions due to noncovalent interactions, but they become viscous on applications of force (Yesilyurt et al. 2016). Gelation can be initiated through UV light or visible light by a polymerization process with free radicals or a photoinitiator. Cross-linking kinetics rely on the reactive functional groups and the hydrogel environment in which the reaction takes place. The preparation of hydrogels depends on the physical and chemical cross-linking (Fig. 2.2). The following are the different techniques for cross-linking hydrogels (Table 2.1).

2.4.1 Self-Assembly

Physically cross-linked hydrogels, often referred to as self-assembled hydrogels, are prepared through the entanglement of polymeric chains via noncovalent interactions (ionic bonding, electrostatic interactions, or hydrogen bonding) (Chao et al.

Table 2.1 Summary of the different mechanisms of cross-linking of injectable hydrogels with their advantages and disadvantages

Mechanisms of cross-linking	Examples	Advantages	Disadvantages	Ref
Chemical cross-linking	Acrylate monomers	Photopolymerization: prevent gelation until light exposed	Require light exposure to the injection site and light penetration to the gel	Sharifi et al. (2009)
	Thiols/acrylates	Self-reactive: Michael-type reaction		Cai et al. (2005)
	Aldehydes	Rapid cross-linking under physiological conditions	Lack of specificity because of the reactivity with amino groups	Lee et al. (2001)
Physical cross-linking	PLGA-PEG-PLGA (ReGel®)	Delivered as one-phase system	Possibility of burst release	Qiao et al. (2005)
	PNIPAAm	Gelation triggered based on stimuli	More viscoelastic nature compared with chemical cross-linking	Alexander et al. (2014) and Ansari et al. (2022)
Ionic cross-linking	Alginate/Ca^{2+}	Gel physiological conditions can entrap the viable cells	Gelation happens rapidly, hence the need for dual delivery	Van Vlierberghe et al. (2011)
Enzyme-mediated cross-linking	Fibrin	High specificity	Concentration-dependent	Stabenfeldt et al. (2012) and Hama et al. (2023)

2020). The physical interactions between the polymeric chains should withstand the conditions for a mechanically stable network that holds the water content of hydrogels. These hydrogels have gained significant interest due to the lack of chemical reactions and no cross-linkers involved (Parhi 2017).

2.4.1.1 Ionic Cross-Linking

The ionizable polymers can be used for ionic cross-linking including pectin, alginate, and proteins such as collagen cross-linked with ions having opposite charges such as Ca^{2+}, Zn^{2+}, etc. The degree of cross-linking, permeability, and drug release kinetics can be modulated by choice of ions (Jay and Saltzman 2009). The short stability of the ionic cross-linking is a limitation in comparison to chemically cross-linked hydrogels. Cross et al. fabricated hybrid hydrogels of HA, poly(2-aminoethyl methacrylate) (PAEM), and PEG (Cross et al. 2015). The authors stated that the composition of the solvent and molecular structure of polycation influence the hydrogel properties.

2.4.1.2 Hydrophobic Interactions

The gelation of hydrogel triggered through variation in pH or temperature could be owing to the alteration in hydrophobic interactions among the polymeric chains. The thermosensitive Pluronic F127 leads to gelation upon temperature change owing to the contribution of hydrophobic interactions (Dumortier et al. 2006). Elevation in temperature causes water loss from the PPO block and aggregation resulting in the self-assembled micellar structure.

2.4.2 Chemical Cross-Linking

The polymer chains are connected through covalent bond formation with cross-linkers. These hydrogels exhibit strong mechanical strength and higher stability. However, cell exposure to cross-linking agents, residual organic solvents, or chemical agents produces cytotoxicity. Further, these hydrogels were classified into three groups.

2.4.2.1 Photo-Initiated Cross-Linking

Photoinitiators and precursors of polymers were implanted into the tumor site, and the gelation was triggered using near-infrared (NIR) light through photo-cross-linking. For photothermal therapy, Zhang et al. utilized TiO_2 @MWCNTs nanocomposite as a photoinitiator and PEG double acrylates (PEGDA) as a matrix component to fabricate hybrid in situ hydrogel (Zhang et al. 2015). However, the photoinitiators are generally activated using UV light; this light penetrates deep tissues leading to cell damage and the generation of free radicals causing alteration in gene expression (Wong et al. 2015; Dimatteo et al. 2018).

2.4.2.2 Precursors Aiding Covalent Cross-Linking

Polymeric precursors functionalized with cross-linkers were employed to fabricate a hydrogel network (Dimatteo et al. 2018). Click chemistry reactions were used to generate covalently linked hydrogels. Diels–Alder reaction, Michael addition reaction, Schiff reaction, and thiol-ene reaction are the commonly known reactions.

2.4.2.3 Enzymatic Cross-Linking

Enzyme-catalyzed cross-linking is one of the approaches for hydrogel preparation. In this approach, the enzyme substrate is added to the molecular polymeric chain to provide enzymatical cross-linker points to produce hydrogel after injection (Mohammadi et al. 2022).

2.5 Stimuli-Responsive Minimally Invasive Injectable Hydrogels

The working principle of stimuli-responsive hydrogels is that they can transform their volume or shape after sensing the environmental stimuli at the targeted region. They can degrade, shrink, or expand in volume upon changes in enzyme activity, electric stimuli, light, temperature, and pH. The injectable hydrogels can be significantly considered for drug delivery for site specificity in the case of various disorders owing to the changes in physiological conditions such as pH, temperature, and glucose level. For the delivery of physically encapsulated drugs, the light-, electric-, temperature-, and pH-responsive systems are considered as most suitable delivery platforms. In contrast, the enzyme- and light-responsive systems are more suitable for the delivery of drug candidates, which are covalently bound to the hydrogels. Delivery of therapeutics through injectable hydrogels will augment the targeted delivery at the specific site and substantially decrease the nonspecific distribution of the drug to the other action. Such delivery systems should offer the prolonged release of therapeutics. In the below section, various types of stimuli-responsive injectable hydrogels were discussed depending on the type of stimuli they respond to (Fig. 2.3).

2.5.1 Photo-Responsive Injectable Hydrogels

Photo-/light-responsive injectable hydrogels are fabricated by the introduction of the photo-responsive material within the hydrogel system (Peng et al. 2022). The photosensitive units undergo different changes after exposure to the light, which include dimerization, isomerization, or cleavage. This ultimately results in various chemical and physical alterations like contraction, degradation, and swelling of the hydrogel network (El-Husseiny et al. 2022). The photo-responsive hydrogels can release the drug in a controlled manner owing to the ability of the hydrogel to shrink or swell after the release or uptake of water upon light irradiation (Li et al. 2019a). The shrinking and swelling of the hydrogel can be modified by the utilization of different photosensitizers such as o-nitrobenzyl ether (o-NB), spiropyranes, and azobenzene during the synthesis of hydrogel. Such polymers undergo the process of reversible photoisomerization that results in alteration in the chain length of the polymer, leading to swelling or shrinkage of the hydrogel network (Iwaso et al. 2016). Recently, Rastogi and coworkers utilized host–guest chemistry between 4-arm PEG and azobenzene to fabricate a photo-responsive hydrogel consisting of Alexa Fluor 750 (fluorescent dye). Azobenzene exhibits reversible photoisomerization among its isomers, i.e., trans (300–380 nm) and cis (>400 nm). Under UV irradiation, the conformation of azobenzene changes to *cis* form which results in the reduction of hydrogel volume that will further permit the release of drug through the hydrogel network (Rastogi et al. 2018). Therefore, photo-responsive minimally invasive injectable hydrogels have significant potential to deliver therapeutics for

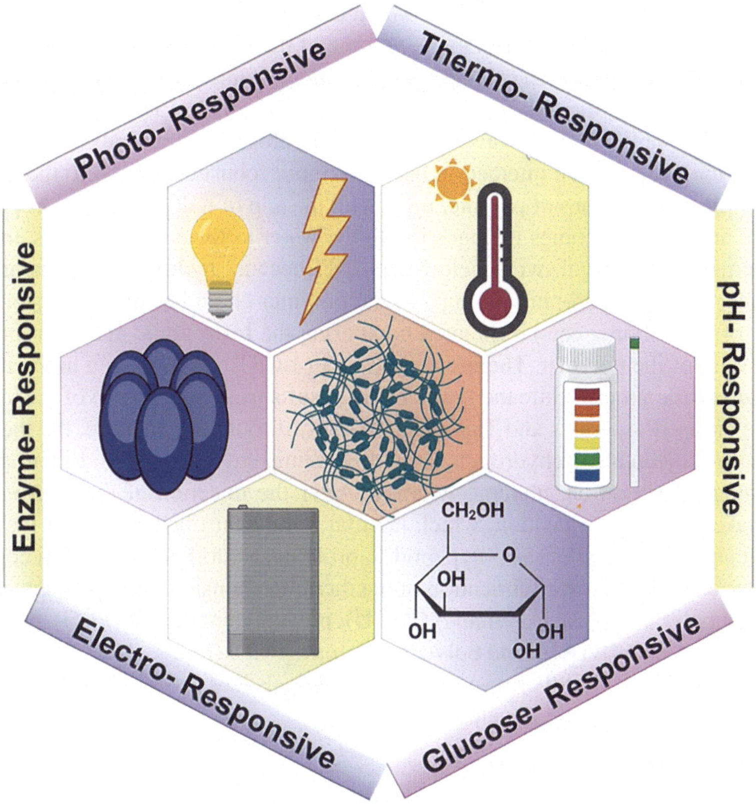

Fig. 2.3 Different types of stimuli-responsive injectable hydrogels

the management of various diseases owing to their biocompatibility, tunable properties, and light sensitivity.

2.5.2 Temperature-Responsive Injectable Hydrogels

The temperature-responsive hydrogels work on the principle that as the environmental temperature changes, the hydrogels change their volume. This type of hydrogel possesses hydrophilic and hydrophobic groups (Huang et al. 2019). When the temperature changes, it affects the hydrogen bonding and the hydrophobic interaction among the polymer chains, which finally results in a change in the hydrogel's volume and structure. Such changes take place at the lower critical solution temperature (LCST) and upper critical solution temperature (UCST) of the hydrogel. Most of the polymers get soluble in water when the temperature increases which is known as the positive temperature dependence. As a result, the positive thermosensitive hydrogels display shrinking and swelling at lower and higher temperatures, respectively. In the case of the polymers which exhibit the LCST, their solubility in the water reduces beyond their LCST which causes shrinkage or contraction of the hydrogels. Such kind of changes in the volume of the hydrogel is called as negative temperature dependence (Li et al. 2023b). In order to fabricate the negative thermo-sensitive hydrogels, the polymers containing more number of hydrophobic groups than the hydrophilic moieties are required. The hydrogel contracts above the LCST because the hydrophobic interaction becomes more robust compared to the hydrogen bonding interactions of polymer units with the aim of reducing the hydrophobic surface area when it comes in contact with the water to lower the system's energy. In contrast, the hydrogel swells below the LCST because hydrogen bonding among the water and the hydrophilic groups in the polymer chain becomes stronger. The LCST of the polymer can be changed by altering the hydrophilic and hydrophobic group ratio of the polymer. Therefore, the higher LCST results from a high ratio of the hydrophilic group, while the lower LCST results from a high ratio of the hydrophobic group (Pasparakis and Tsitsilianis 2020). At room temperature, the thermo-responsive injectable prehydrogels should be semisolid or liquid and then, at the physiological condition, undergo gelation. After the injection of the hydrogel, it undertakes a sol–gel transition, which facilitates the controlled release of the drug (Fan et al. 2022). Most of the polypeptides, proteins, natural polymers, and copolymers are employed for the fabrication of the thermoresponsive hydrogels, which are based on poly(amino ester urethane) (PAEU), polyethylene glycol (PEG), poly(d,l-lactide) (PLA), PNIPAAM, and polycaprolactone (PCL) (Fan et al. 2022; Picchioni and Muljana 2018). The thermo-responsive hydrogels have tremendous potential in the field of controlled and targeted drug delivery.

2.5.3 pH-Responsive Injectable Hydrogels

The pH-responsive injectable hydrogels work on the principle that the volume of hydrogel alters in response to the change in the pH of the tissue environment, which results in the release of the therapeutics from the hydrogel network (Xin and Naficy 2022). In recent decades, many types of pH-responsible hydrogels have been explored in the area of drug delivery. The conventional pH-responsive hydrogels comprise the ionizable side chain, like basic and acidic groups (Singh and Nayak 2023). The degree of dissociation of such moieties changes with the change in the external environment pH. As a result, the cross-linking point of the hydrogel decreases, which further causes alterations in the degree of swelling of the developed hydrogel network (FitzSimons et al. 2021). In the case of polymers, which contain acidic groups like carboxylic acid, they can form new hydrogen bonds in the low pH environment, which leads to a reduction in the volume of hydrogels. Therefore, such hydrogels are not able to absorb more water in an acidic environment, causing a decreased rate of drug release. In contrast, when polymers containing acidic groups come in contact with the alkaline environment, it causes the swelling of hydrogels because of the presence of ionized carboxyl groups, which shows electrostatic repulsion at alkaline pH conditions (Ahmed 2015). In this case, the drug can be released in a faster manner as compared to the acidic environment due to the swelling of hydrogels. Vice versa, the basic group containing polymers expands in the presence of an acidic environment and decreases the volume of hydrogel in basic conditions (Rizwan et al. 2017). Henceforth, the alteration of the pH conditions can cause swelling or shrinkage of the hydrogel network, which can be further utilized to control the release and diffusion rate of the therapeutics (Koetting et al. 2015). The dynamic covalent bonded precursors were used to prepare the hydrogels exhibiting good injectability, self-healing, and shear-thinning properties. Such hydrogels can be prepared through covalent bonding by the utilization of disulfide bonds, complexations, phenylboronate complexations, oxime bonds, hydrazones, and imines. The covalently cross-linked hydrogels are highly stable compared to the hydrogels that are fabricated via noncovalent cross-linking (Picchioni and Muljana 2018).

2.5.4 Electro-responsive Injectable Hydrogels

The electro-responsive hydrogels work on the principle that the hydrogels can swell (on one side), bend, or deswell (on the opposite side) in the existence of an electric field (Kulkarni and Biswanath 2007). The shape and volume of hydrogels can be changed after the application of an electric field, which initiates the release of drugs from the hydrogel network. The experimental setup, type of therapeutics, type of gel, polymer type, and the magnitude of the electric field can affect the release of drugs from the electro-responsive hydrogels (Murdan 2003). The polymers comprising ionizable groups are utilized for the fabrication of the electro-responsive hydrogels, for instance, polyelectrolytes. When an electric field is applied, the hydrogels containing polyelectrolytes deswell and release the drug in a controlled manner. When the electric field is released, the hydrogel swells as it absorbs the aqueous solutions, resulting in the stoppage of drug release. By applying the electric field with a specific intensity for a particular period, the release amount and rate of drug release can be controlled. The mechanism of the drug release via electrically responsive hydrogel includes pH gradient formation resulting from the water electrolysis, stress gradient, and difference in the osmotic pressure after the application of an electric field (Qu et al. 2018; Wang et al. 2016). In a recent study, Huang and group fabricated an electro-responsive nanocarrier hydrogel in a chip-like device by the utilization of the inorganic silica and amphiphilic chitosan for controlled drug delivery. Further, the ethosuximide (an anticonvulsant), which is hydrophilic in nature, was loaded within the nanocarrier hydrogel, and the prepared drug-loaded hydrogel was deposited within a container in between two platinum electrodes. When the different electric voltages were applied then, the drug was released from the nanocarrier hydrogel at different release rates from slow to faster rates. Such kind of release mechanism can be observed owing to the electrostatic interactions among the active molecule and the nanocarrier hydrogel. These interaction mechanisms are based on the electro-osmotic and electrophoretic actions (Huang et al. 2011a). In drug delivery applications, in addition to the conducting polymers, different kinds of conducting additives are widely utilized for the fabrication of electro-responsive hydrogels. Therefore, electro-responsive hydrogels can be extensively employed for the purpose of drug delivery, which offers drug release in a controlled pattern.

2.5.5 Enzyme-Responsive Injectable Hydrogels

The working of enzyme-responsive hydrogels is based on the concept that when the hydrogel network is incorporated with the known sequence of recognition units, it can be particularly identified by the unique set of enzymes (Chandrawati 2016; Dong et al. 2016; Thambi et al. 2016). The recognition units should be approachable to the enzymes. The characteristics of the hydrogel network can influence the interaction among the recognition units and the enzymes. Thus, upon the various enzymatic reactions, the diffusion behavior and the release rate of the hydrogel can be controlled to achieve efficient drug delivery for various applications (Huang et al. 2011b; Skardal et al. 2017; Cirillo et al. 2019). The most commonly explored enzyme-responsive hydrogel delivery systems include the glucosidase- and protease-responsive hydrogels. Glucosidase and the proteases catalyze the hydrolysis of the glycosidic (polysaccharides) and peptide bonds, respectively. Many diseases, such as viral diseases, bacterial infections, inflammatory diseases, cardiovascular diseases, and cancer, are associated with the overexpression of proteases (Turk 2006). Thus, to achieve the controlled release and local delivery for managing such disorders, the protease-responsive hydrogels are gaining major attention. For instance, in the case of cancer patients, the enzyme matrix metalloproteases (MMPs) are upregulated, which leads to the stimulation of angiogenesis, proliferation, and metastasis of cancer cells (Egeblad and Werb 2002). MMPs hydrolyze the peptide bonds in the amino acid under a normal physiological environment, whereas tissue inhibitors regularly control the activity of MMPs in a precise manner. Nevertheless, in disease conditions, the activity of MMP is sustained, which results in malfunctions in tissue activity contributing to the disease progression. Therefore, MMP-responsive injectable hydrogel systems have considerable importance in controlled and targeted cancer therapy (Najafi et al. 2020). Similarly, glucosidase-responsive hydrogels have been employed as a vehicle for the delivery of therapeutics for the management of different diseases, including colon-specific diseases, cardiovascular diseases, and wound healing (Sinha and Kumria 2001; Chourasia and Jain 2004; Vandamme et al. 2002). For instance, the β-galactosidase-responsive injectable hydrogel systems have been explored for the controlled delivery of nitric oxide (NO). The NO plays a major role in various disorders related to the immune, nervous, respiratory, and cardiovascular systems. To maintain normal biological functions and to avoid any kind of toxic effects, it is necessary to maintain the average physiological concentration of the NO. Henceforth, the enzyme-responsive injectable hydrogel can be utilized in drug delivery for the management of various disorders involving abnormal activities related to enzymes.

2.5.6 Glucose-Responsive Injectable Hydrogels

The glucose-responsive hydrogels have been extensively explored for the delivery of insulin in a controlled manner to control the levels of blood glucose (Mansoor et al. 2021). The glucose-responsive hydrogels are fabricated by the utilization of the enzyme glucose oxidase, which is further incorporated into the pH-responsive hydrogels (Gu et al. 2013; Veiseh et al. 2014). When the level of blood glucose rises, the enzyme glucose oxidase will convert the glucose into gluconic acid, which ultimately results in the alteration of the pH in the disease microenvironment. Afterward, the change in the pH conditions causes changes in the volume of the hydrogel, which initiates the release of the active ingredients present within the hydrogel network (Zhao et al. 2017; Traitel et al. 2000). One more approach is there for the synthesis of the glucose-responsive hydrogel that utilizes the specific interactions among glucose and concanavalin A (Con A). This kind of hydrogel is prepared by the complexation of Con A with the glycosylated insulin derivative. The increased concentration of glucose in the diseased condition triggers the glucose-dependent release of the glycosylated insulin derivative due to competitive interaction between the Con A and glucose (Stanca et al. 2024; Makino et al. 1990). Still, there are some major drawbacks associated with this method, which include reaction byproducts (hydrogen peroxide) and immunogenicity due to the use of the enzymes. Therefore, the application of this method is limited. Hence, another method has been developed to fabricate the glucose-responsive hydrogels. In this method, phenylboronic acid (PBA) is used, which can reversibly bind to diols like glucose, thereby forming a stable complex with the five-membered ring. Thus, the glucose-responsive hydrogels can be formulated by using a combination of diol-comprising polymers like PVA- and the PBA-decorated polymers. Then, such hydrogels respond to the altered blood glucose levels, which further results in the controlled release pattern of the therapeutics (Springsteen and Wang 2002). Also, the glucose-responsive hydrogels are not only utilized for the delivery of insulin but also used for therapeutic delivery. Therefore, this type of hydrogel opens the door for new areas of stimuli-responsive drug delivery for the management of various disease modalities.

2.6 Supramolecular Hydrogels for Minimal Invasion

Supramolecular hydrogels are three-dimensional, elastic, and capable of absorbing water. They include inherent mechanical and chemical properties, including toughness, swelling capability, biocompatibility, and biodegradability. Noncovalent interactions such as hydrophobic interaction, metal–ligand coordination, host–guest interactions, hydrogen bonding, electrostatic interactions, and π–π interactions keep them together (Kolipaka et al. 2024). Unlike chemical gels, which are stabilized by chemical cross-linkers, supramolecular hydrogels are supported by these noncovalent interactions, forming macromolecular polymers and a three-dimensional gel network (Pandey et al. 2024; Skopinska-Wisniewska et al. 2021). As mentioned

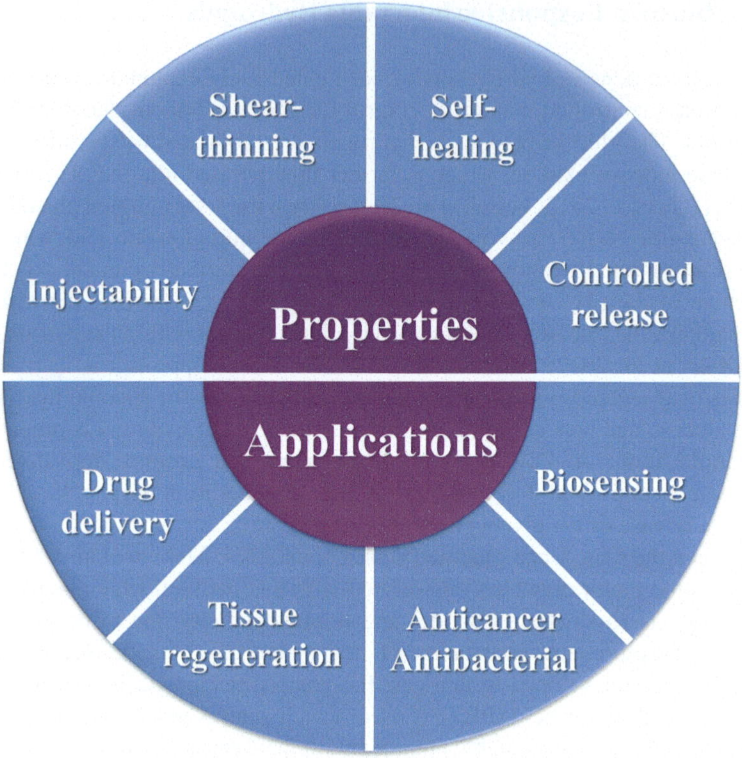

Fig. 2.4 Properties and therapeutic applications of minimally invasive injectable hydrogels

above, rapid association and dissociation provide supramolecular hydrogels with highly dynamic properties. This allows them to continuously reshape, reconfigure, and adapt to environmental changes. Their dynamic nature leads to other unique characteristics such as self-healing, shear-thinning, and responsiveness to external stimuli, which equip them for a range of biological applications such as tissue engineering, controlled administration of medicines and proteins, bioimaging, and biosensing (Fig. 2.4) (Kulkarni et al. 2023). Due to these features, supramolecular hydrogels have attracted much attention as carriers in therapeutic agents and bioactive molecule delivery systems.

Additionally, these minimally invasive supramolecular hydrogels can be categorized based on their composition; a brief description has been given in the following section.

2.6.1 Polymer-Based Hydrogels

These are derived from artificial or natural sources. Natural polymers are especially valued because of their biocompatibility and biodegradability. Polysaccharides, for instance, dextran, alginate, alginate, hyaluronic acid, chitosan, and cellulose, represent a group of hydrophilic natural polymers. On the other hand, hydrogels with certain physical properties can be tailored by adding specific functional groups to synthetic polymers. For instance, because it is easy to functionalize, the synthetic polymer PEG has been investigated extensively in tissue engineering applications (Thang et al. 2023).

2.6.2 Peptide-Based Hydrogels

The development of supramolecular hydrogels has attracted much attention due to the self-assembling properties of peptides. To enable peptides to self-assemble into filaments, many structural motifs can be designed into them, including α-helix, β-sheet, amphiphilic, collagen-like, and elastin-like sequences. These filaments can become entangled at specific concentrations, forming a nanofibrous hydrogel network (Binaymotlagh et al. 2022).

2.6.3 Nucleic Acid-Based Hydrogels

Though less explored than peptides and polymers, nucleic acids, specifically DNA, are becoming increasingly interesting as building blocks for creating supramolecular hydrogels. These hydrogels are created through reversible cross-linking driven by DNA self-assembly. DNA hybridization happens when two complementary single-stranded molecules of DNA bond together to produce a double-stranded structure using Watson–Crick base pairing. These hydrogels can be constructed wholly of DNA or short sequences of DNA grafted on polymer backbones (Li et al. 2016).

2.6.4 Multicomponent/Hybrid Hydrogels

A multicomponent supramolecular hydrogel can be created through physical interactions between peptides or modified peptides and polymers rather than relying on the self-assembly of hybrid lipopeptides. A novel class of hybrid supramolecular hydrogels has been created by fusing peptides with synthetic or natural polymers. These hydrogels improve mechanical characteristics and boost chemical or biological response (Lau and Kiick 2015; Vieira et al. 2017).

2.7 Therapeutic Applications of Injectable Hydrogels as Minimally Invasive Drug Delivery Systems

With an emphasis on oxygen permeability and flexibility, Wichterle and Lim (1960) presented hydrogels as biomaterials for contact lens construction (Wichterle and Lím 1960); since then, hydrogels have been widely used in various industries, including the pharmaceutical and medical sectors. The biomedical community has taken a keen interest in minimally injectable hydrogels because of their potential uses in minimally invasive drug delivery. These hydrogels can be injected into specific body parts and undergo in situ gelation, allowing for targeted implantation and site-specific delivery to tissues or areas where wound healing is prolonged. The ideal injectable hydrogel should ensure a controlled and prolonged release of therapeutic agents at the target site while minimizing drug dispersion to healthy tissues for controlled and localized drug delivery. To achieve this, hydrogels must be capable of detecting changes in the disease environment and releasing the appropriate amount of drug over the required time. Researchers have developed stimuli-responsive injectable hydrogels to meet this need. These hydrogels alter their shape or volume in response to environmental triggers like light, pH, temperature, electrical signals, or enzymatic activity. This preserves healthy tissues while guaranteeing the accurate release of drugs at the intended site. Elisseeff and colleagues demonstrated the first use of invasive hydrogels for tissue engineering and drug delivery in 1999 (Elisseeff et al. 1999). Their study used PEO hydrogels exposed to UV and visible light after chondrocytes or albumins were transdermally injected with PEO dimethacrylate to generate cells or proteins. This achievement came nearly 30 years after Heskins and Guillet first reported the sol–gel transition of PNIPAAM in 1968 (Lanzalaco and Armelin 2017).

2.7.1 Stimuli-Responsive Minimally Invasive Injectable Hydrogel in the Controlled and Local Delivery

Hydrogels with stimuli-responsive properties adjust their volume or morphology in response to environmental conditions. They can either expand, contract, or degrade in response to stimuli. Given that disease conditions in the body often involve low pH, elevated temperatures, and varying glucose levels, injectable hydrogels responsive to these factors are being explored for sustained and targeted drug delivery. Recently, Rasool and colleagues developed a pH-responsive hydrogel employing carrageenan and PVA by physical and chemical cross-linking techniques using (3-aminopropyl)triethoxysilane. The hydrogel's sensitivity to pH was ascribed to the carrageenan's sulfate and -OH functional groups. The –OSO3 groups were protonated at an acidic pH, which improved the connections between molecules through hydrogen bonds. As a result of these groups' ionization, the hydrogel showed the least swelling at acidic pH and the most at basic pH. The drug release from the hydrogel was regulated by this pH-dependent swelling, with a rise in drug release quantity predicted with increasing pH (Rasool et al. 2019). In addition, there has been a lot of interest in thermoresponsive hydrogels created from PEG (hydrophilic) and biodegradable aliphatic polyesters because of their potential for minimally invasive tissue engineering applications and regulated, prolonged drug release. Usually, they are made of copolymers that display thermoresponsive gelation at physiological temperatures, such as PEG–PLA, PEG/PCL, and PEG/PLGA. The hydrophobic properties of the aliphatic polyester blocks and temperature-dependent variations in the hydrogen bonding between polymer and water are the causes of these hydrogels' thermoresponsive activity. In a related study, Singh et al. used a multiblock copolymer poly(ϵ-caprolactone lactide) (PCLA)–PEG–PCLA–PAEU to produce pH- and thermo-sensitive invasive hydrogels. They combined positively charged 2D-layered double hydroxide nanoparticles (LDH) with negatively charged human growth hormone (hGH). The PCLA–PEG–PCLA–PAEU matrix was embedded with the LDH–hGH complex to produce a multistimulus-responsive nanocomposite hydrogel. The pH was changed from 6.0 to 6.8, which caused the sol–gel transition. The ionization of tertiary amines in PAEU caused the hydrogel to become hydrophilic at low pH levels. The creation of micelles and the deionization of tertiary amines caused by an increase in pH started the hydrogelation process. The resultant hydrogel showed prolonged human growth hormone release in vitro and in vivo and was injectable, physiologically stable, and nontoxic to cells. Dual ionic interactions between hGH and the cationic copolymer or the LDH were identified as the cause of the controlled release (Singh et al. 2015). Furthermore, photosensitive functional groups are incorporated into the hydrogel matrix to create photo-responsive injectable hydrogels. When exposed to light, these photosensitive units can cleave, isomerize, or dimerize, changing the hydrogel's chemical and physical properties. These changes can include degradation, swelling, contraction, or chemical alterations within the hydrogel's network. In a work by Rastogi et al., a photo-responsive hydrogel made up of bifunctional azobenzene and 4-arm PEG was used to deliver a model medication, the fluorescent dye Alexa Fluor 750. Reversible photoisomerization occurs between the trans (absorbing 300–380 nm) and cis (absorbing >400 nm) forms of azobenzene. The change to the cis azobenzene conformation caused the hydrogel's volume to drop when exposed to UV light, which made it easier for the model drug to be released from the hydrogel network (Rastogi et al. 2018). Injectable hydrogels offer exciting possibilities for the local delivery of anticancer therapies. Studies proved that hydrogels can provide precise delivery to the tumor site with minimized drug circulation in the blood and reduced dosing frequency. Li et al. (2019b) fabricated self-healing hydrogel using graphene nanoparticles to prevent breast cancer recurrence. These hydrogels demonstrated improved self-healing and mechanical properties with sustained drug delivery. In another work, Xu and coworkers prepared an injectable, NIR/pH-responsive NC hydrogel to manage cancer. This hydrogel was prepared by copolymerizing N-isopropyl acrylamide (NIPAM) with methacrylate poly-β-cyclodextrin (MPCD)-based macromers. The hydrogel was combined with gold nanorods (GNRs) to create a poly(NIPAM-co-MPCD)/GNRs NC hydrogel that reacts to pH and NIR changes. The adamantane-modified drug doxorubicin (DOX) was integrated into the hydrogel, employing host–guest interactions between the adamantane units in DOX and the β-cyclodextrin groups in the polymer. Because the hydrogel broke the acid-sensitive hydrazine bond that bound DOX and adamantane, DOX could be released continuously at an acidic pH. Furthermore, because the photothermal action of the GNRs caused a volume change in the hydrogel network, NIR irradiation increased the release of DOX from the NC hydrogel (Xu et al. 2017).

2.7.2 Minimally Invasive Injectable Hydrogel for Tissue Regeneration

The injectable hydrogels can be used as smart materials due to their advantages over traditional surgical methods, which include reduced healthcare costs and improved patient compliance. Novel injectable hydrogel-based systems with improved characteristics, like ease of use and efficacy in healing defects, have been made possible by advances in tissue engineering. Injectable hydrogels have become valuable biomaterials for delivering various therapeutic agents to different human tissues. For instance, they are increasingly used in cartilage and bone tissue engineering to replace damaged organs. These hydrogels can be synthesized from a variety of synthetic and natural polymers. Natural polymers have been extensively used, including hyaluronic acid, alginate, agarose, chitosan, and fibrin (Vashist et al. 2017). Among these, hyaluronic acid is a top choice for researchers due to its natural occurrence in various tissues, gel-forming capability, biocompatibility, and ease of modification. Presently, several initiatives are being undertaken to increase the mechanical strength of hyaluronic acid-based hydrogels. Various nanofillers, such as carbon nanotubes and cellulose nanocrystals, are being employed to boost their strength. According to a recent study, an injectable hydrogel was created by combining aldehyde-modified nanocrystals with adipic acid dihydrazide-modified hyaluronic acid (ADH-HA) in order to improve it even more. This hydrogel exhibited enhanced mechanical properties, including a more compact network and increased stiffness, with a 0.25 wt% concentration of nanocrystals and a storage modulus (E') of 152.4 kPa. This concentration led to a 135% increase in the elastic modulus compared to hydrogels without the nanocrystals. Furthermore, the hydrogel-encapsulated human adipose-derived stem cells (hASCs) demonstrated significant proliferative capacity (Domingues et al. 2015). Additionally, it is often overlooked that injected hydrogels maintain spatial integrity at the target region. In a related study, Schotman and Dankers examined this issue in the context of hydrogels used for treating myocardial infarction (Schotman and Dankers 2022). They identified key factors influencing hydrogel retention and the delivery of therapeutics, including the hydrogel's formulation (tissue affinity, biodegradability, and mechanical properties such as viscoelasticity, stiffness, and swelling behavior), cardiac pulsation, and therapy strategy (injection timing and volume). A promising method to improve spatial control involves using the pericardial cavity as a natural mold for hydrogel administration (Zhu et al. 2021). According to preliminary clinical trials, injectable hydrogels based on alginate can improve exercise capacity after myocardial infarction and avoid pathological remodeling of the left ventricle. However, most people tolerated these hydrogels well and experienced no serious negative effects (Frey et al. 2014). Another study found a 30-day death rate of 8.6%, indicating the need for more research to verify their safety and efficacy in clinical applications (Anker et al. 2015).

2.8 Conclusion

Hydrogels may absorb and retain large volumes of water or body fluids, providing their porous 3D structure hydrophilic characteristics. They can replicate the physicochemical properties of the ECM or tissues, encapsulating hydrophilic medications, cells, and therapeutic agents. However, because of their hydrophilic characteristics, hydrogels face limitations in encapsulating hydrophobic drugs, as their 3D structures are not well suited for evenly distributing large amounts of these drugs. Additionally, many hydrogels lack sufficient tensile strength, making them unsuitable for load-bearing or stress-bearing tissue applications. Consequently, there are limitations on extensively utilizing hydrogels in biological applications. To overcome these issues, nanocomposite hydrogels were produced by adding different nanoparticles into the hydrogel matrix. Nanocomposite hydrogels have shown improved mechanical strength and enhanced drug loading capacity, enabling encapsulation of hydrophobic drugs and controlled release. Moreover, the shear-thinning and stimuli responsiveness properties of nanocomposite hydrogels can be fine-tuned and optimized by selecting different types of nanomaterials.

However, despite progress, several challenges must be overcome for future advancements and practical applications. Achieving on-demand drug release from these hydrogels in vivo and clinical settings remains challenging, as present systems cannot provide multiple functions simultaneously. For clinical use, particularly in dynamic organs such as the lungs, heart, and arteries, which undergo constant expansion, contraction, and relaxation, and in tissues subjected to high stress, the ideal systems should offer a balance of mechanical stability, tissue adhesiveness, and elasticity. Additionally, managing the immune response is crucial, as minimizing or controlling the inflammation caused by the injected hydrogels is important. Additionally, several aspects of injectable hydrogel systems require further enhancement, including preserving the survival and growth of cells within the hydrogel matrix both during and after injection, shielding drug molecules from deterioration, and enhancing the capacity to load both hydrophilic and hydrophobic drugs. Controlled drug release capabilities are also crucial. These systems must be biodegradable, free of hazardous leftover cross-linkers, and nontoxic breakdown products. Hence, it is essential to promote the development of multifunctional hydrogels that incorporate various functional units to address these challenges. In summary, minimally injectable hydrogels are regarded as highly promising biomaterials for use as drug or bioactive molecule carriers or scaffolds in disease treatment, cancer management, and tissue regeneration. It is anticipated that ongoing research will result in further developments in the creation of injectable hydrogels with the essential characteristics required for various uses.

Acknowledgment The authors would like to acknowledge the research funding support by the Department of Pharmaceuticals (DoP), Ministry of Chemicals and Fertilizers, Govt. of India, to "Pharmaceutical Innovation and Translational Research Lab (PITRL)," Department of Pharmaceutics, National Institute of Pharmaceutical Education and Research (NIPER) Hyderabad, India.

Declaration of Competing Interest Reported none.

References

Ahmed EM (2015) Hydrogel: preparation, characterization, and applications: a review. J Adv Res 6:105–121. https://doi.org/10.1016/J.JARE.2013.07.006

Akash MSH, Rehman K (2015) Recent progress in biomedical applications of Pluronic (PF127): pharmaceutical perspectives. J Control Release 209:120–138. https://doi.org/10.1016/j.jconrel.2015.04.032

Alexander A, Ajazuddin KJ, Saraf S, Saraf S (2014) Polyethylene glycol (PEG)–Poly(N-isopropylacrylamide) (PNIPAAm) based thermosensitive injectable hydrogels for biomedical applications. Eur J Pharm Biopharm 88:575–585. https://doi.org/10.1016/j.ejpb.2014.07.005

Anderson JM, Shive MS (1997) Biodegradation and biocompatibility of PLA and PLGA microspheres. Adv Drug Deliv Rev 28:5–24. https://doi.org/10.1016/S0169-409X(97)00048-3

Anker SD, Coats AJS, Cristian G, Dragomir D, Pusineri E, Piredda M et al (2015) A prospective comparison of alginate-hydrogel with standard medical therapy to determine impact on functional capacity and clinical outcomes in patients with advanced heart failure (AUGMENT-HF trial). Eur Heart J 36:2297–2309. https://doi.org/10.1093/eurheartj/ehv259

Ansari MJ, Rajendran RR, Mohanto S, Agarwal U, Panda K, Dhotre K et al (2022) Poly(N-isopropylacrylamide)-based hydrogels for biomedical applications: a review of the state-of-the-art. Gels 8:454. https://doi.org/10.3390/gels8070454

Binaymotlagh R, Chronopoulou L, Haghighi FH, Fratoddi I, Palocci C (2022) Peptide-based hydrogels: new materials for biosensing and biomedical applications. Materials (Basel) 15(17):5871. https://doi.org/10.3390/ma15175871

Bourges X, Weiss P, Daculsi G, Legeay G (2002) Synthesis and general properties of silated-hydroxypropyl methylcellulose in prospect of biomedical use. Adv Colloid Interface Sci 99:215–228. https://doi.org/10.1016/S0001-8686(02)00035-0

Burdick JA, Prestwich GD (2011) Hyaluronic acid hydrogels for biomedical applications. Adv Mater 23(12):H41–H56. https://doi.org/10.1002/adma.201003963

Cai S, Liu Y, Zheng Shu X, Prestwich GD (2005) Injectable glycosaminoglycan hydrogels for controlled release of human basic fibroblast growth factor. Biomaterials 26:6054–6067. https://doi.org/10.1016/j.biomaterials.2005.03.012

Chandrawati R (2016) Enzyme-responsive polymer hydrogels for therapeutic delivery. Exp Biol Med (Maywood) 241:972–979. https://doi.org/10.1177/1535370216647186

Chao Y, Chen Q, Liu Z (2020) Smart injectable hydrogels for cancer immunotherapy. Adv Funct Mater 30(2):1902785. https://doi.org/10.1002/adfm.201902785

Chenite A, Chaput C, Wang D, Combes C, Buschmann MD, Hoemann CD et al (2000) Novel injectable neutral solutions of chitosan form biodegradable gels in situ. Biomaterials 21:2155–2161. https://doi.org/10.1016/S0142-9612(00)00116-2

Chourasia MK, Jain SK (2004) Polysaccharides for colon targeted drug delivery. Drug Deliv 11:129–148. https://doi.org/10.1080/10717540490280778

Cirillo G, Spizzirri UG, Curcio M, Nicoletta FP, Iemma F (2019) Injectable hydrogels for cancer therapy over the last decade. Pharmaceutics 11:486. https://doi.org/10.3390/PHARMACEUTICS11090486

Cross D, Jiang X, Ji W, Han W, Wang C (2015) Injectable hybrid hydrogels of hyaluronic acid crosslinked by well-defined synthetic polycations: preparation and characterization in vitro and in vivo. Macromol Biosci 15:668–681. https://doi.org/10.1002/mabi.201400491

D'souza AA, Shegokar R (2016) Polyethylene glycol (PEG): a versatile polymer for pharmaceutical applications. Expert Opin Drug Deliv 13:1257–1275. https://doi.org/10.1080/17425247.2016.1182485

Desai RM, Koshy ST, Hilderbrand SA, Mooney DJ, Joshi NS (2015) Versatile click alginate hydrogels crosslinked via tetrazine–norbornene chemistry. Biomaterials 50:30–37. https://doi.org/10.1016/J.BIOMATERIALS.2015.01.048

Dimatteo R, Darling NJ, Segura T (2018) In situ forming injectable hydrogels for drug delivery and wound repair. Adv Drug Deliv Rev 127:167–184. https://doi.org/10.1016/j.addr.2018.03.007

Domingues RMA, Silva M, Gershovich P, Betta S, Babo P, Caridade SG et al (2015) Development of injectable hyaluronic acid/cellulose nanocrystals bionanocomposite hydrogels for tissue engineering applications. Bioconjug Chem 26:1571–1581. https://doi.org/10.1021/acs.bioconjchem.5b00209

Dong R, Pang Y, Su Y, Zhu X (2015) Supramolecular hydrogels: synthesis, properties and their biomedical applications. Biomater Sci 3:937–954. https://doi.org/10.1039/C4BM00448E

Dong Y, Wang W, Veiseh O, Appel EA, Xue K, Webber MJ et al (2016) Injectable and glucose-responsive hydrogels based on boronic acid-glucose complexation. Langmuir 32:8743–8747. https://doi.org/10.1021/ACS.LANGMUIR.5B04755/ASSET/IMAGES/LARGE/LA-2015-04755H_0001.JPEG

Dumortier G, Grossiord JL, Agnely F, Chaumeil JC (2006) A review of Poloxamer 407 pharmaceutical and pharmacological characteristics. Pharm Res 23:2709–2728. https://doi.org/10.1007/s11095-006-9104-4

Egeblad M, Werb Z (2002) New functions for the matrix metalloproteinases in cancer progression. Nat Rev Cancer 3:161–174. https://doi.org/10.1038/nrc745

El-Husseiny HM, Mady EA, Hamabe L, Abugomaa A, Shimada K, Yoshida T et al (2022) Smart/stimuli-responsive hydrogels: cutting-edge platforms for tissue engineering and other biomedical applications. Mater Today Bio 13:100186. https://doi.org/10.1016/J.MTBIO.2021.100186

Elisseeff J, Anseth K, Sims D, McIntosh W, Randolph M, Langer R (1999) Transdermal photopolymerization for minimally invasive implantation. Proc Natl Acad Sci U S A 96:3104–3107. https://doi.org/10.1073/pnas.96.6.3104

Fan R, Cheng Y, Wang R, Zhang T, Zhang H, Li J et al (2022) Thermosensitive hydrogels and advances in their application in disease therapy. Polymers 14:2379. https://doi.org/10.3390/POLYM14122379

FitzSimons TM, Anslyn EV, Rosales AM (2021) Effect of pH on the properties of hydrogels cross-linked via dynamic Thia-Michael addition bonds. ACS Polym Au 2:129–136. https://doi.org/10.1021/ACSPOLYMERSAU.1C00049

Frey N, Linke A, Süselbeck T, Müller-Ehmsen J, Vermeersch P, Schoors D et al (2014) Intracoronary delivery of injectable bioabsorbable scaffold (IK-5001) to treat left ventricular remodeling after ST-elevation myocardial infarction: a first-in-man study. Circ Cardiovasc Interv 7:806–812. https://doi.org/10.1161/CIRCINTERVENTIONS.114.001478

Gajbhiye S, Wairkar S (2022) Collagen fabricated delivery systems for wound healing: a new roadmap. Biomater Adv 142:213152. [CrossRef] - Google Search n.d. https://www.google.com/search?q=Gajbhiye%2C+S.%3B+Wairkar%2C+S.+Collagen+fabricated+delivery+systems+for+wound+healing%3A+A+new+roadmap.+Biomater.+Adv.+2022%2C+142%2C+213152.+%5BCrossRef%5D&oq=Gajbhiye%2C+S.%3B+Wairkar%2C+S.+Collagen+fabricated+delivery+systems+for+wound+healing%3A+A+new+roadmap.+Biomater.+Adv.+2022%2C+142%2C+213152.+%5BCrossRef%5D&gs_lcrp=EgZjaHJvbWUyBggAEEUYOdIBCDEzMjBqMGo3qAIIsAIB&sourceid=chrome&ie=UTF-8. Accessed 13 Aug 2024

Gao C, Cheng K, Li Y, Gong R, Zhao X, Nie G et al (2022) Injectable immunotherapeutic hydrogel containing RNA-loaded lipid nanoparticles reshapes tumor microenvironment for pancreatic cancer therapy. Nano Lett 22:8801–8809. https://doi.org/10.1021/acs.nanolett.2c01994

Gong C, Shi S, Dong P, Kan B, Gou M, Wang X et al (2009) Synthesis and characterization of PEG-PCL-PEG thermosensitive hydrogel. Int J Pharm 365:89–99. https://doi.org/10.1016/j.ijpharm.2008.08.027

Grant GT, Morris ER, Rees DA, Smith PJ, Thom D (1973) Biological interactions between polysaccharides and divalent cations: the egg-box model. FEBS Lett 32:195–198. Google Search n.d. https://www.google.com/search?q=%5B115%5D+G.T.+Grant%2C+E.R.+Morris%2C+D.A.+Rees%2C+P.J.+Smith%2C+D.+Thom%2C+Biological+interactions+between+polysacch arides+and+divalent+cations%3A+the+egg-box+model%2C+FEBS+Lett.+32+(1973)+195%E2%80%93198.&oq=%5B115%5D+G.T.+Grant%2C+E.R.+Morris%2C+D.A.+Rees%2C+P.J.+Smith%2C+D.+Thom%2C+Biological+interactions+between+polysaccharides+and+divalent+cations%3A+the+egg-box+model%2C+FEBS+Lett.+32+(1973)+195%E2%80%93198.&gs_lcrp=EgZjaHJvbWUyBggAEEUYOdIBBzM4M2owajeoAgiwAgE&sourceid=chrome&ie=UTF-8. Accessed 12 Aug 2024

Grover GN, Lam J, Nguyen TH, Segura T, Maynard HD (2012) Biocompatible hydrogels by oxime click chemistry. Biomacromolecules 13:3013–3017. https://doi.org/10.1021/BM301346E/SUPPL_FILE/BM301346E_SI_001.PDF

Gu Z, Aimetti AA, Wang Q, Dang TT, Zhang Y, Veiseh O et al (2013) Injectable nano-network for glucose-mediated insulin delivery. ACS Nano 7:4194–4201. https://doi.org/10.1021/NN400630X

Guillén-Carvajal K, Valdez-Salas B, Beltrán-Partida E, Salomón-Carlos J, Cheng N (2023) Chitosan, gelatin, and collagen hydrogels for bone regeneration. Polymers (Basel) 15:2762. https://doi.org/10.3390/polym15132762

Hama R, Ulziibayar A, Reinhardt JW, Watanabe T, Kelly J, Shinoka T (2023) Recent developments in biopolymer-based hydrogels for tissue engineering applications. Biomolecules 13:280. https://doi.org/10.3390/biom13020280

Hodgson SM, Bakaic E, Stewart SA, Hoare T, Adronov A (2016) Properties of poly(ethylene glycol) hydrogels cross-linked via strain-promoted alkyne-azide cycloaddition (SPAAC). Biomacromolecules 17:1093–1100. https://doi.org/10.1021/ACS.BIOMAC.5B01711/ASSET/IMAGES/MEDIUM/BM-2015-017114_0011.GIF

Hoffman AS (2002) Hydrogels for biomedical applications. Adv Drug Deliv Rev 54:3–12. https://doi.org/10.1016/S0169-409X(01)00239-3

Huang WC, Lee TJ, Hsiao CS, Chen SY, Liu DM (2011a) Characterization and drug release behavior of chip-like amphiphilic chitosan–silica hybrid hydrogel for electrically modulated release of ethosuximide: an in vitro study. J Mater Chem 21:16077–16085. https://doi.org/10.1039/C1JM12376A

Huang R, Qi W, Feng L, Su R, He Z (2011b) Self-assembling peptide–polysaccharide hybrid hydrogel as a potential carrier for drug delivery. Soft Matter 7:6222–6230. https://doi.org/10.1039/C1SM05375B

Huang H, Qi X, Chen Y, Wu Z (2019) Thermo-sensitive hydrogels for delivering biotherapeutic molecules: a review. Saudi Pharm J 27:990–999. https://doi.org/10.1016/J.JSPS.2019.08.001

Iwaso K, Takashima Y, Harada A (2016) Fast response dry-type artificial molecular muscles with [c2]daisy chains. Nat Chem 8(6):625–632. https://doi.org/10.1038/nchem.2513

Jay SM, Saltzman WM (2009) Controlled delivery of VEGF via modulation of alginate microparticle ionic crosslinking. J Control Release 134:26–34. https://doi.org/10.1016/j.jconrel.2008.10.019

Jin R, Teixeira LSM, Krouwels A, Dijkstra PJ, Van Blitterswijk CA, Karperien M et al (2010) Synthesis and characterization of hyaluronic acid–poly(ethylene glycol) hydrogels via Michael addition: an injectable biomaterial for cartilage repair. Acta Biomater 6:1968–1977. https://doi.org/10.1016/J.ACTBIO.2009.12.024

Kiamahalleh MV, Mellati A, Madani SA, Pendleton P, Zhang H, Madani SH (2017) Smart carriers for controlled drug delivery: thermosensitive polymers embedded in ordered mesoporous carbon. J Pharm Sci 106:1545–1552. https://doi.org/10.1016/j.xphs.2017.02.010

Koetting MC, Peters JT, Steichen SD, Peppas NA (2015) Stimulus-responsive hydrogels: theory, modern advances, and applications. Mater Sci Eng R Rep 93:1–49. https://doi.org/10.1016/J.MSER.2015.04.001

Kolipaka T, Pandey G, Abraham N, Srinivasarao DA, Raghuvanshi RS, Rajinikanth PS et al (2024) Stimuli-responsive polysaccharide-based smart hydrogels for diabetic wound healing: design aspects, preparation methods and regulatory perspectives. Carbohydr Polym 324:121537. https://doi.org/10.1016/j.carbpol.2023.121537

Kulkarni RV, Biswanath SA (2007) Electrically responsive smart hydrogels in drug delivery: a review. J Appl Biomater Biomech 5(3):125–139. https://doi.org/10.1177/228080000700500301

Kulkarni N, Rao P, Jadhav GS, Kulkarni B, Kanakavalli N, Kirad S et al (2023) Emerging role of injectable dipeptide hydrogels in biomedical applications. ACS Omega 8:3551–3570. https://doi.org/10.1021/acsomega.2c05601

Lanzalaco S, Armelin E (2017) Poly(N-isopropylacrylamide) and copolymers: a review on recent progresses in biomedical applications. Gels 3(4):36. https://doi.org/10.3390/gels3040036

Lau HK, Kiick KL (2015) Opportunities for multicomponent hybrid hydrogels in biomedical applications. Biomacromolecules 16:28–42. https://doi.org/10.1021/bm501361c

Lee KY, Mooney DJ (2012) Alginate: properties and biomedical applications. Prog Polym Sci 37:106–126. https://doi.org/10.1016/j.progpolymsci.2011.06.003

Lee KY, Alsberg E, Mooney DJ (2001) Degradable and injectable poly(aldehyde guluronate) hydrogels for bone tissue engineering. J Biomed Mater Res 56:228–233. https://doi.org/10.1002/1097-4636(200108)56:2<228::AID-JBM1089>3.0.CO;2-9

Lee SY, Jeon SI, Sim SB, Byun Y, Ahn C-H (2021) A supramolecular host-guest interaction-mediated injectable hydrogel system with enhanced stability and sustained protein release. Acta Biomater 131:286–301. https://doi.org/10.1016/j.actbio.2021.07.004

Li J, Mooney DJ (2016) Designing hydrogels for controlled drug delivery. Nat Rev Mater 1:16071. https://doi.org/10.1038/natrevmats.2016.71

Li J, Mo L, Lu C-H, Fu T, Yang H-H, Tan W (2016) Functional nucleic acid-based hydrogels for bioanalytical and biomedical applications. Chem Soc Rev 45:1410–1431. https://doi.org/10.1039/c5cs00586h

Li L, Scheiger JM, Levkin PA (2019a) Design and applications of photoresponsive hydrogels. Adv Mater 31:1807333. https://doi.org/10.1002/ADMA.201807333

Li Q, Wen J, Liu C, Jia Y, Wu Y, Shan Y et al (2019b) Graphene-nanoparticle-based self-healing hydrogel in preventing postoperative recurrence of breast cancer. ACS Biomater Sci Eng 5:768–779. https://doi.org/10.1021/acsbiomaterials.8b01475

Li Y, Yang HY, Lee DS (2022) Biodegradable and injectable hydrogels in biomedical applications. Biomacromolecules 23:609–618. https://doi.org/10.1021/acs.biomac.1c01552

Li Y, Xu Z, Wang J, Pei X, Chen J, Wan Q (2023a) Alginate-based biomaterial-mediated regulation of macrophages in bone tissue engineering. Int J Biol Macromol 230:123246. https://doi.org/10.1016/J.IJBIOMAC.2023.123246

Li L, Guo J, Kang C (2023b) LCST-UCST transition property of a novel retarding swelling and thermosensitive particle gel. Materials 16:2761. https://doi.org/10.3390/MA16072761

Lim JYC, Lin Q, Xue K, Loh XJ (2019) Recent advances in supramolecular hydrogels for biomedical applications. Mater Today Adv 3:100021. https://doi.org/10.1016/j.mtadv.2019.100021

Lo Y-W, Sheu M-T, Chiang W-H, Chiu Y-L, Tu C-M, Wang W-Y et al (2019) In situ chemically crosslinked injectable hydrogels for the subcutaneous delivery of trastuzumab to treat breast cancer. Acta Biomater 86:280–290. https://doi.org/10.1016/j.actbio.2019.01.003

Makino K, Mack EJ, Okano T, Kim SW (1990) A microcapsule self-regulating delivery system for insulin. J Control Release 12:235–239. https://doi.org/10.1016/0168-3659(90)90104-2

Mansoor S, Kondiah PPD, Choonara YE (2021) Advanced hydrogels for the controlled delivery of insulin. Pharmaceutics 13:2113. https://doi.org/10.3390/PHARMACEUTICS13122113

Mellati A, Kiamahalleh MV, Madani SH, Dai S, Bi J, Jin B et al (2016a) Poly(N-isopropylacrylamide) hydrogel/chitosan scaffold hybrid for three-dimensional stem cell culture and cartilage tissue engineering. J Biomed Mater Res A 104:2764–2774. https://doi.org/10.1002/JBM.A.35810

Mellati A, Valizadeh Kiamahalleh M, Dai S, Bi J, Jin B, Zhang H (2016b) Influence of polymer molecular weight on the in vitro cytotoxicity of poly (N-isopropylacrylamide). Mater Sci Eng C Mater Biol Appl 59:509–513. https://doi.org/10.1016/j.msec.2015.10.043

Mellati A, Hasanzadeh E, Gholipourmalekabadi M, Enderami SE (2021) Injectable nanocomposite hydrogels as an emerging platform for biomedical applications: a review. Mater Sci Eng C Mater Biol Appl 131:112489. https://doi.org/10.1016/j.msec.2021.112489

Mohammadi M, Karimi M, Malaekeh-Nikouei B, Torkashvand M, Alibolandi M (2022) Hybrid in situ- forming injectable hydrogels for local cancer therapy. Int J Pharm 616:121534. https://doi.org/10.1016/j.ijpharm.2022.121534

Montalbano G, Toumpaniari S, Popov A, Duan P, Chen J, Dalgarno K et al (2018) Synthesis of bioinspired collagen/alginate/fibrin based hydrogels for soft tissue engineering. Mater Sci Eng C Mater Biol Appl 91:236–246. https://doi.org/10.1016/j.msec.2018.04.101

Murdan S (2003) Electro-responsive drug delivery from hydrogels. J Control Release 92:1–17. https://doi.org/10.1016/S0168-3659(03)00303-1

Najafi M, Asadi H, Van Den Dikkenberg J, Van Steenbergen MJ, Fens MHAM, Hennink WE et al (2020) Conversion of an injectable MMP-degradable hydrogel into core-cross-linked micelles. Biomacromolecules 21:1739–1751. https://doi.org/10.1021/ACS.BIOMAC.9B01675/ASSET/IMAGES/LARGE/BM9B01675_0008.JPEG

Oprea M, Voicu SI (2020) Recent advances in composites based on cellulose derivatives for biomedical applications. Carbohydr Polym 247:116683. https://doi.org/10.1016/j.carbpol.2020.116683

Ortiz de Solorzano I, Bejagam KK, An Y, Singh SK, Deshmukh SA (2020) Solvation dynamics of *N*-substituted acrylamide polymers and the importance for phase transition behavior. Soft Matter 16:1582–1593. https://doi.org/10.1039/C9SM01798D

Pandey G, Phatale V, Khairnar P, Kolipaka T, Shah S, Famta P et al (2024) Supramolecular self-assembled peptide-engineered nanofibers: a propitious proposition for cancer therapy. Int J Biol Macromol 256:128452. https://doi.org/10.1016/j.ijbiomac.2023.128452

Parhi R (2017) Cross-linked hydrogel for pharmaceutical applications: a review. Adv Pharm Bull 7:515–530. https://doi.org/10.15171/apb.2017.064

Pasparakis G, Tsitsilianis C (2020) LCST polymers: thermoresponsive nanostructured assemblies towards bioapplications. Polymer (Guildf) 211:123146. https://doi.org/10.1016/J.POLYMER.2020.123146

Peng K, Zheng L, Zhou T, Zhang C, Li H (2022) Light manipulation for fabrication of hydrogels and their biological applications. Acta Biomater 137:20–43. https://doi.org/10.1016/J.ACTBIO.2021.10.003

Peppas NA, Hilt JZ, Khademhosseini A, Langer R (2006) Hydrogels in biology and medicine: from molecular principles to bionanotechnology. Adv Mater 18:1345–1360. https://doi.org/10.1002/adma.200501612

Piantanida E, Alonci G, Bertucci A, De Cola L (2019) Design of nanocomposite injectable hydrogels for minimally invasive surgery. Acc Chem Res 52:2101–2112. https://doi.org/10.1021/acs.accounts.9b00114

Picchioni F, Muljana H (2018) Hydrogels based on dynamic covalent and non covalent bonds: a chemistry perspective. Gels 4:21. https://doi.org/10.3390/GELS4010021

Qiao M, Chen D, Ma X, Liu Y (2005) Injectable biodegradable temperature-responsive PLGA–PEG–PLGA copolymers: synthesis and effect of copolymer composition on the drug release from the copolymer-based hydrogels. Int J Pharm 294:103–112. https://doi.org/10.1016/j.ijpharm.2005.01.017

Qiu Y, Park K (2001) Environment-sensitive hydrogels for drug delivery. Adv Drug Deliv Rev 53:321–339. https://doi.org/10.1016/S0169-409X(01)00203-4

Qu J, Zhao X, Ma PX, Guo B (2018) Injectable antibacterial conductive hydrogels with dual response to an electric field and pH for localized "smart" drug release. Acta Biomater 72:55–69. https://doi.org/10.1016/J.ACTBIO.2018.03.018

Rasool A, Ata S, Islam A, Khan RU (2019) Fabrication of novel carrageenan based stimuli responsive injectable hydrogels for controlled release of cephradine. RSC Adv 9:12282–12290. https://doi.org/10.1039/c9ra02130b

Rastogi SK, Anderson HE, Lamas J, Barret S, Cantu T, Zauscher S et al (2018) Enhanced release of molecules upon ultraviolet (UV) light irradiation from photoresponsive hydrogels prepared from bifunctional azobenzene and four-arm poly(ethylene glycol). ACS Appl Mater Interfaces 10:30071–30080. https://doi.org/10.1021/acsami.6b16183

Rizwan M, Yahya R, Hassan A, Yar M, Azzahari AD, Selvanathan V et al (2017) pH sensitive hydrogels in drug delivery: brief history, properties, swelling, and release mechanism, material selection and applications. Polymers 9:137. https://doi.org/10.3390/POLYM9040137

Schotman MJG, Dankers PYW (2022) Factors influencing retention of injected biomaterials to treat myocardial infarction. Adv Mater Interfaces 9(7):2100942. https://doi.org/10.1002/ADMI.202100942

Sharifi S, Mirzadeh H, Imani M, Rong Z, Jamshidi A, Shokrgozar M et al (2009) Injectable in situ forming drug delivery system based on poly(ε-caprolactone fumarate) for tamoxifen citrate delivery: Gelation characteristics, in vitro drug release and anti-cancer evaluation. Acta Biomater 5:1966–1978. https://doi.org/10.1016/j.actbio.2009.02.004

Sharma S, Parmar A, Kori S, Sandhir R (2016) PLGA-based nanoparticles: a new paradigm in biomedical applications. Trends Anal Chem 80:30–40. https://doi.org/10.1016/j.trac.2015.06.014

Shen J, Lin M, Ding M, Yu N, Yang C, Kong D et al (2022) Tumor immunosuppressive microenvironment modulating hydrogels for second near-infrared photothermal-immunotherapy of cancer. Mater Today Bio 16:100416. https://doi.org/10.1016/j.mtbio.2022.100416

Shenoy R, Tibbitt MW, Anseth KS, Bowman CN (2013) Formation of core-shell particles by interfacial radical polymerization initiated by a glucose oxidase-mediated redox system. Chem Mater 25:761–767. https://doi.org/10.1021/CM303913F/ASSET/IMAGES/LARGE/CM-2012-03913F_0007.JPEG

Singh J, Nayak P (2023) pH-responsive polymers for drug delivery: trends and opportunities. J Polym Sci A 61:2828–2850. https://doi.org/10.1002/POL.20230403

Singh NK, Nguyen QV, Kim BS, Lee DS (2015) Nanostructure controlled sustained delivery of human growth hormone using injectable, biodegradable, pH/temperature responsive nanobiohybrid hydrogel. Nanoscale 7:3043–3054. https://doi.org/10.1039/C4NR05897F

Sinha VR, Kumria R (2001) Polysaccharides in colon-specific drug delivery. Int J Pharm 224:19–38. https://doi.org/10.1016/S0378-5173(01)00720-7

Skardal A, Murphy SV, Crowell K, Mack D, Atala A, Soker S (2017) A tunable hydrogel system for long-term release of cell-secreted cytokines and bioprinted in situ wound cell delivery. J Biomed Mater Res B Appl Biomater 105:1986–2000. https://doi.org/10.1002/JBM.B.33736

Skopinska-Wisniewska J, De la Flor S, Kozlowska J (2021) From supramolecular hydrogels to multifunctional carriers for biologically active substances. Int J Mol Sci 22(14):7402. https://doi.org/10.3390/ijms22147402

Springsteen G, Wang B (2002) A detailed examination of boronic acid-diol complexation. Tetrahedron 58:5291–5300. https://doi.org/10.1016/S0040-4020(02)00489-1

Stabenfeldt SE, Gourley M, Krishnan L, Hoying JB, Barker TH (2012) Engineering fibrin polymers through engagement of alternative polymerization mechanisms. Biomaterials 33:535–544. https://doi.org/10.1016/j.biomaterials.2011.09.079

Stanca M, Ławí Nska K, Wnek GE, Bercea M, Lupu A (2024) Recent insights into glucose-responsive Concanavalin A-based smart hydrogels for controlled insulin delivery. Gels 10:260. https://doi.org/10.3390/GELS10040260

Tao F, Cheng Y, Shi X, Zheng H, Du Y, Xiang W et al (2020) Applications of chitin and chitosan nanofibers in bone regenerative engineering. Carbohydr Polym 230:115658. https://doi.org/10.1016/J.CARBPOL.2019.115658

Thambi T, Phan VHG, Lee DS (2016) Stimuli-sensitive injectable hydrogels based on polysaccharides and their biomedical applications. Macromol Rapid Commun 37:1881–1896. https://doi.org/10.1002/MARC.201600371

Thang NH, Chien TB, Cuong DX (2023) Polymer-based hydrogels applied in drug delivery: an overview. Gels 9(7):523. https://doi.org/10.3390/gels9070523

Traitel T, Cohen Y, Kost J (2000) Characterization of glucose-sensitive insulin release systems in simulated in vivo conditions. Biomaterials 21:1679–1687. https://doi.org/10.1016/S0142-9612(00)00050-8

Truong VX, Zhou K, Simon GP, Forsythe JS (2015) Nitrile oxide-norbornene cycloaddition as a bioorthogonal crosslinking reaction for the preparation of hydrogels. Macromol Rapid Commun 36:1729–1734. https://doi.org/10.1002/MARC.201500314

Turk B (2006) Targeting proteases: successes, failures and future prospects. Nat Rev Drug Discov 5(9):785–799. https://doi.org/10.1038/nrd2092

Van Vlierberghe S, Dubruel P, Schacht E (2011) Biopolymer-based hydrogels as scaffolds for tissue engineering applications: a review. Biomacromolecules 12:1387–1408. https://doi.org/10.1021/bm200083n

Vandamme TF, Lenourry A, Charrueau C, Chaumeil JC (2002) The use of polysaccharides to target drugs to the colon. Carbohydr Polym 48:219–231. https://doi.org/10.1016/S0144-8617(01)00263-6

Vashist A, Kaushik A, Alexis K, Dev Jayant R, Sagar V, Vashist A et al (2017) Bioresponsive injectable hydrogels for on-demand drug release and tissue engineering. Curr Pharm Des 23(24):3595–3602. https://doi.org/10.2174/1381612823666170516144914

Veiseh O, Tang BC, Whitehead KA, Anderson DG, Langer R (2014) Managing diabetes with nanomedicine: challenges and opportunities. Nat Rev Drug Discov 14(1):45–57. https://doi.org/10.1038/nrd4477

Vieira VMP, Hay LL, Smith DK (2017) Multi-component hybrid hydrogels—understanding the extent of orthogonal assembly and its impact on controlled release. Chem Sci 8:6981–6990. https://doi.org/10.1039/c7sc03301j

Wan C, Sun Y, Hu Y, Huang J, Lu L, Gao Y et al (2021) Peptide hydrogels loaded with irradiated tumor cell secretions enhance cancer immunotherapy. Nano Today 41:101323. https://doi.org/10.1016/j.nantod.2021.101323

Wang F, Garza LA, Kang S, Varani J, Orringer JS, Fisher GJ et al (2007) In vivo stimulation of de novo collagen production caused by cross-linked hyaluronic acid dermal filler injections in photodamaged human skin. Arch Dermatol 143(2):155–163. https://doi.org/10.1001/archderm.143.2.155

Wang Q, Wang Q, Teng W (2016) Injectable, degradable, electroactive nanocomposite hydrogels containing conductive polymer nanoparticles for biomedical applications. Int J Nanomedicine 11:131–144. https://doi.org/10.2147/IJN.S94777

Wichterle O, Lím D (1960) Hydrophilic gels for biological use. Nature 185(4706):117–118. https://doi.org/10.1038/185117a0

Wong DY, Ranganath T, Kasko AM (2015) Low-dose, long-wave UV light does not affect gene expression of human mesenchymal stem cells. PLoS One 10:e0139307. https://doi.org/10.1371/journal.pone.0139307

Wu X, He C, Wu Y, Chen X (2016) Synergistic therapeutic effects of Schiff's base cross-linked injectable hydrogels for local co-delivery of metformin and 5-fluorouracil in a mouse colon carcinoma model. Biomaterials 75:148–162. https://doi.org/10.1016/J.BIOMATERIALS.2015.10.016

Xi W, Scott TF, Kloxin CJ, Bowman CN (2014) Click chemistry in materials science. Adv Funct Mater 24:2572–2590. https://doi.org/10.1002/ADFM.201302847

Xin H, Naficy S (2022) Drug delivery based on stimuli-responsive injectable hydrogels for breast cancer therapy: a review. Gels 8:45. https://doi.org/10.3390/GELS8010045

Xu X, Huang Z, Huang Z, Zhang X, He S, Sun X et al (2017) Injectable, NIR/pH-responsive nanocomposite hydrogel as long-acting implant for chemophotothermal synergistic cancer therapy. ACS Appl Mater Interfaces 9:20361–20375. https://doi.org/10.1021/acsami.7b02307

Yang Y, Lu Y, Zeng K, Heinze T, Groth T, Zhang K (2021) Recent progress on cellulose-based ionic compounds for biomaterials. Adv Mater 33(28):e2000717. https://doi.org/10.1002/adma.202000717

Yang Y, Cui Y, Cao W, Zhao M, Lin W, Xu R et al (2022) Nanohydroxyapatite stimulates PD-L1 expression to boost melanoma combination immunotherapy. ACS Nano 16:18921–18935. https://doi.org/10.1021/acsnano.2c07818

Yesilyurt V, Webber MJ, Appel EA, Godwin C, Langer R, Anderson DG (2016) Injectable self-healing glucose-responsive hydrogels with pH-regulated mechanical properties. Adv Mater 28:86–91. https://doi.org/10.1002/adma.201502902

Yu F, Cao X, Li Y, Zeng L, Yuan B, Chen X (2013) An injectable hyaluronic acid/PEG hydrogel for cartilage tissue engineering formed by integrating enzymatic crosslinking and Diels–Alder "click chemistry". Polym Chem 5:1082–1090. https://doi.org/10.1039/C3PY00869J

Zarrintaj P, Ramsey JD, Samadi A, Atoufi Z, Yazdi MK, Ganjali MR et al (2020) Poloxamer: a versatile tri-block copolymer for biomedical applications. Acta Biomater 110:37–67. https://doi.org/10.1016/j.actbio.2020.04.028

Zhang H, Zhu X, Ji Y, Jiao X, Chen Q, Hou L et al (2015) Near-infrared-triggered in situ hybrid hydrogel system for synergistic cancer therapy. J Mater Chem B 3:6310–6326. https://doi.org/10.1039/C5TB00904A

Zhao L, Wang L, Zhang Y, Xiao S, Bi F, Zhao J et al (2017) Glucose oxidase-based glucose-sensitive drug delivery for diabetes treatment. Polymers 9:255. https://doi.org/10.3390/POLYM9070255

Zhu D, Li Z, Huang K, Caranasos TG, Rossi JS, Cheng K (2021) Minimally invasive delivery of therapeutic agents by hydrogel injection into the pericardial cavity for cardiac repair. Nat Commun 12:1412. https://doi.org/10.1038/s41467-021-21682-7

Advances in Iontophoresis-Based Drug Delivery

3

Popat S. Kumbhar, Kaustubh Ajit Kolekar, Preeti Karade,
Sanjay B. Lohar, Sukriti Vishwas, Kalpana S. Patil,
Sachin Kumar Singh, Kamal Dua, John Disouza,
and Vandana B. Patravale

Abstract

The localized dermal and transdermal delivery of therapeutics for managing diverse skin diseases (acne, psoriasis, skin cancer), ocular diseases, etc., is challenging. The anatomical features and presence of barriers are key hurdles in efficiently delivering drugs. Iontophoresis has emerged as a promising avenue for enhancing the permeability of drugs by overcoming the challenges mentioned above. It is an innovative drug delivery technique that utilizes low-level electrical currents to enhance the dermal or transdermal transport of charged therapeutic agents. The significant versatility in delivering drug doses in a controlled manner is the key benefit of the iontophoresis technique. This chapter covers recent advances in iontophoretic approaches used for drug delivery, focusing on iontophoresis transport mechanisms and routes of administration. Key factors influencing iontophoresis efficacy, such as current density, electrode design, etc., are

P. S. Kumbhar
Department of Pharmaceutics, Tatyasaheb Kore College of Pharmacy, Warananagar, Maharashtra, India

K. A. Kolekar · S. Vishwas
School of Pharmaceutical Sciences, Lovely Professional University, Phagwara, Punjab, India

P. Karade
Appasaheb Birnale College of Pharmacy, Sangli, Maharashtra, India

S. B. Lohar · K. S. Patil
Department of Pharmacognosy, Tatyasaheb Kore College of Pharmacy, Warananagar, Maharashtra, India

S. Kumar Singh
School of Pharmaceutical Sciences, Lovely Professional University, Jalandhar, Punjab, India

Discipline of Pharmacy, Graduate School of Health, University of Technology Sydney, Sydney, NSW, Australia

M. R. Babu et al. (eds.), *Transdermal Applications of Minimally Invasive Drug Delivery Systems*, https://doi.org/10.1007/978-981-96-6275-3_3

discussed. Significant advancements in iontophoresis including the integration of iontophoresis with nanocarriers and microneedles, which improve drug permeation and provide targeted drug delivery, are also explained. In addition, the future perspectives section provides insight into further research required into combination therapies and real-time monitoring systems to enhance the clinical effectiveness of the iontophoretic drug delivery approach. Overall, iontophoresis represents a promising approach that could revolutionize noninvasive drug administration methods in modern drug delivery systems.

Keywords

Iontophoresis · Noninvasive · Charged therapeutic agents · Low-level electrical currents · Electrode design · Transdermal transport

3.1 Introduction

Iontophoresis is a noninvasive technique wherein a small electric current is utilized to increase the transdermal delivery of cargo. This system is comprised of two electrodes and a power supply. The two electrodes present are the drug delivery electrode and the returning electrode. The drug delivery electrode is situated within or near the drug formulation on the surface of the desired tissue. To cut off the electrical circuit and enable current to pass, the returning electrode is positioned on nearby tissue (Kalia et al. 2004). It is now a good alternative to traditional drug delivery methods including oral and intravenous routes. Iontophoresis works on the basis of applying a tiny electric current to the skin, thereby making it easier for charged drug molecules to pass through the skin. The application of a small electric current across the skin causes the permeation of therapeutics through the epidermis into the blood, avoiding hepatic first-pass metabolism (Lakshmanan et al. 2014). Thus, this technique confers various benefits including noninvasiveness, systemic as well as localized cargo delivery, the capability of delivering a range of therapeutics, controlled cargo release, improved permeation, prevention of hepatic and gastrointestinal metabolism of drugs, and increased patient compliance. However, the chief

K. Dua
Discipline of Pharmacy, Graduate School of Health, University of Technology Sydney, Sydney, NSW, Australia

Faculty of Health, Australian Research Centre in Complementary and Integrative Medicine, University of Technology Sydney, Ultimo, NSW, Australia

J. Disouza (✉)
Department of Pharmaceutics, SYBES's Bombay Institute of Pharmacy and Research, Dombivli (East), Maharashtra, India

V. B. Patravale (✉)
Department of Pharmaceutical Sciences and Technology, Institute of Chemical Technology, Matunga, Mumbai, Maharashtra, India
e-mail: vb.patravale@ictmumbai.edu.in

drawbacks of this approach include suitability for delivering drug molecules that undergo ionization, skin irritation on prolonged use due to the electric current, and lack of uniform drug distribution at the affected site (Guy et al. 2000).

The iontophoretic drug transport mainly occurs via electroosmosis, electropermeabilization, and direct-field effect mechanisms. The iontophoretic drug delivery is affected by a range of parameters involving drug physicochemical properties, formulation attributes, experimental factors, and device parameters. The polarity and ionization constant of the drug determines the kind of iontophoresis, which can be either cathodal for anionic drugs or anodal for cationic drugs. The primary device variables that significantly determine the extent of drug penetration across biological barriers are the length of time the current is delivered, the amount of current utilized, and the electrode size (Murdan 2008).

This system has shown significant promise in treating various diseases owing to its capability of controlling cargo release and delivering therapeutics in a localized manner. Iontophoresis system is utilized for the management of pain, diabetes, skin disorders (acne, psoriasis, atopic dermatitis), neurodegenerative diseases such Alzheimer's and Parkinson's, and ocular disorders involving glaucoma and age-related macular degeneration. In addition, it demonstrated significant applications in cosmetics, hormonal therapy, and delivery of large molecules such as vaccines and proteins (Kigasawa et al. 2010).

The latest advances in both nanotechnology and knowledge of skin physiology have contributed to advancements in iontophoresis-based cargo delivery. The use of different nanocarriers and microneedles in iontophoresis delivery has expanded its potential for delivering an array of therapeutics such as biologics, peptides, and vaccines in managing diseases via different routes of administration (Dragicevic and Maibach 2018). Thus, iontophoresis and advanced iontophoresis could be a promising way to deliver therapeutics via improving their permeability through different barriers. This chapter covers transport mechanisms and routes of administration of iontophoresis and factors affecting iontophoresis-based drug delivery. Further, applications of iontophoresis and advances in iontophoresis drug delivery are discussed.

3.2 Transport Mechanisms of Iontophoresis

Iontophoresis is a noninvasive strategy wherein the drug molecules are permeated through anatomic barriers via a small electric current. Commonly two sorts of voltage supply such as direct and alternating current are employed in iontophoresis (Karpiński 2018). The applied electric field causes the permeation of drug molecules through media, cell membranes, or tissues, respectively. The permeation rate of these therapeutic molecules is 10–2000 times higher than conventional methods. This iontophoresis-based drug delivery was effective in ophthalmology, dentistry, dermatology, and otorhinolaryngology (Karpiński 2018). The key working principle of iontophoresis involves the attraction of oppositely charged ions and the repulsion of ions with the same charge (Eljarrat-Binstock and Domb 2006). The ocular permeation of therapeutics across biological membranes can be increased using iontophoresis through electroosmosis, electropermeabilization, and direct-field effect (electrophoresis).

One of the key mechanisms for the improved transport of large monovalent ions throughout iontophoresis is electroosmosis (electroosmotic flow). The bulk fluid flow that happens when a voltage difference is applied over a charged membrane is recognized as electroosmosis. It is the main mechanism for the transport of neutral molecules (Eljarrat-Binstock and Domb 2006). Both ionic and neutral medications can be delivered more effectively by the solvent's motion. In the electropermeabilization mechanism, the permeability of tissue both during and after iontophoresis is enhanced by modulating the tissue barrier through the application of an electric field. The application of an electric field can result in a change in membrane porosity and membrane transport pathways. The direct-field effect (Nernst-Planck effect) relies on the movement of ions due to an applied electrical potential gradient. This effect is chiefly responsible for the increase in the flux of small ions. Based on the charge, the ionized substances are drawn to the anode or the cathode by direct-field effect (Li and Hao 2018). The transport mechanisms of iontophoresis are depicted in Fig. 3.1.

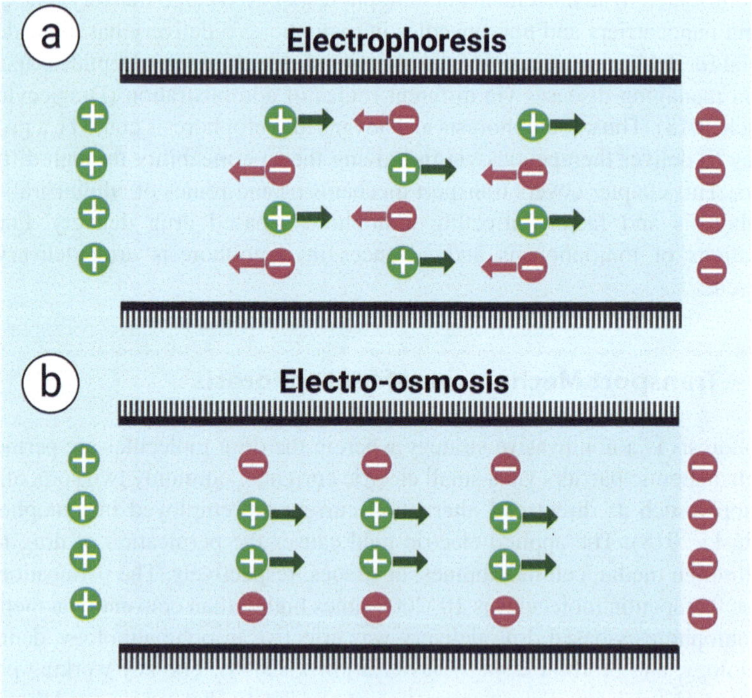

Fig. 3.1 The transport mechanisms of iontophoresis

3.3 Routes of Iontophoresis Administration

A different route of administration of iontophoresis varies with its clinical applications. Routes like transdermal, ocular, intranasal, and intrathecal have been reported for iontophoresis administration. Transdermal is a commonly employed route for iontophoresis administration wherein the drug is kept on the skin beneath an electrode that generates current. Drugs like corticosteroids and local anesthetics for pain treatment are administered transdermally using the iontophoresis technique (Perez et al. 2020; Khan et al. 2011).

In ocular or transocular iontophoresis, therapeutics are delivered locally to the eye. The transocular route for iontophoresis administration is divided into transcorneal and transscleral. This approach is used for the delivery of therapeutics in managing posterior and anterior eye disorders. Antibiotics and steroid drugs can be given through ocular iontophoresis. Nasal iontophoresis is used to increase drug permeation across nasal mucosa via avoiding blood-brain barriers. Iontophoresis delivery through the nasal route is capable of delivering drugs locally as well as systemically. The intrathecal route of iontophoresis administration is employed for delivering drugs to the cerebrospinal fluid through the spinal cord or to targeted areas of the central nervous system. Local anesthetics and therapeutics for managing neurodegenerative diseases (Alzheimer's and Parkinson's) can be delivered using iontophoresis via the intrathecal route. Thus, iontophoresis is a flexible technique with the capability to deliver a number of therapeutics via multiple routes of administration. It can effectively administer drugs through the skin or mucous membranes, thereby rendering it helpful for targeted drug delivery, controlling pain, inflammatory management, and diagnostic purposes (Perez et al. 2020; Khan et al. 2011).

3.4 Factors Affecting Iontophoresis

According to the Faraday's law, the iontophoretic flux (J_{drug}) of a drug molecule is the sum of the fluxes due to electroosmosis (J_{EO}), electromigration (J_{EM}), and passive delivery (J_{p}) (Gratieri et al. 2013; Kalaria et al. 2018):

$$J_{\mathrm{drug}} = J_{\mathrm{EM}} + J_{\mathrm{EO}} + J_{\mathrm{p}} = \left(id^{*}\, T_{\mathrm{drug}}\right) / \left(Z_{\mathrm{drug}}{}^{*}\, F\right) + V_{\mathrm{w}}{}^{*} C_{\mathrm{drug}} + K_{\mathrm{drug}}{}^{*} C_{\mathrm{drug}}$$

where

id is the applied current density

T_{drug} and Z_{drug} are the transport number and charge of the drug, respectively

F is Faraday's constant

V_{w} is volume of the solvent flow

C_{drug} is concentration of the drug

K_{drug} is permeability coefficient of the drug.

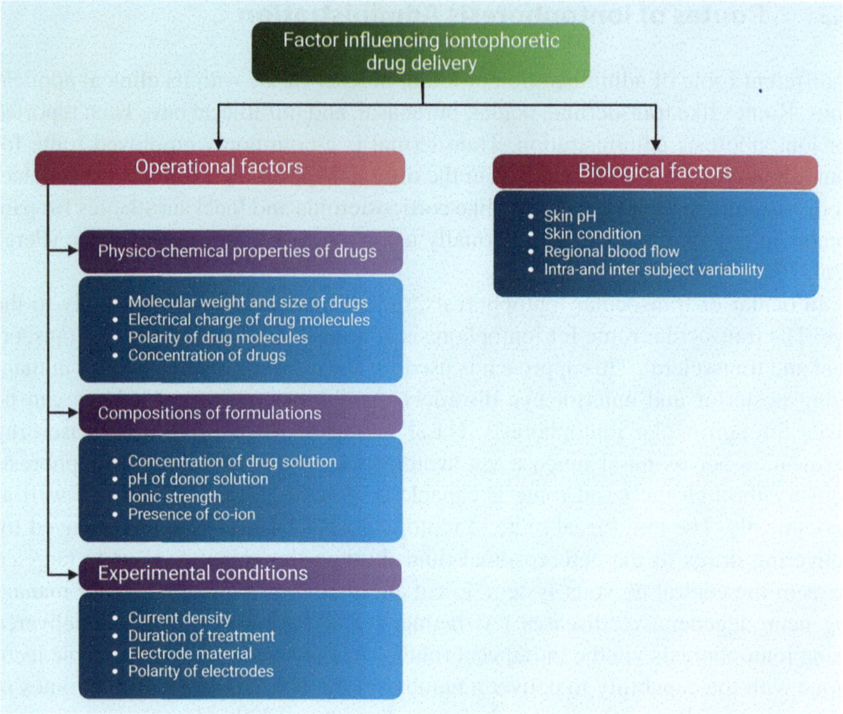

Fig. 3.2 Various factors influencing iontophoresis drug delivery

Various factors influencing iontophoretic drug delivery can be categorized into four classes, including physicochemical properties of drugs, composition of formulation, experimental factors, and biological factors (Fig. 3.2).

3.4.1 Physicochemical Properties of Drugs

The molecular size and molecular weight of the drug is a major element concerning the efficiency of iontophoretic drug delivery applications. The small-sized and more hydrophilic ions are transported at a faster rate as compared to the larger ions (Miller et al. 1987; Miller and Smith 1989). A number of research correlating flux as a characteristic of molecular weight were conducted. These studies concluded that for electro-repulsive iontophoresis when all other conditions were kept constant, the transport of compounds decreased with an increase in molecular weight. However, modern techniques like iontophoresis, electroporation, and phonophoresis help transport large molecule like peptides (Khan et al. 2011).

The electrical charge of the drug is one of the most important factors affecting the iontophoretic process. The nature of the electrical charge on the drug molecules plays an important role in the determination of mechanisms (e.g., electroosmosis and electro-repulsion) of drug transport (Khan et al. 2011).

Electro-repulsion is based on the principle that different charges attract each other and equal charges repel each other and show the movement of charged substances through the skin, under the impact of an electric field (Noh et al. 2018). Electroosmosis is relevant in the permeation of substances with higher molecular weight. In this mechanism, solvent flow is induced by an electric field, which subsequently causes the movement of neutral or charged molecules through the skin (Pikal 2001). It has been reported that the transport of the cations occurs in a comparatively better manner as compared to the anions (Kalia et al. 2004). Furthermore, an amplified positive electrical charge on the drug molecule reduces the permeation rate of the peptide drugs. It may be due to the formation of strong binding of the drug with the membrane (Priya et al. 2006). The influence of polarity of several drug molecules has already been studied. Mostly hydrophilic drugs show the optimal permeation flux (nalbuphine and its ester derivatives) and are considered ideal candidates (Li and Hao 2018; Khan et al. 2011).

3.4.2 Compositions of Formulations

The compositions of formulation were observed to be significant factors influencing iontophoresis. Increasing drug concentration results in greater drug delivery but only to some extent as the effect of iontophoresis depends on the tissue on which (e.g., thickness, permeability, the presence of pores) the electrodes are applied. Research has shown that diclofenac concentration in plasma does not increase linearly and iontophoresis failed to achieve more effective local concentrations of diclofenac. However, researchers presented the effect of geraniol and l-menthol on enhancing the delivery of diclofenac and decreasing the percentage of inflammation (Arunkumar et al. 2018). Another research team studied electrical, physicochemical, and device-related factors acting on the permeation kinetics of in vitro iontophoresis and found that iontophoresis increased the transdermal permeation flux of ketorolac as compared to the diffusion (Tiwari and Udupa 2003). Researchers also conducted an iontophoresis study for the delivery of 5-aminolevulinic acid as a function of pH and investigated the mechanisms accountable for its electro-transport. Their study concluded that significant 5-aminolevulinic acid delivery at pH 7.4 can be achieved by increasing the drug concentration in the anodal formulation. The pH is an important factor governing the iontophoretic delivery of drugs. The pH of the formulation should be optimized to ensure maximum ionization of the compound. Moreover, it also influences the skin permeability as well as the chemical stability of the drug involved. With the decrease in pH, the concentration of hydrogen ions increases, resulting in vasodilation. Therefore, when working with vasodilators, it is important to keep the pH as close to 7. The research team investigated various factors influencing the skin permeability of glycyrrhizin. These included the donor pH levels ranging from 4 to 7, drug concentration, current intensity, external chloride ion concentration, and the use of permeation enhancers such as urea and Tween 80.

The steady-state skin permeation flux of glycyrrhizin did not differ between pH 4 and 5 but increased significantly at pH 6 ($p < 0.05$) and nearly significantly at pH 7. In addition, researchers increased the positive charges of pramipexole dihydrochloride by decreasing the pH of the formulation and studied its effect on iontophoretic transport across the skin. Their study revealed that skin permselectivity is an important component for the iontophoretic transport of drugs. The results suggest that the decrease in pramipexole dihydrochloride delivery at higher valence at pH 3 was mainly due to pH-dependent selectivity of pramipexole ion permeation across the skin and not electroosmosis.

Extensive research has been carried out to analyze the effect of pH on the penetration of various drug molecules, such as enalaprilat, lidocaine, rotigotine, insulin, and acetate ions (Khan et al. 2011; Miller and Smith 1989; Siddiqui et al. 1987; Nayak et al. 2019). Some studies indicated that with an increase in the ionic strength of the system, the permeation rate of the drug decreases as extraneous ions compete with the drug ions (Thysman et al. 1992). One research study investigated the effects of ionic strength upon transungual passive and iontophoretic transport. Results demonstrated significant ionic strength effects upon the partitioning and transport of a cationic ion in transungual transport, which may be instrumental in the development of transungual delivery systems. A study was carried out for the effect of ionic strength on many peptides, and drugs like ketorolac showed an increased flux occurring at low electrolyte concentrations (Khan et al. 2011). Co-ions are an ion that contains an equal electrical charge of different kinds. These co-ions are more mobile and smaller in size than the drug ions (Khan et al. 2011). The buffering agents used to maintain the pH of the solution are the principal source of co-ions.

3.4.3 Experimental Factors

Current density is the amount of current transported per unit of electrode surface area. Higher applied current facilitates the drug permeation/transportation rate (Khan et al. 2011). However, it should not produce any kinds of adverse effects (like skin irritations and burns) on the skin. A study showed a linear relationship between the iontophoretic flux of a number of compounds (verapamil, GRH, diclofenac, and ketorolac) and applied current. In general, 0.5 mA/cm^2 is considered the maximum iontophoretic current that should be used on human beings. The applied current can be easily controlled by the electronics. A linear relationship between the strength of the applied current and the achieved permeation flux of drug molecules has been reported. In the in vitro studies carried out using animals, the current profile is kept constant and a very low voltage of about 10 V. Their study indicated that different current profiles significantly regulate iontophoretic transport and constant direct current is most efficient in terms of total transmembrane drug delivery. However, few studies indicated that the use of direct current can result in skin polarization and irritation. Moreover, the amount of the drug permeated linearly increases with the increase in the period of injection of the current (Nayak et al. 2019). Electrode materials should be nontoxic. The most frequently used electrodes are zinc/zinc chloride (Zn/ZnCl2) and Ag/AgCl electrodes. Ag/AgCl electrodes are considered the most ideal electrode for the iontophoretic delivery of drugs as Ag/AgCl electrodes can resist the alterations in the pH. Other electrode materials include platinum and aluminum (Khan et al. 2011).

3.4.4 Biological Factors

The skin condition is a crucial biological factor that plays an important role in determining the permeation properties of the drug molecules during iontophoresis (Khan et al. 2011). In a study carried out by researchers, the skin of different areas of the human body was selected to study passive diffusion of methyl salicylate. Their study indicated passive diffusion of methyl salicylate took place maximally from the abdomen for all subjects (Dixit et al. 2007). In another study carried out by the research team, it was observed that maximum passive diffusion of hydrocortisone occurred from areas with numerous hair follicles. During iontophoresis delivery, systemic and underlying tissue drug absorption is influenced by dermal blood supply. However, during iontophoresis, blood supply does not affect the drug penetration fluxes through the epidermis. Experiments conducted in vivo *supported the finding that the* permeation flux of the drugs *was* not greatly influenced in males and females, as well as between hairy and hairless skin.

It has been suggested by the researchers that the barrier properties of stratum corneum can be overcome by oleic acid and propylene glycol. Moreover, the study demonstrated enhanced drug delivery using co-application of oleic acid with iontophoresis (Jiang et al. 2000). In vivo experiments demonstrated that a minor difference in the flux rate was observed following transdermal iontophoresis between males and females. These experiments indicate that local anesthetics, vasoconstrictors, some corticosteroids, several anticancer drugs, and several antiviral agents are suitable for iontophoresis (Gangarosa et al. 1978).

3.5 Applications of Iontophoresis

Iontophoresis is a noninvasive and advanced drug delivery technique. This technique required a very low level of electric current to deliver the ionic drug from the biological membrane. Therefore, this technique improves the penetration of the cargo for the biological barrier such as transdermal, ocular, buccal, and transbuccal, leading to an increase in the bioavailability of the drug in the site of action. Various drugs delivered using iontophoresis via transdermal, ocular, buccal, and transbuccal routes in treating different diseases are summarized in Table 3.1.

Table 3.1 Drugs delivered using iontophoresis via transdermal, ocular, buccal, and transbuccal routes in treating different diseases and their outcomes

Name of drug	Type of disease targeted	Outcomes	References
Transdermal-based iontophoresis drug delivery			
Dexamethasone sodium phosphate	Psoriasis	• Controlled cargo release • Significantly ($p < 0.005$) improved permeability of dexamethasone sodium phosphate	Zhou et al. (2023)
Terazosin hydrochloride	Hypertension	• Substantial improvement in percutaneous penetration of terazosin hydrochloride	Jiang et al. (2021)
Salicylic acid	Onychomycosis	• Improved trans-nail delivery flux	Murthy et al. (2007)
Ketoprofen	Cutaneous vein ipsilateral	• Substantially enhanced local absorption of ketoprofen	Tashiro et al. (2000)
Ocular-based iontophoresis drug delivery			
Dexamethasone phosphate	Noninfectious anterior uveitis	• Significant ($p < 0.005$) improvement in ocular permeation of dexamethasone phosphate	Perez et al. (2020)
Latanoprost	Glaucoma	• Controlled latanoprost release for 7 days • Substantial enhancement in ocular permeability of latanoprost	Kim et al. (2022a)
Bevacizumab	Conjunctivitis	• 14-fold improvement in the ocular permeation than conventional delivery of bevacizumab	Molokhia et al. (2020)
Buccal and transbuccal-based iontophoresis drug delivery			
Sumatriptan succinate	Acute migraine	• Significantly ($p < 0.005$) improved permeation of sumatriptan succinate through pig esophageal epithelium	Telò et al. (2016)
Buflomedil hydrochloride and dexamethasone phosphate	Oral submucous fibrosis	• Controlled release of both drugs • Substantially improved permeability of both drugs from porcine esophageal mucosa	Tyagi et al. (2021)
Salmon calcitonin	Hypercalcemia	• Significant ($p < 0.005$) improvement in permeability of salmon calcitonin through buccal cavity	Oh et al. (2011)
Prilocaine hydrochloride and lidocaine hydrochloride	Buccal anesthesia	• 12-fold enhanced permeation of prilocaine hydrochloride and lidocaine hydrochloride through the mucosa	Cubayachi et al. (2015)

(continued)

Table 3.1 (continued)

Name of drug	Type of disease targeted	Outcomes	References
Transungual iontophoresis			
Terbinafine	Onychomycosis	• 21- and 37-fold increased permeation through finger and toenails, respectively	Nair et al. (2011)
Efinaconazole	Onychomycosis	• Significantly ($p < 0.0001$) enhanced permeation through nail	Nair et al. (2023)
Transpapillary iontophoresis			
Lapatinib and 5-fluorouracil	Breast cancer	• Significantly ($p < 0.005$) transpapillary permeability of drugs	Fernandes et al. (2024)
Resveratrol	Breast cancer	• Enhanced transpapillary permeability of drug through porcine mammary papilla using ion iontophoresis than passive diffusion • Increased C_{max} and AUC of resveratrol administered transpapillary using iontophoresis when compared to its oral administration	Gadag et al. (2022)

3.5.1 Transdermal Iontophoresis

In transdermal iontophoresis, a low electric current is applied to the skin to drive charged drug molecules, improving their penetration through the stratum corneum. This approach is prominently effective for delivering hydrophilic and macromolecules, like protein and peptides, which otherwise have low permeability through the skin layers. This technique is commonly applied in treating pain and skin diseases and in diagnostics. However, the transdermal iontophoresis may cause skin irritation due to the electric current generated in this technique upon prolonged use.

Transdermal iontophoresis has shown a promising role in the control of pain and inflammation, orthopedic diseases, and dermatological disorders (acne, psoriasis, eczema, etc.). Additionally, it is found to be potential in cosmetic applications like vitamin supplements and antiaging and in managing hypocalcaemia. The researchers have delivered terpene with luteinizing hormone-releasing hormone using iontophoresis. They reported that the significantly increasing therapeutic agent's permeability in porcine skin (Bhatia and Singh 1998) was due to the electric current generated in iontophoresis. Moreover, the research team investigated the potential of iontophoresis for pain management with nonsteroidal anti-inflammatory drugs. They observed significantly ($p < 0.05$) improved permeation of the drug through the skin and therefore enhanced pain relief compared to traditional therapy (Zaciewski et al. 2007). Further, in diagnostics, iontophoresis is also employed, notably with pilocarpine to induce sweating for cystic fibrosis testing. Here, sweat is collected and analyzed, allowing for reliable diagnosis (Dhote et al. 2012). Thus, transdermal iontophoresis represents a novel, effective, and safe approach for drug delivery and diagnostic applications.

3.5.2 Ocular Iontophoresis

This is a targeted drug delivery method designed to administer therapeutic agents directly to the eye, enhancing treatment efficacy while reducing systemic side effects. This approach is especially valuable for treating ocular disease. These diseases include glaucoma, retinal disorders, ocular infections, and many more. It has also shown potential applications in the delivery of gene and topical anesthetics. The application of ocular iontophoresis enables charged drug molecules to penetrate the ocular barrier, achieving localized drug delivery with precision (Dhote et al. 2012). However, the key hurdles associated with ocular iontophoresis are eye or surrounding tissue irritation following long-time exposure, suitable for delivery of only charged drug molecules, and the necessity of a particular device for iontophoresis.

The researchers evaluated the effectiveness of ocular iontophoresis for the delivery of dexamethasone with the objective of enhancing the intraocular concentration of dexamethasone. The iontophoretic delivery of dexamethasone resulted in a 50- to 100-fold increase in intraocular drug delivery in the rabbit models, compared to the traditional drug delivery technique. This marked improvement in drug bioavailability underscores the potential of iontophoresis to optimize ocular drug delivery by overcoming biological barriers and achieving therapeutic drug levels at the specific target site (Perez et al. 2020). Another research group investigated the iontophoresis potential for delivery of negatively charged nanoparticles to suprachoroidal space in treating posterior eye disorders. They noticed a significant accumulation of the nanoparticles at the posterior segment of the eye, signifying the capability of iontophoresis in localized delivery of drugs to the eye regions.

3.5.3 Buccal and Transbuccal Iontophoresis

The buccal route offers a convenient method for systemic drug delivery, effectively bypassing first-pass metabolism. Iontophoresis through buccal mucosa enhances drug permeation and is beneficial for hormone therapies and local anesthesia. It is also found to be effective in controlling pain and in the delivery of large molecules (vaccines and biologics). This approach is allied with several downsides like irritation at the site of application due to the long-time use, suitable for delivery of only ionizable drug molecules, and less convenience with the use of electric devices, impacting patient compliance.

The research team evaluated the combined effects of iontophoresis and chemical enhancer on the buccal delivery of lidocaine and nicotine. They reported that the iontophoresis, when combined with chemical enhancers, significantly improved drug permeation by eight times for lidocaine and 450 times for nicotine hydrogen tartrate compared to passive diffusion (Wei et al. 2012). Therefore, buccal iontophoresis holds significant potential for enhancing therapeutic outcomes and improving the bioavailability of drugs for systemic delivery. Similarly, other researchers explored the delivery of 0.5% ondansetron through the transbuccal route. They observed the significantly improved transbuccal delivery of drug by 3.3-, 5.2-, and 7.1-fold with 0.1 mA, 0.2 mA, and 0.3 mA, respectively, compared to the control group. Moreover, iontophoresis techniques are safe and do not cause major morphological changes in the buccal tissue (Hu et al. 2011). Therefore, buccal iontophoresis holds significant potential for enhancing therapeutic outcomes and improving the bioavailability of drugs for systemic delivery.

3.5.4 Transungual Iontophoresis

Transungual iontophoresis involves the penetration of therapeutics into the nail. The nail serves as an important barrier due to its thickness and innate nature. The application of iontophoresis is effective in overcoming this barrier and assists in enhancing the permeation of the drug through the nail. This approach can be promising to treat nail diseases such as nail psoriasis, onychomycosis, etc. The researchers explored the potential of transungual delivery of terbinafine using iontophoresis in the treatment of onychomycotic nails. They noticed 21-fold and 37-fold enhanced permeation of terbinafine through finger and toenails via iontophoresis. In addition, a more than 12-fold increase in terbinafine concentration was observed in the onychomycotic due to iontophoresis (Nair et al. 2011). Similarly, the research team studied the effect of iontophoresis on delivery of efinaconazole by transungual route. The application of iontophoresis at 10.5 V demonstrated enhanced permeation (79 μg/cm^2) of efinaconazole through transungual route (Nair et al. 2023).

3.5.5 Transtympanic Iontophoresis

The delivery of therapeutic to the ear is challenging due to the presence of blood-cochlear barriers. Therefore, transtympanic administration can be effective to overcome the challenges associated with barriers. This technique involves the delivery of therapeutics through the eardrum (tympanic membrane) into the inner ear by the application of tiny currents. This approach can directly deliver therapeutic into the affected area of the ear, thereby minimizing the negative effects. The transtympanic iontophoresis devices generally used possess bare-like anode that contacts the ear canal, resulting in discomfort. The formulations in the form of solutions or gel are delivered through transtympanic iontophoresis. It is employed to treat various ear conditions like inflammation, hearing loss (Meniere's disease), ear infections such as labyrinthitis, and many more. The key limitations of this approach are the threat of injury to the eardrum, skin irritation, and limited effectiveness. One research evaluated the effectiveness of delivering gadopentetate through transtympanic iontophoresis for treating inner and middle ear diseases. The gadopentetate was administered to eight rabbits via iontophoresis at a dose of 469 mg/mL. The application of iontophoresis caused a significant accumulation of gadopentetate in the middle and inner parts of the ear (Christodoulou et al. 2003).

3.5.6 Transpapillary Iontophoresis

The delivery of therapeutics to the breast tissues is required for treating various breast-related diseases. However, cargo delivery to the breast by intravenous route is limited by mammary duct basement membrane. Transpapillary administration of drugs can be effective to overcome the challenges in the path of intravenous administration. Transpapillary iontophoresis is the approach wherein drug molecules are delivered through the papillary layer which is located under the epidermis of the skin using a tiny current. The papillary layer is a crucial region for drug absorption because it is comprised of flexible connective tissue and possesses an ample supply of blood. This technique is frequently employed to administer drugs precisely to specific locations in ocular and dermatological disorders. However, the limitations of this technique include unsuitability for delivery of large-molecule drugs, skin irritation, and suitability for delivery of ionizable drugs only (Fernandes et al. 2024; Gadag et al. 2022). The research team designed transferosomes loaded with lapatinib and 5-fluorouracil and investigated their potential against breast cancer through transpapillary iontophoresis. They noticed a substantial enhancement in the transpapillary permeation of both drugs upon application of iontophoresis (Fernandes et al. 2024).

3.5.7 Trans-Intestinal Iontophoresis

The oral delivery of large-molecule drugs (biologics) is limited by the epithelial cells of the intestine. Additionally, oral delivery of these biologics is associated with poor stability in the gastrointestinal tract. The challenges mentioned above necessitate the delivery of biologics via intravenous routes. However, intravenous delivery of biologics is painful, causing contempt for the dosage schedule and patient noncompliance (Kumbhar et al. 2024). Trans-intestinal iontophoresis can be a promising avenue in the intestinal delivery of therapeutics. This approach enhances the permeation of therapeutics across intestinal membranes using the application of tiny currents. Thus, this strategy could be effective in enhancing the oral bioavailability of poorly absorbed drugs. The researchers studied the delivery of insulin using iontophoresis. They fabricated a mucoadhesive patch laden with insulin. These patches were applied to the intestine with a current of 0.33 mA/cm^2 for 18 m. They noticed a substantial drop in glucose levels due to increased paracellular absorption of insulin by iontophoresis (Banerjee et al. 2019).

3.6 Advances in Iontophoretic Drug Delivery

Nanotechnology is demonstrating a significant role in increasing the performance of iontophoretic systems. Recent advancements in iontophoretic drug delivery have focused on enhancing the efficacy and versatility of this technique through innovative approaches. Two significant developments include the integration of nanocarriers and microneedles with iontophoresis. Combining these nanotechnological approaches like nanoparticles and microneedles with iontophoresis can be promising to remarkably enhance permeation of drug molecules (Moura et al. 2022).

3.6.1 Use of Iontophoresis with Nanocarriers

The combination of iontophoresis with nanocarriers such as polymeric, lipidic, metallic, and exosome-based systems has shown promising results in optimizing drug delivery. The lipid-based nanocarriers such as nanostructured lipid carriers (NLCs) loaded with therapeutic agents exhibit enhanced stability and controlled release during transdermal iontophoresis. The application of a low-level electric current enables the nanocarrier-drug complex to penetrate the skin barrier more effectively. Many studies demonstrated that the iontophoresis can significantly increase therapeutic agent concentrations in systemic circulation through various mechanisms. These combined approaches facilitate quicker and higher drug delivery rates, which is helpful in rapid drug action studies. Furthermore, nanocarriers enhance the drug solubilization and controlled release profile, which leads to improving the bioavailability of cargo. Thus, iontophoresis with nanocarrier loaded with therapeutic agents increases the pharmacokinetic and pharmacodynamic efficacy of the drug. Moreover, this approach represents significant steps toward designing a noninvasive, efficient, and patient-friendly therapeutic solution (Moura et al. 2022).

Given this, the researchers have investigated the potential of delivering latanoprost-loaded nanoparticles using iontophoresis in the treatment of glaucoma. They developed poly(lactic-co-glycolic acid) (PLGA) nanoparticles encapsulated with latanoprost and delivered topically. This nanoparticulate-based iontophoretic delivery demonstrated substantially higher permeation and controlled cargo release. Moreover, this system displayed a 23-fold increase in effectiveness when compared to the marketed eye drops (Kim et al. 2022a). Another research team studied the iontophoretic delivery of dexamethasone encapsulated into PLGA nanoparticles to the anterior segment of the eye. This iontophoretic-based delivery of nanoparticles demonstrated tenfold more permeation from the rabbit eye than plain PLGA nanoparticles. Furthermore, the release of dexamethasone was controlled for 10 days in comparison to the release of the drug from the marketed eye drop (Kim et al. 2022b).

3.6.2 Use of Iontophoresis with Microneedles

Microneedles offer an innovative approach when combined with iontophoresis to improve drug delivery. This technique involves the use of tiny needles that create microchannels in the skin, promoting deeper penetration of therapeutic agents while the electrical currents drive the therapeutic agent further into the tissue. This combination significantly improves drug delivery efficiency, particularly for larger molecules and peptides that typically struggle to cross the biological barrier.

Recent advancements have led to the development of wearable iontophoresis patches equipped with built-in batteries, simplifying the drug delivery process and enhancing patient comfort by removing the need for bulky external power sources. These integrated systems allow for precise control over drug dosages and controlled release profiles, making them particularly suitable for chronic disease conditions that require regular administration of medication such as pain relievers. The combination of microneedles and iontophoresis can not only improve the bioavailability of macromolecules but also reduce action of onset, making it a better option for various applications in transdermal drug delivery (Li et al. 2021a; Kusama et al. 2021; Yang et al. 2020).

The researchers have designed an iontophoresis microneedle array patch for regulated transdermal delivery of insulin in managing diabetes. This patch displayed controlled insulin delivery for a long duration. Moreover, a significant antidiabetic effect was noticed in the diabetes-induced rat following treatment with a microneedle array patch (Yang et al. 2020). Similarly, the research team has explored the delivery of insulin-loaded microneedle array patch using iontophoresis in managing diabetes. A remarkable (99%) sin permeation of insulin was observed from the developed systems. Notably, this system lowered and controlled blood sugar levels for more than 5.4 h in diabetes-induced rats (Li et al. 2021b). Thus, iontophoresis with microneedles represents a promising advancement in drug delivery technology, with the potential to significantly enhance therapeutic outcomes while improving patient experience.

3.7 Conclusion and Future Perspectives

Iontophoresis-based drug delivery is a potential nonintrusive way to enhance the transdermal administration of pharmaceuticals. This approach is capable of overcoming the limitations associated with oral and injectable delivery of therapeutics through the application of electric current. Moreover, this system is capable of improving permeability and controlling cargo release, bypassing the hepatic and gastrointestinal metabolism of drugs, and enhancing patient compliance in the management of multiple diseases. Despite the significant benefits of an iontophoresis-based drug delivery system, the capability to load and transport only ionized forms of drugs, irritation on prolonged use, and drug stability are key limitations. Therefore, more research is crucial to addressing the aforementioned difficulties.

Various factors are influencing iontophoresis-based drug delivery; this system seems to have an optimistic future as long as research and technological advancements persist to support its likelihood for widespread clinical application. Iontophoresis has the potential to be a paradigm shift in the treatment of a variety of diseases with further development and improvement, enabling patients with a more convenient, efficient, and successful way to take treatments. However, further advances in electrical current regulation, electrode construction, and skin pre-treatment approaches are essential to achieve better efficacy with iontophoresis.

The combination of nanocarriers (polymeric and lipidic nanocarriers) and microneedles, which improve the effectiveness, reliability, and flexibility of transdermal drug administration, has greatly advanced iontophoresis-based drug delivery. The advances in technologies have the potential to enhance drug stability, deliver a wide range of therapeutics, and improve the penetrability of drugs. Nevertheless, further research is essential for increasing the scalability of nanocarrier and microneedle fabrication process. Moreover, constraints on the marketplace and regulations need to be overcome to introduce iontophoresis-based treatments into clinical practice. Further, consistency, safety, and efficacy confirmations are crucial for advanced iontophoresis-based drug delivery technologies.

References

Arunkumar S, Shivakumar HN, Narasimha Murthy S (2018) Effect of terpenes on transdermal iontophoretic delivery of diclofenac potassium under constant voltage. Pharm Dev Technol 23(8):806–814

Banerjee A, Chen R, Arafin S, Mitragotri S (2019) Intestinal iontophoresis from mucoadhesive patches: a strategy for oral delivery. J Control Release 297:71–78

Bhatia KS, Singh J (1998) Mechanism of transport enhancement of LHRH through porcine epidermis by terpenes and iontophoresis: permeability and lipid extraction studies. Pharm Res 15:1857–1862

Christodoulou P, Doxas PG, Papadakis CE, Prassopoulos P, Maris T, Helidonis ES (2003) Transtympanic iontophoresis of gadopentetate dimeglumine: preliminary results. Otolaryngol Head Neck Surg 129(4):408–413

Cubayachi C, do Couto RO, de Gaitani CM, Pedrazzi V, de Freitas O, Lopez RF (2015) Needle-free buccal anesthesia using iontophoresis and amino amide salts combined in a mucoadhesive formulation. Colloids Surf B Biointerfaces 136:1193–1201

Dhote V, Bhatnagar P, Mishra PK, Mahajan SC, Mishra DK (2012) Iontophoresis: a potential emergence of a transdermal drug delivery system. Sci Pharm 80(1):1

Dixit N, Bali V, Baboota S, Ahuja A, Ali J (2007) Iontophoresis—an approach for controlled drug delivery: a review. Curr Drug Deliv 4(1):1–10

Dragicevic N, Maibach H (2018) Combined use of nanocarriers and physical methods for percutaneous penetration enhancement. Adv Drug Deliv Rev 127:58–84

Eljarrat-Binstock E, Domb AJ (2006) Iontophoresis: a non-invasive ocular drug delivery. J Control Release 110(3):479–489

Fernandes NB, Velagacherla V, Spandana KJ, Bhagya N, Mehta CH, Gadag S, Sabhahit JN, Nayak UY (2024) Co-delivery of lapatinib and 5-fluorouracil transfersomes using transpapillary iontophoresis for breast cancer therapy. Int J Pharm 650:123686

Gadag S, Narayan R, Sabhahit JN, Hari G, Nayak Y, Pai KS, Garg S, Nayak UY (2022) Transpapillary iontophoretic delivery of resveratrol loaded transfersomes for localized delivery to breast cancer. Biomater Adv 140:213085

Gangarosa LP, Park NH, Fong BC, Scott DF, Hill JM (1978) Conductivity of drugs used for iontophoresis. J Pharm Sci 67(10):1439–1443

Gratieri T, Alberti I, Lapteva M, Kalia YN (2013) Next generation intra-and transdermal therapeutic systems: Using non-and minimally-invasive technologies to increase drug delivery into and across the skin. Eur J Pharm Sci 50(5):609–622

Guy RH, Kalia YN, Delgado-Charro MB, Merino V, López A, Marro D (2000) Iontophoresis: electrorepulsion and electroosmosis. J Control Release 64(1–3):129–132

Hu L, Damaj BB, Martin R, Michniak-Kohn BB (2011) Enhanced in vitro transbuccal drug delivery of ondansetron HCl. Int J Pharm 404(1–2):66–74

Jiang SJ, Hwang SM, Choi EH, Ahn SK, Elias PM, Lee SH (2000) Structural and functional effects of oleic acid and iontophoresis on hairless mouse stratum corneum. J Invest Dermatol 114(1):64–70

Jiang C, Jiang X, Wang X, Shen J, Zhang M, Jiang L, Ma R, Gan T, Gong Y, Ye J, Gao W (2021) Transdermal iontophoresis delivery system for terazosin hydrochloride: an in vitro and in vivo study. Drug Deliv 28(1):454–462

Kalia YN, Naik A, Garrison J, Guy RH (2004) Iontophoretic drug delivery. Adv Drug Deliv Rev 56(5):619–658

Kalaria DR, Singhal M, Patravale V, Merino V, Kalia YN (2018) Simultaneous controlled iontophoretic delivery of pramipexole and rasagiline in vitro and in vivo: transdermal polypharmacy to treat Parkinson's disease. Eur J Pharm Biopharm 127:204–212

Karpiński TM (2018) Selected medicines used in iontophoresis. Pharmaceutics 10(4):204

Khan A, Yasir M, Asif M, Chauhan I, Singh AP, Sharma R, Singh P, Rai S (2011) Iontophoretic drug delivery: history and applications. J Appl Pharm Sci 1(3):11–24

Kigasawa K, Kajimoto K, Hama S, Saito A, Kanamura K, Kogure K (2010) Noninvasive delivery of siRNA into the epidermis by iontophoresis using an atopic dermatitis-like model rat. Int J Pharm 383(1–2):157–160

Kim SN, Min CH, Kim YK, Ha A, Park CG, Lee SH, Park KH, Choy YB (2022a) Iontophoretic ocular delivery of latanoprost-loaded nanoparticles via skin-attached electrodes. Acta Biomater 144:32–41

Kim SN, Min CH, Kim BH, Lee S, Ji HB, Kim CR, Han JH, Im CH, Yu HG, Choy YB (2022b) Iontophoretic delivery of dexamethasone-loaded nanoparticles to the anterior segment of the eye. J Ind Eng Chem 116:199–206

Kumbhar PS, Kolekar K, Nadaf S, Vishwas S, Kamble V, Bhagwat D, Gupta G, Deshannavar UB, Singh SK, Dua K, Disouza J (2024) Expanding the arsenal against diabetes mellitus via transdermal application of drug-loaded nanoparticles and microneedles: recent advances and future perspectives. J Drug Deliv Sci Technol 98:105930

Kusama S, Sato K, Matsui Y, Kimura N, Abe H, Yoshida S, Nishizawa M (2021) Transdermal electroosmotic flow generated by a porous microneedle array patch. Nat Commun 12(1):658

Lakshmanan S, Gupta GK, Avci P, Chandran R, Sadasivam M, Jorge AE, Hamblin MR (2014) Physical energy for drug delivery; poration, concentration and activation. Adv Drug Deliv Rev 71:98–114

Li SK, Hao J (2018) Transscleral passive and iontophoretic transport: theory and analysis. Expert Opin Drug Deliv 15(3):283–299

Li X, Huang X, Mo J, Wang H, Huang Q, Yang C, Zhang T, Chen HJ, Hang T, Liu F, Jiang L (2021a) A fully integrated closed-loop system based on mesoporous microneedles-iontophoresis for diabetes treatment. Adv Sci 8(16):e2100827

Li Y, Yang J, Zheng Y, Ye R, Liu B, Huang Y, Zhou W, Jiang L (2021b) Iontophoresis-driven porous microneedle array patch for active transdermal drug delivery. Acta Biomater 121:349–358

Miller LL, Smith GA (1989) Iontophoretic transport of acetate and carboxylate ions through hairless mouse skin. A cation exchange membrane model. Int J Pharm 49(1):15–22

Miller LL, Smith GA, Chang AC, Zhou QX (1987) Electrochemically controlled release. J Control Release 6(1):293–296

Molokhia S, Papangkorn K, Butler C, Higuchi JW, Brar B, Ambati B, Li SK, Higuchi WI (2020) Transscleral iontophoresis for noninvasive ocular drug delivery of macromolecules. J Ocul Pharmacol Ther 36(4):247–256

Moura RB, Andrade LM, Alonso L, Alonso A, Marreto RN, Taveira SF (2022) Combination of lipid nanoparticles and iontophoresis for enhanced lopinavir skin permeation: impact of electric current on lipid dynamics. Eur J Pharm Sci 168:106048

Murdan S (2008) Enhancing the nail permeability of topically applied drugs. Expert Opin Drug Deliv 5(11):1267–1282

Murthy SN, Wiskirchen DE, Bowers CP (2007) Iontophoretic drug delivery across human nail. J Pharm Sci 96(2):305–311

Nair AB, Kiran Vaka SR, Murthy SN (2011) Transungual delivery of terbinafine by iontophoresis in onychomycotic nails. Drug Dev Ind Pharm 37(10):1253–1258

Nair AB, Aldhubiab B, Shah J, Jacob S, Attimarad M, Sreeharsha N, Venugopala KN, Joseph A, Morsy MA (2023) Design, development, and evaluation of constant voltage iontophoresis for the transungual delivery of Efinaconazole. Pharmaceutics 15(5):1422

Nayak AK, Dey S, Pal K, Banerjee I (2019) Iontophoretic drug delivery systems. In: Bioelectronics and medical devices. Woodhead Publishing, Sawston, pp 393–420

Noh G, Keum T, Seo JE, Bashyal S, Eum NS, Kweon MJ, Lee S, Sohn DH, Lee S (2018) Iontophoretic transdermal delivery of human growth hormone (hGH) and the combination effect of a new type microneedle, Tappy Tok Tok®. Pharmaceutics 10(3):153

Oh DH, Chun KH, Jeon SO, Kang JW, Lee S (2011) Enhanced transbuccal salmon calcitonin (sCT) delivery: effect of chemical enhancers and electrical assistance on in vitro sCT buccal permeation. Eur J Pharm Biopharm 79(2):357–363

Perez VL, Wirostko B, Korenfeld M, From S, Raizman M (2020) Ophthalmic drug delivery using iontophoresis: recent clinical applications. J Ocul Pharmacol Ther 36(2):75–87

Pikal MJ (2001) The role of electroosmotic flow in transdermal iontophoresis. Adv Drug Deliv Rev 46(1–3):281–305

Priya B, Rashmi T, Bozena M (2006) Transdermal iontophoresis. Expert Opin Drug Deliv 3(1):127–138

Siddiqui O, Sun Y, Liu JC, Chien YW (1987) Facilitated transdermal transport of insulin. J Pharm Sci 76(4):341–345

Tashiro Y, Kato Y, Hayakawa E, Ito K (2000) Iontophoretic transdermal delivery of ketoprofen: effect of iontophoresis on drug transfer from skin to cutaneous blood. Biol Pharm Bull 23(12):1486–1490

Telò I, Tratta E, Guasconi B, Nicoli S, Pescina S, Govoni P, Santi P, Padula C (2016) In-vitro characterization of buccal iontophoresis: the case of sumatriptan succinate. Int J Pharm 506(1–2):420–428

Thysman S, Préat V, Roland M (1992) Factors affecting iontophoretic mobility of metoprolol. J Pharm Sci 81(7):670–675

Tiwari SB, Udupa N (2003) Investigation into the potential of iontophoresis facilitated delivery of ketorolac. Int J Pharm 260(1):93–103

Tyagi V, Serna-Jiménez CE, Kalia YN (2021) Controlled simultaneous iontophoresis of buflomedil hydrochloride and dexamethasone sodium phosphate to the mucosa for oral submucous fibrosis. Int J Pharm 607:121009

Wei R, Simon L, Hu L, Michniak-Kohn B (2012) Effects of iontophoresis and chemical enhancers on the transport of lidocaine and nicotine across the oral mucosa. Pharm Res 29:961–971

Yang J, Li Y, Ye R, Zheng Y, Li X, Chen Y, Xie X, Jiang L (2020) Smartphone-powered iontophoresis-microneedle array patch for controlled transdermal delivery. Microsyst Nanoeng 6(1):112

Zaciewski T, Brickman K, Wellock AR, Adams DZ (2007) The use of iontophoresis in the pain management of soft tissue injuries in the emergency department: a pilot study. J Emerg Med 33(3):330–331

Zhou Y, Jia X, Pang D, Jiang S, Zhu M, Lu G, Tian Y, Wang C, Chao D, Wallace G (2023) An integrated mg battery-powered iontophoresis patch for efficient and controllable transdermal drug delivery. Nat Commun 14(1):297

Modern Perspectives of Ultrasound-Based Drug Delivery

4

Kaustubh Ajit Kolekar, Popat S. Kumbhar, Vaishali Yadav, Sukriti Vishwas, Kalpana S. Patil, Sachin Kumar Singh, Kamal Dua, John Disouza, and Vandana B. Patravale

Abstract

Ultrasound-based drug delivery is a transformative approach that enhances the efficacy and precision of therapeutic agents. It improves the delivery of drugs through diverse mechanisms including cavitations, acoustic streaming, sonoporation, and thermal effects. This method enables targeted delivery, reduces side effects, and offers versatile, noninvasive applications. This chapter discusses ultrasound technology fundamentals, including hyperthermia and acoustic streaming, which facilitate drug transport. Additionally, the chapter focuses on recent advancements including smart drug delivery systems like liposomes, polymers, and microbubbles that respond to ultrasound for modified release.

K. A. Kolekar · S. Vishwas
School of Pharmaceutical Sciences, Lovely Professional University, Phagwara, Punjab, India

P. S. Kumbhar
Department of Pharmaceutics, Tatyasaheb Kore College of Pharmacy, Warananagar, Maharashtra, India

V. Yadav
New College of Pharmacy, Kolhapur, Maharashtra, India

K. S. Patil
Department of Pharmacognosy, Tatyasaheb Kore College of Pharmacy, Warananagar, Maharashtra, India

S. Kumar Singh (✉)
School of Pharmaceutical Sciences, Lovely Professional University, Jalandhar, Punjab, India

Discipline of Pharmacy, Graduate School of Health, University of Technology Sydney, Sydney, NSW, Australia

M. R. Babu et al. (eds.), *Transdermal Applications of Minimally Invasive Drug Delivery Systems*, https://doi.org/10.1007/978-981-96-6275-3_4

Moreover, potential applications of ultrasound-mediated drug delivery systems for managing various diseases are also discussed. Further, challenges regarding safety, optimal ultrasound parameters, and delivery efficiency are briefed. Overall, ultrasound technology has the potential to revolutionize drug delivery, paving the way for innovative therapies. Further investigation is needed to address present constraints and optimize its clinical application.

Keywords

Ultrasound-based drug delivery · Mechanisms · Sonophoresis · Smart drug delivery systems · Combination therapies · Diseases

4.1 Introduction

Ultrasound-based drug delivery is a novel approach in therapeutics and biomedical engineering that offers a nonintrusive means of improving the specific distribution of therapeutics. It is reported to be a potential approach for overcoming the bottlenecks associated with conventional delivery of therapeutics. In this technique, high-frequency sound waves modify drugs, drug delivery carriers, and tissue behavior that causes enhanced effectiveness and precision of therapy. This ultrasound stimulates delivery of the drugs across cell membranes by enhancing cellular permeability.

Ultrasound waves with low energy (0.1–100 mW/cm^2) are commonly employed for diagnostic purposes. On the other hand, ultrasound waves with high energy varying between 100 and 10,000 mW/cm^2 are utilized for achieving noninvasive delivery of therapeutics at specific sites. Sonophoresis is the key mechanism observed in ultrasound-based drug delivery. Other mechanisms include acoustic waves, acoustic streaming, cavitation effects, thermal effects, microbubble technology, and many more. Further, ultrasound-based drug delivery is influenced by various factors such as frequency and intensity of ultrasound, exposure duration of ultrasound, and pulse duration and duty cycle (Gourevich et al. 2013; Delaney et al. 2022).

K. Dua
Discipline of Pharmacy, Graduate School of Health, University of Technology Sydney, Sydney, NSW, Australia

Faculty of Health, Australian Research Centre in Complementary and Integrative Medicine, University of Technology Sydney, Ultimo, NSW, Australia

J. Disouza (✉)
Department of Pharmaceutics, SYBES's Bombay Institute of Pharmacy and Research, Dombivli (East), Maharashtra, India

V. B. Patravale (✉)
Department of Pharmaceutical Sciences and Technology, Institute of Chemical Technology, Matunga, Mumbai, Maharashtra, India
e-mail: vb.patravale@ictmumbai.edu.in

Recent developments in ultrasound technology have demonstrated the way ultrasound reacts with biological tissues, creating new avenues for a range of medical applications, including gene and cardiovascular therapies, cancer treatment, and neurological problems. The use of ultrasound in amalgamation with different smart drug delivery strategies (nanoparticles, hydrogels, exosomes, microbubbles, etc.) offers a very site-specific and controlled release of cargo (Seah and Teo 2018). Moreover, the integration of ultrasound with other combination therapies like chemotherapies, immunotherapies, and targeted drug delivery systems is capable of enhancing the uptake and localization of drugs at target sites (Ren et al. 2022). This strategy has shown a wide range of applications in the treatment of cancers and neurodegenerative, cardiovascular, infectious, and autoimmune diseases.

With advancements like concentrated ultrasound-mediated rupture of the blood-brain barrier, high-intensity focused ultrasound, and real-time drug delivery evaluation, ultrasound is set to evolve into a key component of precision medicine in the years to come (Low et al. 2021). Regardless of the significant benefits of ultrasound-based drug delivery, it is associated with hurdles like optimization, safety, and regulating concerns, and continued study and trial to fully reap the benefits of ultrasound in drug delivery systems are crucial. This chapter covers the fundamental of ultrasound technology and the implications and recent advances of ultrasound-mediated drug delivery. Further, the applications of ultrasound drug delivery for treating various diseases are explained.

4.2 Fundamentals of Ultrasound Technology

Ultrasound (sonography) technology has become an important approach in drug delivery due to its unique properties allowing for more precise and effective therapeutic application. It is a safe and noninvasive technique that demonstrated significant potential in both therapy and diagnosis. It works on the principle of generation of high-frequency ultrasound waves. The key elements of ultrasound technology include sound wave frequency, cavitations, and sonoporation. All these elements of ultrasound technology have shown a significant role in improving cargo absorption and targeted drug delivery.

4.2.1 Principle of Ultrasound

The ultrasound-based drug delivery approach operates on sound waves above the range of human hearing (above 20 kHz), which penetrate biological tissues. This principle enables both therapeutic and imaging uses. These waves can be focused on targeted areas to improve drug delivery without any invasiveness. In drug delivery, ultrasound technology induces mechanical as well as thermal effects within the body tissues that help in transport and the release of the cargo more effectively (Fig. 4.1). This technology works by several mechanisms such as acoustic waves, cavitation effects, sonoporation, thermal effects, microbubble technology, and

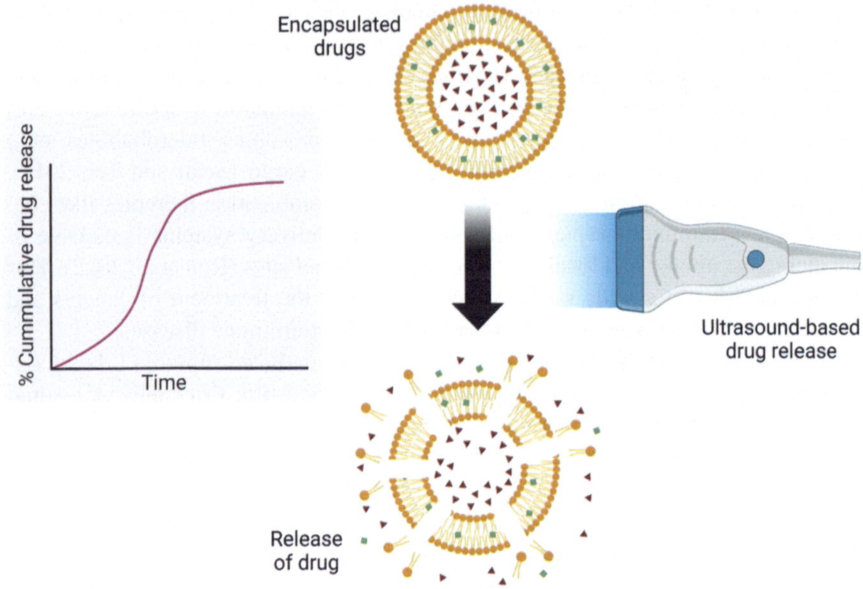

Fig. 4.1 Ultrasound-based release of drug from the formulation

acoustic streaming (Delaney et al. 2022). The interventional (therapeutic) ultrasound is associated with the generation of high amounts of acoustic output, which can be directed toward particular targets to heat, ablate, or lyse tissues.

4.2.2 Mechanisms of Ultrasound-Induced Drug Delivery

Ultrasound-based drug delivery approach improves the targeted drug delivery through various mechanisms.

4.2.2.1 Acoustic Waves

These waves are produced with the frequencies above 20 kHz. At this frequency, the drug can easily penetrate the biological barriers. The acoustic waves are primarily focused on the specific body part which allows for targeted drug delivery without any invasiveness (Sirsi and Borden 2014).

4.2.2.2 Cavitation Effects

In the mechanism of cavitation, the ultrasound-induced formation and collapse of microbubbles generate significant mechanical forces. These forces can easily disrupt cell membranes, leading to increased drug penetration and cavitation. In inertial cavitation, the rapid collapse of bubbles forms shock waves that cause localized tissue disruption, while in stable cavitation, oscillating bubbles without collapse and structural damage increase membrane permeability (Sirsi and Borden 2014).

4.2.2.3 Sonoporation

This phenomenon refers to an increase in the drug permeability particularly macromolecules such as proteins and peptides by temporarily altering the cell membrane (Delaney et al. 2022). The ultrasound waves are responsible for creating momentary openings in the cell membrane.

4.2.2.4 Thermal Effects

Ultrasound can also generate localized heat within tissue, which leads to an increase in blood flow and enhances drug distribution. Thermal effects work particularly well with thermosensitive carriers that release their payload upon heating (Seah and Teo 2018).

4.2.2.5 Microbubble Technology

This technology utilizes small gas-filled bubbles of the size 1–100 μM floated in liquid. Microbubble-loaded drugs are exposed to ultrasound and oscillation and eventually rupture to release their contents directly into targeted tissues. This approach is effective in delivering chemotherapeutic agents and gene therapies to specific sites (Sirsi and Borden 2014).

4.2.2.6 Acoustic Streaming

It refers to the consistent, average time movement of fluid induced when sound waves interact with a fluid mixture. Ultrasound creates fluid motion, increasing drug diffusion by raising the effective penetration of cargo within the tissue that supports drug distribution in targeted sites (Sirsi and Borden 2014).

4.2.3 Ultrasound Parameters and their Effects on Drug Delivery

The effectiveness of ultrasound-based drug delivery is significantly influenced by several parameters.

4.2.3.1 Frequency

Ultrasound frequencies determine both penetration depth and cavitation intensity. Lower frequencies penetrate deeper into tissue but are also more likely to produce heat, which may require careful management to avoid undesirable thermal effects (Sirsi and Borden 2014).

4.2.3.2 Intensity

Higher ultrasound intensities tend to enhance cavitation effects, which can improve cargo transport across cell membranes. However, excessive intensity may risk tissue damage, necessitating precise control to balance efficiency with safety (Sirsi and Borden 2014).

4.2.3.3 Pulse Duration and Duty Cycle

The timing of ultrasound pulses plays an important role in drug delivery efficiency. Continuous ultrasound waves create sustained effects, whereas pulsed waves allow tissue to recover between pulses, which can reduce the risk of thermal damage and optimize drug uptake (Sirsi and Borden 2014).

4.2.3.4 Exposure Duration

The duration of ultrasound exposure directly affects drug uptake levels. The longer exposure times generally improve drug delivery as well as increase the risk of tissue heating. Therefore, the optimal exposure time must be balanced to maximize drug uptake while minimizing thermal injury (Wilson et al. 2024; Tharkar et al. 2019).

4.3 Implications of Ultrasound-Mediated Drug Delivery System

4.3.1 Hyperthermia

In an ultrasound-mediated drug delivery system, leverage the thermal effects of the ultrasound waves to enhance drug delivery, particularly through a process called hyperthermia. The hyperthermia approach has significant implications for facilitating the efficacy of the treatment in various diseases (Couture et al. 2014). Hyperthermia works by using high-intensity focused ultrasound waves to localized heating of tissues. This heating can stimulate the release of the cargo from the temperature-sensitive carrier. These carriers are designed to release their payload at slightly higher temperature (42–43 °C) than normal body temperature (Santos et al. 2017). The use of hyperthermia in conjunction with ultrasound-mediated drug delivery presents several clinical advantages. These include targeted drug delivery, enhanced efficacy, and real-time monitoring (Gasselhuber et al. 2012; de Maar et al. 2021). However, the implication of ultrasound-mediated drug delivery with hyperthermia is promising, with several challenges such as clinical translation, patient variability, safety concerns, etc.

4.3.2 Cavitation

It is a critical phenomenon in ultrasound-mediated drug delivery systems. This approach significantly improves the delivery and efficacy of the therapeutic agents. The cavitation process involves the formation, oscillation, and collapse of microbubbles when subjected to ultrasound, leading to various mechanical effects that promote targeted drug delivery. The two types of cavitation are stable cavitation and inertial cavitation and it works via several mechanisms. These include temporary pore formation at the cell membrane, increase in the permeability of blood vessels, and micro-streaming that promotes drug retention by influencing the extracellular matrix (Delaney et al. 2022). Cavitation is prominently used in cancer therapies as

(Chowdhury et al. 2017) well as gene therapies. However, there are few challenges related to the optimization of parameters like ultrasound frequency, intensity, duration, and microbubble characteristics. Moreover, cavitation effects are reversible and are associated with the risk of unintended damage to surrounding healthy tissues (de Maar et al. 2021; He et al. 2022).

4.3.3 Sonophoresis

It is also known as phonophoresis. This technique utilizes ultrasound to improve the delivery of therapeutic agents through the transdermal route. This method exploits the mechanical and thermal effects of ultrasound to increase skin permeability, thereby facilitating the transport of drugs through the stratum corneum, the outermost layer of the skin (Fig. 4.2) (Bommannan et al. 1992). The dual-frequency ultrasound waves are used to create bubbles with an inner cavity in the skin which oscillate and collapse to form microchannels in the skin that increase its permeability to reach the drug in deeper tissue and then into the systemic circulation (Mitragotri 2017; Rao and Nanda 2009). In combination with other techniques like chemical enhancers, iontophoresis, and electroporation, sonophoresis shows enhanced drug delivery (Seah and Teo 2018). It has promising applications in transdermal drug delivery, ocular drug delivery, and gene therapy. However, there are few challenges related to the effectiveness of sonophoresis due to the variability in results, risk of healthy tissue damage, and regulatory-related challenges (Chowdhury et al. 2017).

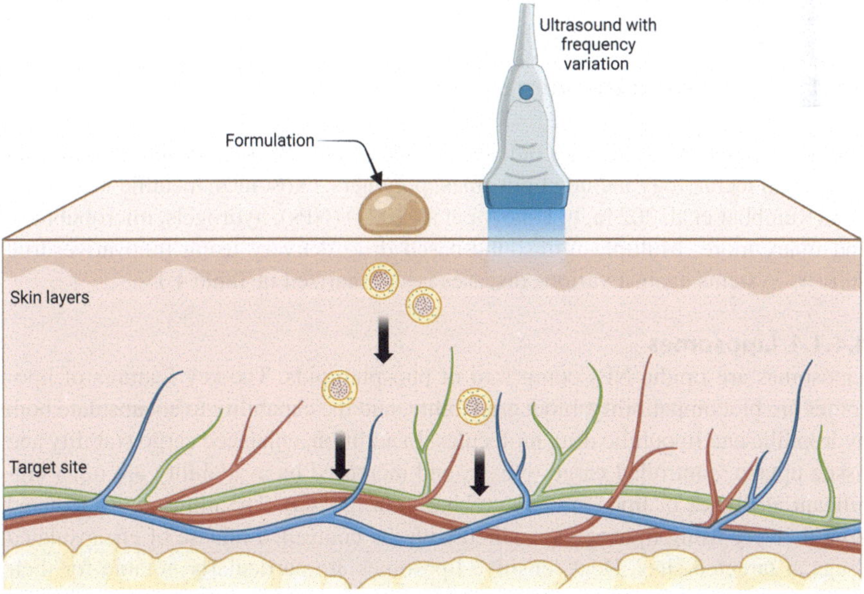

Fig. 4.2 Targeted delivery of drug formulation through different skin layers using ultrasound

4.3.4 Acoustic Streaming

It is a significant phenomenon in ultrasound-mediated drug delivery systems that enhances the transport and distribution of therapeutic agents within biological tissues. This process involves the steady flow of fluid induced by the oscillation of ultrasound waves, which can effectively facilitate drug delivery and improve treatment outcomes (Sirsi and Borden 2014). It works by absorption of ultrasound energy by tissues, leading to localized pressure changes and fluid movement, which enhances drug penetration and microbubble interaction. It is commonly used in cancer therapies, gene therapies, and transdermal drug delivery approaches. However, there are challenges related to the optimization of various parameters and safety concerns. Thus, clinical translation is the major challenge of acoustic streaming (Delaney et al. 2022; Honari and Sirsi 2023).

4.4 Recent Advances and Innovations

Stimuli-sensitive drug delivery systems have demonstrated promising applications in the treatment of variety of diseases. Among these, ultrasound-mediated drug delivery systems have gained significant attention for their ability to increase the efficacy of therapeutic agent delivery while minimizing side effects. This approach utilizes ultrasound waves to facilitate the targeted release of therapeutic agents, often in conjunction with various smart drug delivery systems, combination therapy, and real-time monitoring and imaging. Below are recent innovations and advancements in this field (Kashkooli et al. 2023).

4.4.1 Smart Drug Delivery Systems

Different kinds of smart drug delivery systems focused used in integration with ultrasound technology include liposomes, polymers, exosomes, metallic nanoparticles (Kumbhar et al. 2024a, b; Umeyor et al. 2024) (NPs), hydrogels, microbubbles, and many more. Multiple ultrasound-based drug delivery using innovative drug delivery systems against various diseases is summarized in Table 4.1.

4.4.1.1 Liposomes

Liposomes are lipidic NPs comprised of phospholipids. The key features of liposomes are biocompatibility, biodegradability, and the capability to encapsulate both hydrophilic and lipophilic drug molecules. In addition, enhanced cargo stability and tissue uptake, controlled cargo release, and increased bioavailability are other significant attributes of liposomes (Kumbhar et al. 2024a). The ultrasound-mediated drug delivery using liposomes is allied with the controlled release of encapsulated drugs at targeted sites. Heat-sensitive liposomes are particularly notable for their ability to release drugs upon heating induced by ultrasound (Couture et al. 2014).

Table 4.1 Ultrasound-based drug delivery using innovative drug delivery against diverse diseases

Drug name	Type of nanocarrier	Diseases targeted	Performance	References
Doxorubicin	Liposomes	Breast cancer	• Significant release of doxorubicin from ultrasound-mediated liposomes than conventional liposomes • Significant ($p < 0.005$) reduction in tumor size	Kim et al. (2022)
Calcein	Liposomes	Cervical cancer	• Significant ($p < 0.005$) uptake of calcein in HeLa cells	AlSawaftah et al. (2021)
Camptothecin and doxorubicin	Polymeric NPs	Glioblastoma	• Enhanced permeation of drugs into brain	Sun et al. (2023)
–	Exosomes	Alzheimer's	• Substantial enhancement in amyloid-beta degradation by exosomes following stimulation by ultrasound	Deng et al. (2021)
Bleomycin	Microbubbles	Melanoma	• Enhanced antitumor effects • Reduced required doses of chemotherapeutics	Sonoda et al. (2007)
Gemcitabine	Microbubbles	Pancreatic cancer	• Significantly ($p < 0.05$) inhibited the tumor growth • Elevated survival rate up to 10%	Kotopoulis et al. (2014)
Doxorubicin	Microbubbles	Pancreatic cancer	• 12 times more drug concentration at tumor cells • Significantly reduced tumor growth • Reduced systemic side effects	Tinkov et al. (2010)

(continued)

Table 4.1 (continued)

Drug name	Type of nanocarrier	Diseases targeted	Performance	References
Doxorubicin	Microbubbles	Liver cancer	• Sevenfold increased in drug delivery at tumor site • Dramatically prevented tumor growth Reduced systematic side effect	(2020)
10-Hydroxycamptothecin	Microbubbles	Cancer	• Ultrasound-based drug delivery significantly prevented cell growth (71%) compared to free drug (47%) • Reduced systemic exposure	Li et al. (2012)
Paclitaxel	Aptamer-modified PLGA nanobubbles	Prostate tumor	• Significant cell apoptosis in prostate cancer cells • Prolonged survival of treated mice • No significant systemic toxicity	Wu et al. (2017a)
siRNA	Nanobubbles	Glioma	• Significantly enhanced siRNA transfection and gene silencing efficiency • Achieved superior glioma inhibition	Cai et al. (2018)
10-Hydroxycamptothecin	Nanodroplets	Tumor theranostics	• Enhanced accuracy of diagnosis and monitoring	Liu et al. (2018)
Paclitaxel	Nanoemulsions	Breast and pancreatic cancer	• Enhanced tumor regression and suppressed metastasis	Rapoport et al. (2011)
Curcumin	Mixed micelles	Cancer	• Enhanced tumor targeting and micelle penetration due to ultrasound-induced permeabilization • 6.5-fold greater tumor weight reduction compared to controls	Wu et al. (2017b)

Researchers have studied ultrasound-mediated delivery of doxorubicin using liposomes. This system displayed more than 60% release of doxorubicin in response to ultrasound from liposomes in the treatment of breast cancer. A significant reduction (56%) in growth of the tumor was noticed in vivo in animals treated with ultrasound-mediated liposomes than conventional liposomes (Kim et al. 2022). Furthermore, in vivo a pharmacokinetic study in mice displayed remarkable AUC and C_{max} of doxorubicin from liposomes when compared to plain doxorubicin. In another study, the research team investigated the potential of delivering calcein using transferrin-decorated liposomes via ultrasound technology in managing cervical cancer. They noticed a significantly higher uptake of calcein into cancer cells from targeted liposomes used with ultrasound technology than non-targeted liposomes used without ultrasound technology (AlSawaftah et al. 2021).

4.4.1.2 Polymeric Nanocarriers

Polymeric nanocarriers fabricated using both natural and synthetic polymers are found to be promising for drug delivery against a variety of diseases. These types of nanocarriers have the capability to load both hydrophilic and lipophilic therapeutics, control payload release, and improve therapeutic biopharmaceutical attributes, pharmacokinetic performance, and stability of drugs. Moreover, these nanocarriers have been reported to be biocompatible and biodegradable (Kumbhar et al. 2022). The ultrasound-responsive polymeric nanocarriers can be designed to release drugs in a controlled manner. These systems can be tailored for specific applications, enhancing the therapeutic index of drugs by ensuring localized delivery (Rana et al. 2023). Researchers have fabricated polymeric NPs incorporated with camptothecin and doxorubicin for treatment of glioblastoma using ultrasound technology. The researhers developed camptothecin and doxorubicin hyaluronic acid polymer conjugated with NPs. The considerable improvement in the permeation of both drugs was noticed in the brain studied in murine GBM model from developed system. This enhanced permeation of therapeutics could be attributed to the use of ultrasound (Sun et al. 2023).

4.4.1.3 Exosomes

The membrane-bound vesicles within the size range of 30–150 nm are recognized as exosomes. They offer multiple benefits involving immunogenicity, biodegradability, prolonged circulation and regulated payload release, and significant stability to loaded drug molecules (Kumbhar et al. 2024b). They are responsible for lowering the rapid clearance of therapeutics. Exosomes as carriers in conjunction with ultrasound can facilitate targeted drug delivery due to their natural ability to fuse with cell membranes. Present research is exploring how ultrasound can augment the uptake of exosome-encapsulated therapeutics (Sridharan and Lim 2023). The study by researchers explored the potential of using ultrasound-mediated exosomes in managing Alzheimer's disease. They generated exosomes using glial cells present in the brain. The ultrasound-mediated exosomes caused enhanced breakdown of amyloid-beta, providing significant benefits in Alzheimer's. Thus, they concluded that stimulation of exosomes using ultrasound can be effective in controlling Alzheimer's disease (Deng et al. 2021).

4.4.1.4 Metallic NPs

The metallic NPs such as gold, silver, copper, iron, and zinc have displayed potential roles in drug delivery and treatment of a range of diseases. Augment in solubility, stability, circulation time, and therapeutic effectiveness are the noteworthy features of metallic NPs. These NPs can be used alongside ultrasound to improve imaging and therapy. Their unique properties allow them to act as both contrast agents and drug carriers, enhancing the precision of drug delivery while providing real-time imaging capabilities (Liu et al. 2016). The ultrasound-based metallic NPs have demonstrated significant promise in drug delivery to tumors and efficiently treat cancer. For instance, the research team designed an ultrasound-mediated acoustic-actuated tumor-targeted gold nanoparticulate system for enhancing cancer therapy. They concluded that this technology can be efficient in cancer treatment through the destruction of tumors (Tang et al. 2023).

4.4.1.5 Hydrogel

Hydrogels are comprised of a polymeric network with the capability of absorbing water. Both natural and synthetic polymers can be employed for the fabrication of hydrogel. The biomimetic nature of hydrogel makes it suitable for the delivery of therapeutics to manage a variety of diseases. Ultrasound-responsive hydrogels are being developed for applications such as tissue engineering and wound healing. These hydrogels can release drugs in response to ultrasound stimulation, providing a controlled release profile that is beneficial for localized treatment (Rana et al. 2023). Thus, the ultrasound-responsive hydrogel can enhance patient compliance by minimizing the challenges associated with conventional administration of drugs. The research study investigated the potential of protein delivery using ultrasound-responsive hydrogel. They observed controlled and stimuli-responsive release of the protein from hydrogel. Similarly, another research team successfully prepared an ultrasound-responsive component to manage the release of biomolecules using beta-cyclodextrin and adamantine. They noticed controlled and site-specific cargo release of encapsulated biomolecules. These results suggested that ultrasound-based biomolecule delivery might be an effective and innovative approach for treatment of various diseases (Yamaguchi et al. 2019).

4.4.1.6 Microbubbles

Microbubble shells are prepared using polymer, lipids, proteins, and sugar. The safety, stability, lifetime, and mechanical elasticity of microbubble depend on its components and their concentration and thickness. Microbubbles are perhaps the most widely studied component in ultrasound-based drug delivery. They serve as contrast agents in ultrasound imaging and enhance drug delivery through sonoporation. Recent studies indicate that microbubbles can significantly increase the local uptake of chemotherapeutic agents when used in combination with ultrasound (Wasielewska and White 2022). In view of this, the research team has designed an ultrasound-responsive microbubble incorporated with piperacillin and Fe_3O_4 NPs in the treatment of infection caused by *Pseudomonas aeruginosa*. The microbubbles demonstrated substantially enhanced penetration of piperacillin and Fe_3O_4 NPs into biofilm, leading to efficient treatment of infection caused by *Pseudomonas aeruginosa* (Xiu et al. 2023). Moreover, the researchers fabricated doxorubicin-loaded microbubbles for the treatment of cancer using ultrasound-based drug delivery. They evaluated various parameters in a cancer-induced animal model and compared it with nanoformulation and free drug solution. They noticed that the plasma concentration of cargo was significantly lower in microbubble-treated animals (0.5%/mL) compared to free doxorubicin (3%/mL). Moreover, in myocardial cells, doxorubicin concentration was noticeably lowered in microbubble-treated animals (0.17%/g tissue) compared to free cargo (0.320%/g tissue). Importantly, the ultrasound-treated group had a significantly higher concentration of the drug (2.5%/g) compared to free doxorubicin (0.38%/g) and prevented tumor growth remarkably. Thus, authors suggested that the ultrasound-based drug delivery is a promising approach for improving drug delivery at tumor sites, reducing systemic side effects of the cargo (Cochran et al. 2011).

4.4.2 Combination Therapies

In combination therapy, ultrasound technology integrates with other treatment modalities, including chemotherapies, immunotherapies, and targeted drug delivery systems, to improve drug uptake and localization at target sites. Chemotherapeutics like doxorubicin and bleomycin can significantly reach into the tumor cell by sonoporation mechanism. Moreover, ultrasound can enhance the delivery of immune checkpoint inhibitors or other immunotherapeutic agents by increasing local drug concentrations at tumor sites while reducing systemic exposure. This combination aims to improve the overall response rates and reduce adverse effects associated with systemic immunotherapy. Furthermore, combination therapies enhance the penetration of NPs through physiological barriers, improving their accumulation at target tissues. This synergy has been shown to increase the effectiveness of therapies while minimizing systemic side effects (Delaney et al. 2022).

4.4.3 Real-Time Monitoring and Imaging

Ultrasound technology has advanced significantly, particularly in real-time monitoring and imaging. These innovations enhance the precision and efficacy of drug delivery by allowing clinicians to visualize and control the delivery process dynamically. It is mainly categorized into passive cavitation imaging and photoacoustic imaging.

4.4.3.1 Passive Cavitation Imaging (PCI)

This method is mostly implemented to monitor and evaluate cavitation events in the engineering and medical domains. Recent advancements include the integration of PCI with focused ultrasound therapy. This technique enables real-time tracking of microbubble behavior during drug delivery, providing feedback on the location and concentration of drugs administered across the blood-brain barrier (BBB). The "cavitation dose painting" approach allows for spatially targeted drug delivery, optimizing treatment while avoiding damage to healthy tissues. This method represents a significant improvement over traditional imaging techniques like MRI or PET, which do not provide real-time feedback during procedures (Yang et al. 2019).

4.4.3.2 Photoacoustic Imaging

This noninvasive imaging method integrates the detail of ultrasonography resolution with the molecular imaging features of optical imaging. Photoacoustic imaging combined with ultrasound is being explored for monitoring drug release from nanocarriers. In this approach, near-infrared (NIR)-triggered nanodroplets can be visualized using ultrasound and photoacoustic signals, allowing us to monitor the behavior of these carriers in vivo. This method enables precise control over drug release timing and location, enhancing treatment efficacy while minimizing side effects associated with systemic drug administration (Cao et al. 2023).

4.5 Applications of Ultrasound-Mediated Drug Delivery Systems

The ultrasound-mediated drug delivery approach represents a promising advancement in therapeutic delivery, using ultrasound frequencies to improve targeted drug delivery. This enhances treatment efficacy and minimizes systemic side effects. Below are the key applications across various disease contexts.

4.5.1 Cancer

The ultrasound-mediated drug delivery systems have shown significant potential in cancer treatment by enhancing the localized delivery of chemotherapeutic agents. This method primarily involves sonoporation, where ultrasound induces the formation of temporary pores in cell membranes, facilitating drug uptake and improving bio-distribution to the cancer site. The researchers have explored ultrasound-based delivery using curcumin-loaded pluronic polymer micelles for the treatment of breast cancer. Their study observed a 6.5-fold reduction in tumor growth compared to the control group in the animal model. Therefore, the ultrasound-based drug delivery approach improves the therapeutic efficacy in cancer treatment (Wu et al. 2017b). In another interesting study, the researchers investigated a combination of chemotherapy approach associated with ultrasound for pancreatic cancer. Their findings reported a 40% reduction in tumor growth relative to the traditional drug delivery method (Logan et al. 2021). Similarly, the research team constructed a magneto-liposome-ligated microbubble complex for pancreatic cancer using an ultrasound approach. This combination demonstrated an 80% reduction in cancer cell volume in animals. Thus, the ultrasound technology is attributing the efficacy to improved permeability and targeted drug release (Dwivedi et al. 2020).

4.5.2 Neurodegenerative Diseases

In neurodegenerative diseases, ultrasound technology is gaining attention for its ability to facilitate cargo transport across the blood-brain barrier (BBB). The BBB presents a significant challenge for drug delivery in conditions such as Alzheimer's and Parkinson's diseases. Ultrasound can temporarily disrupt the BBB, allowing larger therapeutic molecules, including antibodies and gene therapies, to penetrate brain tissues effectively. This technique not only enhances targeted drug delivery but also minimizes systemic exposure and associated side effects (Delaney et al. 2022). Preliminary studies indicate that ultrasound technology may improve the delivery of neuroprotective agents and facilitate the administration of gene therapies aimed at reversing neurodegenerative processes.

For instance, researchers have studied the safety, feasibility, and reversibility of ultrasound technology in the treatment of Alzheimer's disease. They reported that the ultrasound technology significantly reduces amyloid plaque formation by 5% in Alzheimer's disease progression. Therefore, ultrasound technology is a promising approach for targeting amyloid plaques in neurodegenerative diseases (D'Haese et al. 2020). Similarly, the research team demonstrated a promising neuroprotective effect using lipid-coated glial cell-derived neurotrophic factor (GDNF) microspheres for treating Parkinson's disease via an ultrasound delivery approach. Their findings suggest that low-frequency ultrasound can significantly increase the level of GDNF in the brain, elevating the bioavailability of GDNF in the targeted area. Therefore, an increase in localized concentration of therapeutic agent reduces Parkinson's disease progression in animal models (Wang et al. 2014).

4.5.3 Cardiovascular Diseases

In cardiovascular diseases, ultrasound-based approaches are showing promise in improving the delivery of therapeutic agents to cardiac tissues. This approach confers a range of benefits such as enhanced therapeutic effectiveness, reduced negative effects, and improved patient compliance. The researchers have demonstrated the feasibility and efficacy of ultrasound technology in a study involving ten participants with severe symptomatic aortic valve stenosis. The researchers noticed significant results regarding time efficiency and energy delivery when using ultrasound for the therapy (Messas et al. 2021). Another research team investigated the effectiveness of ultrasound-based therapy in the treatment of stenosis in 40 participants. They noticed sustained improvements in aortic valve function for up to 180 days following disease treatment. These findings underscore the potential of ultrasound technology to improve treatment outcomes in cardiac conditions by enabling targeted and controlled cargo delivery (Messas et al. 2023).

4.5.4 Other Diseases

Beyond cancer, neurodegenerative diseases, and cardiovascular disease, ultrasound-based delivery approaches are being investigated for a variety of other diseases such as infectious diseases, autoimmune disorders, and diabetes. The ultrasound-based delivery is capable of enhancing the delivery of antibiotics directly to the infected tissues, improving treatment outcomes while reducing systemic toxicity (Díez-Vidal et al. 2024). Moreover, targeted delivery of immunomodulatory drugs can help manage conditions like rheumatoid arthritis more effectively (Di Matteo et al. 2020). The research team studied the effectiveness of ultrasound in the delivery of microbubbles incorporated with tocilizumab to treat rheumatoid arthritis. They achieved the destruction of microbubbles using ultrasound and therefore efficient release of loaded drug (Liu et al. 2023).

The ultrasound technology has also demonstrated the localized delivery of insulin or other antidiabetic medications, potentially improving glycemic control without systemic side effects (Vandemergel 2021). For instance, one study achieved prolonged and on-demand antidiabetic effects of insulin loaded into nanocapsule-based microgel using ultrasound technology. The application of ultrasound caused quick release of insulin to control blood glucose levels. Overall, ultrasound-mediated drug delivery systems represent a versatile platform with broad applications across various medical fields, offering significant improvements in targeted therapy and patient outcomes. Continued research and clinical trials will be essential to fully realize their potential and address challenges related to clinical translation.

4.6 Conclusion and Future Concern

Ultrasound-based drug delivery is a noninvasive technique employed for localized and site-specific delivery of therapeutics. This technique utilizes ultrasound waves of low or high intensity, creating a cavitation effect that causes cell membrane disruption, resulting in enhanced permeability of drugs. This system offers various advantages such as nonintrusive nature, targeted drug delivery, compatibility with diverse drug formulations, and controlled payload release. The efficiency of ultrasound-based drug delivery is impacted by the frequency, intensity, and exposure duration of ultrasound waves. Therefore, the optimization of these parameters is crucial for achieving considerable effectiveness.

The use of ultrasound drug delivery in combination with smart drug delivery is found to be promising in enhancing its effectiveness. Various innovative drug delivery approaches comprising polymeric NPs, lipidic NPs, metallic NPs, and microbubbles are employed in integration with ultrasound in treating different types of cancer, glioblastoma, Alzheimer's disease, etc. However, a selection of suitable and simple fabrication techniques is required. In addition, more research is needed to overcome challenges related to scalability and regulatory aspects.

Although ultrasound-based medication delivery is still in its infancy, it represents an attractive medical prospect due to its potential to improve treatment efficacy and minimize side effects. Thus, further advances in technology, nanoparticulate systems, and regulatory aspects are essential to serve ultrasound as a mainstay of drug delivery in the nearby future.

References

AlSawaftah NM, Awad NS, Paul V, Kawak PS, Al-Sayah MH, Husseini GA (2021) Transferrin-modified liposomes triggered with ultrasound to treat HeLa cells. Sci Rep 11(1):11589

Bommannan D, Okuyama H, Stauffer P, Guy RH, Sonophoresis I (1992) The use of high-frequency ultrasound to enhance transdermal drug delivery. Pharm Res 9:559–564

Cai W, Lv W, Feng Y, Yang H, Zhang Y, Yang G, Duan Y, Wang J (2018) The therapeutic effect in gliomas of nanobubbles carrying siRNA combined with ultrasound-targeted destruction. Int J Nanomedicine 13:6791–6807

Cao Y, Dumani DS, Hallam KA, Emelianov SY, Ran H (2023) Real-time monitoring of NIR-triggered drug release from phase-changeable nanodroplets by photoacoustic/ultrasound imaging. Photo-Dermatology 30:100474

Chowdhury SM, Lee T, Willmann JK (2017) Ultrasound-guided drug delivery in cancer. Ultrasonography 36(3):171

Cochran MC, Eisenbrey JR, Soulen MC, Schultz SM, Ouma RO, White SB, Furth EE, Wheatley MA (2011) Disposition of ultrasound sensitive polymeric drug carrier in a rat hepatocellular carcinoma model. Acad Radiol 18(11):1341–1348

Couture O, Foley J, Kassell NF, Larrat B, Aubry JF (2014) Review of ultrasound mediated drug delivery for cancer treatment: updates from pre-clinical studies. Transl Cancer Res 3(5):494–511

D'Haese PF, Ranjan M, Song A, Haut MW, Carpenter J, Dieb G, Najib U, Wang P, Mehta RI, Chazen JL, Hodder S (2020) B-Amyloid plaque reduction in the hippocampus after focused ultrasound-induced blood–brain barrier opening in Alzheimer's disease. Front Hum Neurosci 14:593672

de Maar JS, Rousou C, van Elburg B, Vos HJ, Lajoinie GP, Bos C, Moonen CT, Deckers R (2021) Ultrasound-mediated drug delivery with a clinical ultrasound system: in vitro evaluation. Front Pharmacol 12:768436

Delaney LJ, Isguven S, Eisenbrey JR, Hickok NJ, Forsberg F (2022) Making waves: how ultrasound-targeted drug delivery is changing pharmaceutical approaches. Mater Adv 3(7):3023–3040

Deng Z, Wang J, Xiao Y, Li F, Niu L, Liu X, Meng L, Zheng H (2021) Ultrasound-mediated augmented exosome release from astrocytes alleviates amyloid-β-induced neurotoxicity. Theranostics 11(9):4351

Di Matteo A, Mankia K, Azukizawa M, Wakefield RJ (2020) The role of musculoskeletal ultrasound in the rheumatoid arthritis continuum. Curr Rheumatol Rep 22:41

Díez-Vidal A, Martínez-Martín P, González-Muñoz B, Tung-Chen Y (2024) Point-of-care ultrasound in infectious diseases: Current insights and future perspectives. Clin Infect Dis 79(2):420–429

Dwivedi P, Kiran S, Han S, Dwivedi M, Khatik R, Fan R, Mangrio FA, Du K, Zhu Z, Yang C, Huang F (2020) Magnetic targeting and ultrasound activation of liposome–microbubble conjugate for enhanced delivery of anticancer therapies. ACS Appl Mater Interfaces 12(21):23737–23751

Gasselhuber A, Dreher MR, Partanen A, Yarmolenko PS, Woods D, Wood BJ, Haemmerich D (2012) Targeted drug delivery by high intensity focused ultrasound mediated hyperthermia combined with temperature-sensitive liposomes: computational modelling and preliminary in vivo validation. Int J Hyperth 28(4):337–348

Gourevich D, Dogadkin O, Volovick A, Wang L, Gnaim J, Cochran S, Melzer A (2013) Ultrasound-mediated targeted drug delivery with a novel cyclodextrin-based drug carrier by mechanical and thermal mechanisms. J Control Release 170(3):316–324

He J, Liu Z, Zhu X, Xia H, Gao H, Lu J (2022) Ultrasonic microbubble cavitation enhanced tissue permeability and drug diffusion in solid tumor therapy. Pharmaceutics 14(8):1642

Honari A, Sirsi SR (2023) The evolution and recent trends in acoustic targeting of encapsulated drugs to solid tumors: strategies beyond Sonoporation. Pharmaceutics 15(6):1705

Kashkooli FM, Jakhmola A, Hornsby TK, Tavakkoli JJ, Kolios MC (2023 Mar) Ultrasound-mediated nano drug delivery for treating cancer: fundamental physics to future directions. J Control Release 1(355):552–578

Kim YS, Ko MJ, Moon H, Sim W, Cho AS, Gil G, Kim HR (2022) Ultrasound-responsive liposomes for targeted drug delivery combined with focused ultrasound. Pharmaceutics 14(7):1314

Kotopoulis S, Delalande A, Popa M, Mamaeva V, Dimcevski G, Gilja OH, Postema M, Gjertsen BT, McCormack E (2014) Sonoporation-enhanced chemotherapy significantly reduces primary tumour burden in an orthotopic pancreatic cancer xenograft. Mol Imaging Biol 16:53–62

Kumbhar P, Manjappa A, Shah R, Jha NK, Singh SK, Dua K, Disouza J, Patravale V (2022) Inhalation delivery of repurposed drugs for lung cancer: approaches, benefits and challenges. J Control Release 341:1–5

Kumbhar PS, Kolekar K, Nadaf S, Vishwas S, Kamble V, Bhagwat D, Gupta G, Deshannavar UB, Singh SK, Dua K, Disouza J (2024a) Expanding the arsenal against diabetes mellitus via transdermal application of drug-loaded nanoparticles and microneedles: recent advances and future perspectives. J Drug Deliv Sci Technol 98:105930

Kumbhar P, Kolekar K, Vishwas S, Shetti P, Kumbar V, Pinto TD, Paiva-Santos AC, Veiga F, Gupta G, Singh SK, Dua K (2024b) Treatment avenues for age-related macular degeneration: breakthroughs and bottlenecks. Ageing Res Rev 98:102322

Li P, Zheng Y, Ran H, Tan J, Lin Y, Zhang Q, Ren J, Wang Z (2012) Ultrasound triggered drug release from 10-hydroxycamptothecin-loaded phospholipid microbubbles for targeted tumor therapy in mice. J Control Release 162(2):349–354

Liu D, Yang F, Xiong F, Gu N (2016) The smart drug delivery system and its clinical potential. Theranostics 6(9):1306

Liu J, Xu F, Huang J, Xu J, Liu Y, Yao Y, Ao M, Li A, Hao L, Cao Y, Hu Z (2018) Low-intensity focused ultrasound (LIFU)-activated nanodroplets as a theranostic agent for noninvasive cancer molecular imaging and drug delivery. Biomater Sci 6(11):2838–2849

Liu S, Zhao C, Zhang R, Wang Q, Wang M, Jiang Y, Yang M (2023) Role of ultrasonic microbubbles in treating rheumatoid arthritis: enhancing the efficacy of tocilizumab via contrast-enhanced ultrasound-monitored, ultrasound-targeted microbubble destruction. Ultrasound Med Biol 49(11):2407–2412

Logan KA, Nesbitt H, Callan B, Gao J, McKaig T, Taylor M, Love M, McHale AP, Callan JF (2021) Synthesis of a gemcitabine-modified phospholipid and its subsequent incorporation into a single microbubble formulation loaded with paclitaxel for the treatment of pancreatic cancer using ultrasound-targeted microbubble destruction. Eur J Pharm Biopharm 165:374–382

Low SS, Lim CN, Yew M, Chai WS, Low LE, Manickam S, Tey BT, Show PL (2021) Recent ultrasound advancements for the manipulation of nanobiomaterials and nanoformulations for drug delivery. Ultrason Sonochem 80:105805

Messas E, IJsselmuiden A, Goudot G, Vlieger S, Zarka S, Puymirat E, Cholley B, Spaulding C, Hagège AA, Marijon E, Tanter M (2021) Feasibility and performance of noninvasive ultrasound therapy in patients with severe symptomatic aortic valve stenosis: a first-in-human study. Circulation 143(9):968–970

Messas E, Ijsselmuiden A, Trifunović-Zamaklar D, Cholley B, Puymirat E, Halim J, Karan R, van Gameren M, Terzić D, Milićević V, Tanter M (2023) Treatment of severe symptomatic aortic valve stenosis using non-invasive ultrasound therapy: a cohort study. Lancet 402(10419):2317–2325

Mitragotri S (2017) Sonophoresis: ultrasound-mediated transdermal drug delivery. In: Percutaneous penetration enhancers physical methods in penetration enhancement. Springer, Berlin, Heidelberg, pp 3–14

Rana A, Adhikary M, Singh PK, Das BC, Bhatnagar S (2023) "Smart" drug delivery: a window to future of translational medicine. Front Chem 10:1095598

Rao R, Nanda S (2009) Sonophoresis: recent advancements and future trends. J Pharm Pharmacol 61(6):689–705

Rapoport N, Nam KH, Gupta R, Gao Z, Mohan P, Payne A, Todd N, Liu X, Kim T, Shea J, Scaife C (2011) Ultrasound-mediated tumor imaging and nanotherapy using drug loaded, block copolymer stabilized perfluorocarbon nanoemulsions. J Control Release 153(1):4–15

Ren WW, Xu SH, Sun LP, Zhang K (2022) Ultrasound-based drug delivery system. Curr Med Chem 29(8):1342–1351

Santos MA, Goertz DE, Hynynen K (2017) Focused ultrasound hyperthermia mediated drug delivery using thermosensitive liposomes and visualized with in vivo two-photon microscopy. Theranostics 7(10):2718

Seah BC, Teo BM (2018) Recent advances in ultrasound-based transdermal drug delivery. Int J Nanomedicine 13:7749–7763

Sirsi SR, Borden MA (2014) State-of-the-art materials for ultrasound-triggered drug delivery. Adv Drug Deliv Rev 72:3–14

Sonoda S, Tachibana K, Uchino E, Yamashita T, Sakoda K, Sonoda KH, Hisatomi T, Izumi Y, Sakamoto T (2007) Inhibition of melanoma by ultrasound-microbubble-aided drug delivery suggests membrane permeabilization. Cancer Biol Ther 6(8):1282–1289

Sridharan B, Lim HG (2023) Exosomes and ultrasound: the future of theranostic applications. Mater Today Bio 19:100556

Sun T, Krishnan V, Pan DC, Filippov SK, Ravid S, Sarode A, Kim J, Zhang Y, Power C, Aday S, Guo J (2023) Ultrasound-mediated delivery of flexibility-tunable polymer drug conjugates for treating glioblastoma. Bioeng Transl Med 8(2):e10408

Tang Q, Zhang F, Luo L, Duan Y, Zhu T, Ni Y, Wang Y, Qi H, Jiang S, Zhou J, Ma X (2023) Ultrasound-induced gold nanoparticle united with acoustic reprogramming of macrophages for enhanced cancer therapy. ACS Appl Mater Interfaces 15(44):50926–50939

Tharkar P, Varanasi R, Wong WS, Jin CT, Chrzanowski W (2019) Nano-enhanced drug delivery and therapeutic ultrasound for cancer treatment and beyond. Front Bioeng Biotechnol 7:324

Tinkov S, Coester C, Serba S, Geis NA, Katus HA, Winter G, Bekeredjian R (2010) New doxorubicin-loaded phospholipid microbubbles for targeted tumor therapy: in-vivo characterization. J Control Release 148(3):368–372

Umeyor CE, Kumbhar P, Trivedi R, Kolekar P, Kolekar K, Bhalgat G, Patravale VB (2024) Physicochemical properties of carriers for targeting lymphatic system. In: Advanced targeting of the lymphatic system. Springer Nature, Cham, pp 205–228

Vandemergel X (2021) Point-of-care ultrasound (POCUS) in the field of Diabetology. Int J Chronic Dis 2021(1):8857016

Wang X, Cui G, Yang X, Zhang Z, Shi H, Zu J, Hua F, Shen X (2014) Intracerebral administration of ultrasound-induced dissolution of lipid-coated GDNF microbubbles provides neuroprotection in a rat model of Parkinson's disease. Brain Res Bull 103:60–65

Wasielewska JM, White AR (2022) Focused ultrasound-mediated drug delivery in humans—a path towards translation in neurodegenerative diseases. Pharm Res 39(3):427–439

Wilson MG, Parikh A, Dara A, Beaver AS, Kubanek J (2024) Targeted drug release from stable and safe ultrasound-sensitive nanocarriers. Front Mol Biosci 11:1408767

Wu M, Wang Y, Wang Y, Zhang M, Luo Y, Tang J, Wang Z, Wang D, Hao L, Wang Z (2017a) Paclitaxel-loaded and A10-3.2 aptamer-targeted poly (lactide-co-glycolic acid) nanobubbles for ultrasound imaging and therapy of prostate cancer. Int J Nanomedicine 12:5313–5330

Wu P, Jia Y, Qu F, Sun Y, Wang P, Zhang K, Xu C, Liu Q, Wang X (2017b) Ultrasound-responsive polymeric micelles for sonoporation-assisted site-specific therapeutic action. ACS Appl Mater Interfaces 9(31):25706–25716

Wu Y, Sun T, Tang J, Liu Y, Li F (2020) Ultrasound-targeted microbubble destruction enhances the antitumor efficacy of doxorubicin in a mouse hepatocellular carcinoma model. Ultrasound Med Biol 46(3):679–689

Xiu W, Ren L, Xiao H, Zhang Y, Wang D, Yang K, Wang S, Yuwen L, Li X, Dong H, Li Q (2023) Ultrasound-responsive catalytic microbubbles enhance biofilm elimination and immune activation to treat chronic lung infections. Sci Adv 9(4):eade5446

Yamaguchi S, Higashi K, Azuma T, Okamoto A (2019) Supramolecular polymeric hydrogels for ultrasound-guided protein release. Biotechnol J 14(5):1800530

Yang Y, Zhang X, Ye D, Laforest R, Williamson J, Liu Y, Chen H (2019) Cavitation dose painting for focused ultrasound-induced blood-brain barrier disruption. Sci Rep 9(1):2840

Thermal Ablation-Based Drug Delivery System

5

Saloni Sharma, K. Gowthamarajan,
Karri V. V. S. Narayana Reddy, C. J. Sanjay, and M. P. Gowrav

Abbreviations

DNA	Deoxyribonucleic acid
G-apparatus	Golgi apparatus
KHz	Kilohertz
MRI	Magnetic resonance imaging
MW	Microwave ablation
NIR	Near infrared
RNA	Ribonucleic acid
RF	Radiofrequency ablation
SPIONs	Superparamagnetic ionised nanoparticles
TME	Tumour microenvironment
TRAIL	Tumour necrosis factor-related apoptosis-inducing ligand
US	Ultrasound

S. Sharma · K. Gowthamarajan (✉) · K. V. V. S. Narayana Reddy · C. J. Sanjay
M. P. Gowrav
Department of Pharmaceutics, JSS College of Pharmacy, Ooty JSS Academy of Higher Education and Research, Nilgiris, Tamil Nadu, India

Department of Oral Medicine and Radiology, JSS Dental College and Hospital, JSS Academy of Higher Education and Research, Mysuru, Karnataka, India

Department of Pharmaceutics, JSS College of Pharmacy, JSSAHER,
Mysuru, Karnataka, India
e-mail: gowthamsang@jssuni.edu.in

M. R. Babu et al. (eds.), *Transdermal Applications of Minimally Invasive Drug Delivery Systems*, https://doi.org/10.1007/978-981-96-6275-3_5

5.1 Introduction

External stimuli-based drug delivery system is the most upcoming prominent approach of drug delivery system for the treatment of cancers in the form of transdermal drug delivery system, hyperthermia-based drug delivery mechanisms and theranostic approaches of thermal ablation for the deposition of drug delivery (Mi et al. 2020). This mechanism involves the local application of external stimuli maintained either at high or low temperatures for better efficacy towards tumour reduction by apoptosis and reducing the spread of cancerous tissues throughout the system (Shi et al. 2017).

The application of this technology introduced lateral imaging systems of the cancerous tissues, fluorescence-guided procedures and other imaging modalities towards the emerging ways of tumour therapy (Cherukula et al. 2019). Thermo-ablation has shown several advantages over the present conventional systems of some highly invasive surgical procedures to increased survival time of the surrounding tumours, reduction in cost, better patient compliance by withstanding the thermal therapy-based procedures and reduced treatment time. However, in these advanced technologies, they are prone to certain disadvantages, such as reduced frequency of ablation, inconsistent efficacy, mortality rates and randomised trials of efficacy between the patients (Li et al. 2019). Due to these reasons, the detection of stages of cancer is delayed, for which better therapeutic tools using radiofrequency thermal ablation method, cryoablation method of thermal ablation and microwave ablation techniques pave a way for advancements towards thermal ablation-based drug delivery systems (Wang et al. 2016).

5.2 Mechanisms of Energy-Based Cell Death

Thermal based methods are used because they have the tendency to ablate the infected cells at the tumour microenvironment at the biological and cellular levels, thus paving way to pathways for tumour destruction (Lu et al. 2017). External heat stimuli therapy is divided into three zones: (Mi et al. 2020) central zone that initiates the coagulation of the tumour cells, (Shi et al. 2017) known as the peripheral phase that has direct contact of the tumour cells with the external stimuli that leads to apoptosis and (Cherukula et al. 2019) conduction zone where the conversion thermal energy occurs in the tumorous cells that surrounds the thermal region (Wang et al. 2014).

Present understanding of cells penetrated to heat needs to be maintained at uniformly reduced temperatures over a long period of time in order to prevent from rupturing the cells; usually, at 45 °C, the cells are prone to irreversible cell damage with an elevated exposure with external stimuli for around 30 m, and when the cells are exposed to temperatures of above 60 °C, the effect of damage decreased, thus resulting to protein denaturation, which leads to cytotoxicity and coagulation (Min et al. 2015).

Due to heat-based therapy, the change in temperatures when in contact with the cells may lead to loss permeability of the cell membranes that will lead to membrane fluidity, thus leading to impairment within the cells, malfunction of filaments and other tubules responsible to cause cytolysis, thus altering the sub-cellular mechanisms and functions, but these changes do not have any correlation with the change in temperatures, due to which these heat variations have direct impact on the mitochondria that leads to heat-induced injury (Jin et al. 2017). During the process of NIR exposure, there is leakage of protons within the cells that leads to DNA damage, thus leading mitochondria swelling that leads to heat-mediated cell death that is caused due to DNA depletion. DNA polymerase enzymes are released that causes repair synthesis that is suppressed due to heat conversion on the cells by chromatin fibres (Papa et al. 2017). An alternative mechanism of cell death is by the RNA disruption by the release of lysosomal enzymes. In this case, it is important to correlate and understand the thermo-sensitivity between the healthy and tumour cells based on the metabolic stress rendered on the cells (Mi et al. 2010). The acidic environment of the tumour site delays the cellular damage by thermal external stimuli by interacting with Golgi apparatus. This mechanism is illustrated by the RNA induction of cellular apoptosis that causes ischaemia and releases lysosomes responsible for invasion of granulocytes, cytokine release and stimulation of immune response (Chu and Dupuy 2014).

5.3 Types of Thermal Ablation

5.3.1 Radiofrequency

The radiofrequency method of thermal ablation is the most widely used therapy towards cancer treatments targeting the main organs and bone cancers too. This method works by creating an electrical network of the body using an (RF)infrared fluoresence probe of about 20An thus connected to an image-guided system that is used in the application towards ultrasound, computed tomography system, magnetic resonance imaging system and surgery. The radiation frequency generated is about 380–550 kHz, thus creating a voltage gap between the body and the probe inducing the radiation (Chauhan et al. 2020). The heating chamber in this system generates a uniform ablation frequency that dissipates, thus forming a thermal conduction. Thus, leading to thermal conversion of the tissue by voltage frequency irradiation thus producing friction to act between the cells and the tissue (Mohan and Rapoport 2010). This method illustrates the inversion proportionality between the thermal conductivity rate and heat produced from the system, thus leading to reduced temperatures too to act upon the cells. The required temperature in this method is to be maintained at 55 °C in order to cause the cancerous cell death (Kim et al. 2008). If the temperatures rise up to 10,060 °C, this will lead to cellular death by necrosis, blood coagulation and boiling and charring of the tissues, thus reducing the thermal resistance and reducing the mechanism of thermal ablation by altering the properties. Hence, in order to maintain the optimum temperature, proper thermal procedures need to be maintained throughout the radiofrequency mechanism (You et al. 2012). The ablation phase is directly dependent upon the electrical conductivity capacity of the target tissue in turn increasing the ablation zone. The RF efficacy can be improved by saline induction into the cells prior to the therapy, thus improving the electrical conductivity of the cells and large surface active heating area based on the probe design and nature of the external stimuli (Bikram et al. 2007).

The RF method of therapy proves its way to be safe and effective treatment of tumour ablation which on the contrary has some potential side effects, such as electrical circuit malfunctions that may lead to serious burns, skin inflammation, fits, nausea, vomiting, post-ablation syndrome, fever and flu-like symptoms; other factors are as follows: ablation size limit needs to be monitored, image stability needs to be guided, accuracy in treatment strategies also needs to be monitored at regular intervals during treatment days, etc. (Lee et al. 2010).

Usually, infr-red fluoresence is characterised as a non-specific treatment since this method can be controlled and maintained at the desired temperatures, ablation size can be monitored and direct targeting the tumour site, thus upon dissipation the cells and tissues get easily dissipated into the region of tumour microenvironment, thus preventing from thermal damage without affecting the healthy cells (Mok et al. 2010). This passive method of tumour thermal therapy leads to a concept of heat-sink effect that helps in the conversion of light radiation into thermal convection by cooling down the nearby vessels. Due to rapid conversion of heat and light in this mechanism, it is difficult to target and irradiate the tumour completely as the treated area of tumour should be within 0.5–1 mm at the peripheral membrane so that retreatment is not required (Das et al. 2008).

5.3.2 Microwaves

Microwave ablation technique involves cellular death by coagulation through necrosis by heat generation. This method uses electromagnetic waves to emit ablation technique to generate the desired heat required for therapy (Winter et al. 2006). This method is regulated by an antenna of 14.5 gauge placed directly at the tumour site to begin with therapy. A set of desired electromagnetic waves is allowed to pass through the cells and penetrate into the tissues by polar water molecules with respect to the magnetic field exerted at the time of magnetic wave generation (Hildebrandt et al. 2002). For the accurate efficacy, the nature of the antenna is responsible for effective therapy results. The ablation zones should be within 10 mm of the tumour size to produce larger coagulation sizes for better accuracy of treatment procedures (Hak et al. 2009). The advantages of this method are that this method focuses on passive transport and only creates the necessary tumour region for effective therapy on the tumorous tissues without affecting the healthy cell. This method does not rely much on heat and temperature intensity and conduction procedures and hence allows uniform heating exposure for a large infected zone (Laginha et al. 2005).

This method focuses on local treatment with side effects like post-ablation inflammation, effusions within the cells and improper heat sink effects within the tissues. Due to the elimination of heat sink effect in this method, the microwave therapy is more consistent and uniform to completely kill the tumour cells (Mabrey and Sturtevant 1976).

5.3.3 Cryoablation

This method is used in the treatment of benign and malignant lessions targeting liver, kidney, lung, bone and breast. The procedure involved in this mechanism involves freeze-thaw cycles with argon and helium gas. Argon gas is used to reduce or maintain the low temperatures of heat at the tumour site for better targeting specificity of the tumours (Burdinski et al. 2010). Gradually, due to reduced temperatures the tissue freezes, osmosis is produced and thus causes imbalance between the cells and tissues in the internal and external tumour environment. In this method, the cell death mechanism involves the dehydration process of the cells and protein imbalance between the tissues, thus leading to the thawing process that leads to cell death by vascular damage (Staruch et al. 2012). The temperature maintained in this process is 30 °C that needs to be reached at 60 °C by adjusting the time and rate to improve the efficacy of the treatment on the cells. The most desired advantage of this process is that the ice formation of cells and tissues can be directly images using different systems like MRI, X-ray CT and fluorescence images (Al-Zu'bi and Mohan 2020). This helps to illustrate the targeted drug delivery system by reducing the risk of healthy tissue damage, and application of cryoablation can also be used for the treatment of large tumours of size >3 cm, which has reduced surgery and cancer treatments (Gasselhuber et al. 2012).

5.4 Drug Delivery Systems

5.4.1 Hyperthermia Drug Delivery System

The hyperthermia drug delivery occurs in the form of drug composites such as liposomes, superparamagnetic nano-conjugates, iron-oxide nanoparticles, magnetic nanocarriers and metallic nanostructures. The structurally engineered molecules have the tendency to trigger the drug release by an external stimuli and external tropism features in the form of magnetic field that helps in tumour ablation-based therapy (Turcotte and Averill-Bates 2001). Generally, nanocarriers and other nanostructures have the tendency to incorporate magnetic agents in the form of iron oxide nanoparticles that can be substituted with polymeric nanoparticles, solid lipid carriers and metallic nanoparticles coated with gold, ferric chloride, etc., conjugated for targeting tumour by magnetic field using the property of magnetic resonance imaging system for tumour imaging. Surface functionalisation of magnetic nanoparticles with aptamers, antibodies and ligand for actively targeting tumour helps in the applications of targeted drug delivery systems by active and passive route of drug delivery (Sharpe et al. 2002).

Different studies have been performed in order to illustrate the hyperthermia drug delivery that includes Chiang et al. (Jiang et al. 2017) that illustrated the synergistic mechanism of nanocarriers in the form of superparamagnetic nanoparticles with the effective release of doxorubicin in the presence of magnetic external stimuli using ligand targeting drug delivery system for enhanced cancer therapy. Followed by another study, Cazares-Cortes et al. (2019) formulated albumin nanoparticles in the form of a photosensitiser agent for targeting cervical cancer by the release of iron oxide drug from albumin nanoparticles using hyperthermia-based drug delivery for anti-cancer efficacy using magneto-fluorescent imaging by enhancing targeted drug delivery system.

Zhang et al. (Kopp et al. 2003) formulated multi-nanostructures in the form of TRAIL conjugation polymeric hydrogels and SPIONs nano-complexes that are moulded into thermosensitive agents for combined cancer therapy. Chen et al. (Scanlan et al. 2002) encapsulated graphene nano-sack, a novel form of drug conjugates in the form of multifunctional particles that can initiate the targeted drug delivery system. Yin PT et al. developed a novel approach of stem cell therapy activated by magnetic hyperthermia for the treatment of cancer. Yang et al. (2018a) synthesised a sono-chemical novel approach in the form of a hydrophilic ferric oxide nano-capsule surface modified with albumin for targeting cervical cancer by the presence of an external triggered hyperthermia mechanism for improving its efficacy. Su et al. (Gao et al. 2020) formulated iron oxide nanoparticles that facilitated the drug release by burst gas generation method in the presence of magnetic field exposure loaded with paclitaxel targeting tumour therapy by facilitative thermosensitive agents and chemotherapy simultaneously for better tumour penetration and improved therapeutic applications. Thorat et al. (Bianchi et al. 2009) illustrated a novel approach of superparamagnetic nanoparticles enhancing liposomal drug delivery systems maintaining a controlled release by application of external magnetic field with the effect of rhodamine that acts as a photosensitive agent to achieve the ultimate hyperthermia application for tumour ablation using photodynamic therapy. Di Corato et al. (Figuerola et al. 2010) formulated rhodamine-loaded magnetic liposomes as a novel approach to illustrate the local hyperthermia magnetic field application towards the hyperthermia treatment therapy of tumour, using a dual modality of photodynamic therapy and hyperthermia for an effective targeted drug delivery system. Wang et al. (Yang et al. 2018b) formulated a NIR-mediated magnetic nanostructures for a multipurpose of cell imaging and improved photothermal therapy in the form of metallic gold nanoparticles for effective application of hyperthermia tumour ablation. Stocke et al. (Spirnak 2018) formulated a magnetic nanoparticles loaded with iron oxide to target localised hyperthermia effect by remote activation of the external stimuli at the tumour site of breast cancer.

5.4.2 Theranostic Drug Delivery Systems

Nanoparticles have been advanced both by enhancement in drug delivery systems and imaging modalities for targeting cancer. In recent applications, there have been potential functions of nanoparticles to act as a dual modality for bioimaging as well as therapeutic drug delivery at the cellular level (Farzanegan and Tahmasbi 2023).

This mechanism has paved way for nanotheranostic approach to enhance the ability for simultaneously imaging the target site as well as treat the disease. Theranostics has been an emerging area of research for better disease treatment strategies as well as to improve their ability to target specific tumour environment at the diseased states for application towards noninvasive drug delivery of particle localisation. Novel research has brought about by a wide range of based on their structural morphology, surface functionality, mechanism of action by the activation of pathways and imaging modals for tumour specificity (Hauser et al. 2015).

The main components of theranostic moiety are as follows:

1. *Detection of noninvasive imaging system*: Imaging moieties are usually in the form of metallic, polymeric and magnetic agents that can be moulded towards the bioimaging applications like magnetic resonance imaging, fluorescence imaging, photoacoustic imaging and X-ray CT imaging (Mi 2020).
2. *Polymer coating* is surface modified in order to improve the colloidal stability and for a strong barrier of functional groups for bio-conjugated surface-modified complexes. Magnetic iron oxide particles are highly effective when dissolved in a nanoemulsion with a surface modifier in the form of ligand, antibody, aptamer and other bio-conjugated functional groups for improving the efficacy using targeted drug delivery system (Seo et al. 2022).
3. *Drug-loading capabilities:* Physicochemical encapsulation into polymeric nanoemulsion provides electrostatic forces, covalent bonds and pi-pi interaction with the drug and polymeric nanocarrier to improve the biocompatibility and drug-loading efficiency (Burke et al. 2017).

Many research works have been performed in the recent years towards the advancement of nanotheranostic in thermal ablation-based drug delivery. For instance, Kim et al. (2022) formulated nanoparticles in the form of capsules for efficient drug release characteristics with the conjugation of disulphide bonds by redox potential with GSH to facilitate drug release by targeted drug delivery system. Followed by another work, Luo et al. (Ferreira et al. 2023) encapsulated mesoporous silica-functionalised nano-polymers that are prone to redox potential for effective binding of the drug with the polymeric matrix by disulphide bonds, targeted by a fluorescence dye for enhancing the applications of targeted drug delivery system. Li et al. (2024) formulated peptide-functionalised silica nanoparticle-loaded doxorubicin for effective drug release using peptides for directly targeting the tumour site in the form of redox-responsive drug delivery system for active tumour target drug delivery system.

In another study, Mi et al. (Choi and Jung 2020) formulated oxidative-responsive polymeric vehicles that improve the efficacy towards neutron therapy for solid tumours by the surface modification of copolymer dissolved in DOX drug matrix by disulphide bonds that activates the oxidative reaction (H_2O_2) oxidative hydrogen per-oxidase catalytic reaction catalyse action by targeting hypoxic cancer with light-triggered drug release characteristics. Followed by another study, Zou et al. (van Zon-Rabelink et al. 2003) formulated polymersomes with surface modification of aptamers to bind with di-dulphide bonds for effective lung cancer chemotherapy that was a success due to dual response redox reactions between the molecules and surface construction of di-selenide conjugates. Yet in another study, Christie et al. (Blanchard et al. 1997) formulated polymeric micelles with disulphide bond conjugation upon surface-modified GSH-triggered drug release to enhance cancer tumour-specific therapy by targeting bio-mimetically siRNA at the tumour target site. Wang et al. (Zhu et al. 2007) synthesised an effective chemotherapeutic agent to act as a dual modality of chemotherapy and photothermal therapy, with the surface modification of phosphazene for actively targeting cancer. In another study, Farshad Moradi et al. (Tehrani et al. 2024) focussed on ultrasound-triggered nanoisomers as smart nanocarriers for localised drug delivery with the help of external triggered drug delivery to form a thermosensitive agent towards the application of nanotheranostics in thermal ablation. Hannah et al. (Moon et al. 2023) formulated gold nanoshells that illustrate the application of gold structures towards nanotheranostic application for cancer. This study uses chemotherapy with a bimodal of optical imaging using a fluorescence tag facilitated by NIR-triggered drug release.

5.4.3 Transdermal Drug Delivery System

For the management of skin diseases using drug therapy-mediated drug delivery system, it reduces the systemic toxicity at the disease site and reduces the systemic toxicity that in turn becomes a noninvasive application. Successful topical drug delivery is achieved based on the physicochemical properties of the drug-loaded molecules, surface structures and sizes for better deep penetration into the skin, better drug release efficacy and better solubility in the aqueous as well as oil phase (Lee and Yeo 2015). There is incorporation of surface enhancers that leads to better drug delivery system that is either physical modifiers or chemical substitutes. Physical enhancers include permeation-enhancing agents, prodrug incorporation for better drug delivery and nanoparticle-modified drug delivery systems (Hlady et al. 2012). These methods have paved a way for targeting sites through external stimuli in the form of microneedles, radiofrequency method, micro-abrasion techniques and light-mediated drug delivery systems. The mechanism of action of thermal ablation through skin includes the following: when an intense beam is triggered at the site, the light waves get converted into monochromatic radiation when it hits the molecules, thus converting them into emission moieties (Paul et al. 2024). This concept of laser-derived drug delivery helps in reversible ablation on the skin barrier, thus improving the absorption intensity and deeper penetration into the skin. To enhance the research aspect, many studies have been performed by forming micropores about 30 nm of diameter and width of about 70 nm; with the reduced particle size, the penetration at the site is easier, thus improving the efficacy of ablation technique by Bramson et al. (Lee et al. 2011). In this study, for the synthesis of microchannel spheres in the form of elliptical microstructures to enhance electrically stable transdermal drug delivery of 80 nm diameter 40 nm of width for better penetration into the skin for a successful ablation process, illustrated by Sintov et al. (Tebbenjohanns et al. 2016). In another approach, the control of heating parameters was achieved by the formulation of 2D nanowires in micron scale range for better ablation network. During this process, there is the tendency for the lasers to heat up with ohmic resistance, due to which the laser is irradiated only for 3–5 s. Followed by another approach that activates the electrodes by radiofrequency mechanism within the stratum corneum range, thus resulting to selective ablation procedures based on the RF energy given for the localised treatment. This method used micro-electrodes of 100 nm length and 40 nm width surrounded by a poly-carbon body for better light and heat conversion efficiency upon pressure on the site for about 4 ns with a voltage of 250 V for effective biomedical applications. This method was depicted by Levin et al. (Bond et al. 1958). Followed by another approach by Mikstza et al. (Coulman et al. 2006) synthesised a gene derived drug delivery for topical transfer of drug with micro-flux R device that couples with electrical transport to maintain a controlled drug delivery system for efficient in vivo delivery as well by the action of oligodeoxynucleotides. Yuzhakov et al. (D'Ajello et al. 2015) depicted a microneedle strategy to enhance skin ablation process for drug delivery where the drug conjugates are dissolved in reservoir medium and upon excitation with laser probes of about 200 V for 5 ns; without any skin damage, this therapy rendered successful results with good sampling procedures and improved the efficacy rate of drug delivery. A novel dual approach was noted by Trautman et al. (Lata et al. 2023) using microneedle system as a porting method for electrically transporting controlled drug delivery by the application of micro-electromechanical systemic devices.

5.5 Safety Concerns

Among the recent research works, the novel drugs and other medicines invented are still into a dilemma whether which drug is most suitable for thermal ablation-based drug delivery. In this study, we have illustrated recent research experiments on thermal ablation-based drug delivery based on their penetration capacities, pharmacokinetic parameters, metabolism rates, systemic drug release, controlled release aspects and targeting drug delivery systems. The channels created by ablation procedures are maintained at a specific site of action to emit into the cell membranes, tissue cultures and other thermosensitive molecules. These channels have empty regions on the surface that when penetrated with laser have the ability to erode the skin surface, as a result leading to sterility of the skin, systemic toxicity at the abrasion regions and skin erosion upon heat; for better patient compliance, it is necessary to formulate in such a way that they are intact on the skin and render better drug delivery at the target site.

5.6 Conclusion

This book chapter focuses on different thermal ablation techniques for better drug delivery systems. The thermal mechanisms of different methods are explained in this chapter which includes radiofrequency ablation method, cryoablation, microwave ablation with enhanced drug delivery systems in the form of transdermal drug delivery, hyperthermia drug delivery and theranostic drug delivery for thermal ablation mechanisms. These novel concepts improve localised drug delivery at the target site by application of external stimuli as they provide better drug delivery when compared to the present conventional systems. As a result, it has paved a way for research in the field of therapeutic tools using proteins, antibodies, ligands, aptamers, etc., for efficient targeting of site and diagnostic tool like MRI, ultrasound, X-ray and CT mechanism for localised targets. Triggered drug release with thermal ablation technique increases the bioavailability of the encapsulated drug by interacting through vascular perfusion at the target site bio-mimetically into the cells and tissues for better drug intake and penetration. Such targeted dual functions of molecules enhance the concept of targeted thermal ablation system to provide better therapeutic efficacy. The ultimate goal for a successful thermal ablation process is to tailor surface modifiers on the molecules to improve treatment strategies at the pathogenic site that leads to better patient compliance and increase the span of survival rates for a better quality of life by optimised treatment procedures to reduce side effects.

Acknowledgements The authors would like to thank ICMR-SRF funding with sanction no. 3/2/2/15/2022-NCD-III division for funding, the Department of Science and Technology—Fund for Improvement of Science and Technology Infrastructure in Universities and Higher Educational Institutions (DST-FIST), and Promotion of University Research and Scientific Excellence (DST-PURSE) for the facilities provided.

References

Al-Zu'bi M, Mohan A (2020) Modelling of combination therapy using implantable anticancer drug delivery with thermal ablation in solid tumor. Sci Rep 10(1):19366

Bianchi D, Nasuti F, Martelli E (2009) Coupled analysis of flow and surface ablation in carbon-carbon rocket nozzles. J Spacecr Rockets 46(3):492–500

Bikram M, Gobin AM, Whitmire RE, West JL (2007) Temperature-sensitive hydrogels with SiO2–Au nanoshells for controlled drug delivery. J Control Release 123(3):219–227

Blanchard EM, Iizuka K, Christe M, Conner DA, Geisterfer-Lowrance A, Schoen FJ, Maughan DW, Seidman CE, Seidman JG (1997) Targeted ablation of the murine α-tropomyosin gene. Circ Res 81(6):1005–1010

Bond AC, Rashis B, Levin LR (1958) Experimental ablation cooling. In: NACA conference on high-speed aerodynamics: a compilation of the papers presented. Amers Aeronautical Laboratory, Moffett Field, CA, p 267

Burdinski D, Pikkemaat JA, Emrullahoglu M, Costantini F, Verboom W, Langereis S, Grüll H, Huskens J (2010) Targeted lipoCEST contrast agents for magnetic resonance imaging: alignment of aspherical liposomes on a capillary surface. Angew Chem Int Ed Engl 49(12):2227–2229

Burke BP, Cawthorne C, Archibald SJ (2017) Multimodal nanoparticle imaging agents: design and applications. Phil Trans R Soc A 375(2107):20170261

Cazares-Cortes E, Cabana S, Boitard C, Nehlig E, Griffete N, Fresnais J, Wilhelm C, Abou-Hassan A, Ménager C (2019) Recent insights in magnetic hyperthermia: from the "hot-spot" effect for local delivery to combined magneto-photo-thermia using magneto-plasmonic hybrids. Adv Drug Deliv Rev 138:233–246

Chauhan VP, Stylianopoulos T, Martin JD, Popović Z, Chen O, Kamoun WS, Bawendi MG, Fukumura D, Jain RK (2020) Normalization of tumour blood vessels improves the delivery of nanomedicines in a size-dependent manner. In: Nano-enabled medical applications. Jenny Stanford Publishing, New Delhi, pp 279–311

Cherukula K, Uthaman S, Park IK (2019) "Navigate-dock-activate" anti-tumor strategy: tumor micromilieu charge-switchable, hierarchically activated nanoplatform with ultrarapid tumor-tropic accumulation for trackable photothermal/chemotherapy. Theranostics 9(9):2505

Choi Y, Jung SL (2020) Efficacy and safety of thermal ablation techniques for the treatment of primary papillary thyroid microcarcinoma: a systematic review and meta-analysis. Thyroid 30(5):720–731

Chu KF, Dupuy DE (2014) Thermal ablation of tumours: biological mechanisms and advances in therapy. Nat Rev Cancer 14(3):199–208

Coulman S, Allender C, Birchall J (2006) Microneedles and other physical methods for overcoming the stratum corneum barrier for cutaneous gene therapy. Crit Rev Ther Drug Carrier Syst 23(3):205–258

D'Ajello F, Maglie C, Yuzakov AA, Baydarov AA (2015) Effects of electromagnetic fields on patients. Preliminary results of double blind study and review of the literature. Perm Med J 32(2):52–57

Das M, Mishra D, Maiti TK, Basak A, Pramanik P (2008) Bio-functionalization of magnetite nanoparticles using an aminophosphonic acid coupling agent: new, ultradispersed, iron-oxide folate nanoconjugates for cancer-specific targeting. Nanotechnology 19(41):415101

Farzanegan Z, Tahmasbi M (2023) Evaluating the applications and effectiveness of magnetic nanoparticle-based hyperthermia for cancer treatment: a systematic review. Appl Radiat Isot 198:110873

Ferreira LF, Bermeo Varon LA, Orlande HR, Lamien B (2023) Design under uncertainties of the thermal ablation treatment of skin cancer. ASME J Heat Mass Transf 145(3):031202

Figuerola A, Di Corato R, Manna L, Pellegrino T (2010) From iron oxide nanoparticles towards advanced iron-based inorganic materials designed for biomedical applications. Pharmacol Res 62(2):126–143

Gao H, Yan C, Wu W, Li J (2020) Application of microfluidic chip technology in food safety sensing. Sensors 20(6):1792

Gasselhuber A, Dreher MR, Rattay F, Wood BJ, Haemmerich D (2012) Comparison of conventional chemotherapy, stealth liposomes and temperature-sensitive liposomes in a mathematical model. PLoS One 7(10):e47453

Hak S, Sanders HM, Agrawal P, Langereis S, Grüll H, Keizer HM, Arena F, Terreno E, Strijkers GJ, Nicolay K (2009) A high relaxivity Gd (III) DOTA-DSPE-based liposomal contrast agent for magnetic resonance imaging. Eur J Pharm Biopharm 72(2):397–404

Hauser AK, Wydra RJ, Stocke NA, Anderson KW, Hilt JZ (2015) Magnetic nanoparticles and nanocomposites for remote controlled therapies. J Control Release 219:76–94

Hildebrandt B, Wust P, Ahlers O, Dieing A, Sreenivasa G, Kerner T, Felix R, Riess H (2002) The cellular and molecular basis of hyperthermia. Crit Rev Oncol Hematol 43(1):33–56

Hlady RA, Novakova S, Opavska J, Klinkebiel D, Peters SL, Bies J, Hannah J, Iqbal J, Anderson KM, Siebler HM, Smith LM (2012) Loss of Dnmt3b function upregulates the tumor modifier Ment and accelerates mouse lymphomagenesis. J Clin Invest 122(1):163–177

Jiang PS, Tsai HY, Drake P, Wang FN, Chiang CS (2017) Gadolinium-doped iron oxide nanoparticles induced magnetic field hyperthermia combined with radiotherapy increases tumour response by vascular disruption and improved oxygenation. Int J Hyperthermia 33(7):770–778

Jin Z, Wen Y, Hu Y, Chen W, Zheng X, Guo W, Wang T, Qian Z, Su BL, He Q (2017) MRI-guided and ultrasound-triggered release of NO by advanced nanomedicine. Nanoscale 9(10):3637–3645

Kim JE, Lee JE, Lee SH, Yu JH, Lee JH, Park TG, Hyeon T (2008) Designed fabrication of a multifunctional polymer nanomedical platform for simultaneous cancer-targeted imaging and magnetically guided drug delivery. Adv Mater 20(3):478–483

Kim GH, Kim PH, Kim JH, Kim PN, Won HJ, Shin YM, Choi SH (2022) Thermal ablation in the treatment of intrahepatic cholangiocarcinoma: a systematic review and meta-analysis. Eur Radiol 32(2):1205–1215

Kopp VI, Zhang ZQ, Genack AZ (2003) Lasing in chiral photonic structures. Prog Quantum Electron 27(6):369–416

Laginha KM, Verwoert S, Charrois GJ, Allen TM (2005) Determination of doxorubicin levels in whole tumor and tumor nuclei in murine breast cancer tumors. Clin Cancer Res 11(19):6944–6949

Lata T, Trautman J, Townend P, Wilson RB (2023) Current management of gastro-oesophageal reflux disease—treatment costs, safety profile, and effectiveness: a narrative review. Gastroenterol Rep 11:goad008

Lee JH, Yeo Y (2015) Controlled drug release from pharmaceutical nanocarriers. Chem Eng Sci 125:75–84

Lee JH, Yigit MV, Mazumdar D, Lu Y (2010) Molecular diagnostic and drug delivery agents based on aptamer-nanomaterial conjugates. Adv Drug Deliv Rev 62(6):592–605

Lee JW, Gadiraju P, Park JH, Allen MG, Prausnitz MR (2011) Microsecond thermal ablation of skin for transdermal drug delivery. J Control Release 154(1):58–68

Li Z, Song N, Yang YW (2019) Stimuli-responsive drug-delivery systems based on supramolecular nanovalves. Matter 1(2):345–368

Li B, Qian Y, Huang Y, Li Z (2024) Efficacy and safety of thermal ablation modalities for the treatment of papillary thyroid microcarcinoma: systematic review and network meta-analysis. Heliyon 10(3):e25536

Lu Y, Aimetti AA, Langer R, Gu Z (2017) Bioresponsive materials. Nat Rev Mater 2(1):16075

Mabrey S, Sturtevant JM (1976) Investigation of phase transitions of lipids and lipid mixtures by sensitivity differential scanning calorimetry. Proc Natl Acad Sci U S A 73(11):3862–3866

Mi P (2020) Stimuli-responsive nanocarriers for drug delivery, tumor imaging, therapy and theranostics. Theranostics 10(10):4557

Mi P, Ju XJ, Xie R, Wu HG, Ma J, Chu LY (2010) A novel stimuli-responsive hydrogel for K+-induced controlled-release. Polymer 51(7):1648–1653

Mi P, Cabral H, Kataoka K (2020) Ligand-installed nanocarriers toward precision therapy. Adv Mater 32(13):1902604

Min KH, Min HS, Lee HJ, Park DJ, Yhee JY, Kim K, Kwon IC, Jeong SY, Silvestre OF, Chen X, Hwang YS (2015) pH-controlled gas-generating mineralized nanoparticles: a theranostic agent for ultrasound imaging and therapy of cancers. ACS Nano 9(1):624–645

Mohan P, Rapoport N (2010) Doxorubicin as a molecular nanotheranostic agent: effect of doxorubicin encapsulation in micelles or nanoemulsions on the ultrasound-mediated intracellular delivery and nuclear trafficking. Mol Pharm 7(6):1959–1973

Mok H, Veiseh O, Fang C, Kievit FM, Wang FY, Park JO, Zhang M (2010) pH-sensitive siRNA nanovector for targeted gene silencing and cytotoxic effect in cancer cells. Mol Pharm 7(6):1930–1939

Moon AM, Kim HP, Singal AG, Owen D, Mendiratta-Lala M, Parikh ND, Rose SC, McGinty KA, Agala CB, Burke LM, Abate A (2023) Thermal ablation compared to stereotactic body radiation therapy for hepatocellular carcinoma: a multicenter retrospective comparative study. Hepatol Commun 7(7):e00184

Papa AL, Korin N, Kanapathipillai M, Mammoto A, Mammoto T, Jiang A, Mannix R, Uzun O, Johnson C, Bhatta D, Cuneo G (2017) Ultrasound-sensitive nanoparticle aggregates for targeted drug delivery. Biomaterials 139:187–194

Paul T, Imran KM, Gannon J, Trusiano B, Eden K, Ivester H, Mott M, Powar M, Edwards M, Byron C, Clark-Deener S (2024) First successful engraftment of human liver cancer cell line in highly robust immunocompromised porcine model to test the tumor ablation efficacy by histotripsy. Cancer Res 84(6_Supplement):1456

Scanlan MJ, Gure AO, Jungbluth AA, Old LJ, Chen YT (2002) Cancer/testis antigens: an expanding family of targets for cancer immunotherapy. Immunol Rev 188(1):22–32

Seo B, Kanematsu H, Nakamoto M, Miyabayashi Y, Tanaka T (2022) Corrosion resistance and antibacterial activity of a surface coating created by super-spread wetting of liquid copper on laser-ablated carbon steel. Surf Coat Technol 445:128706

Sharpe M, Easthope SE, Keating GM, Lamb HM (2002) Polyethylene glycol-liposomal doxorubicin: a review of its use in the management of solid and haematological malignancies and AIDS-related Kaposi's sarcoma. Drugs 62(14):2089–2126

Shi J, Kantoff PW, Wooster R, Farokhzad OC (2017) Cancer nanomedicine: progress, challenges and opportunities. Nat Rev Cancer 17(1):20–37

Spirnak JR (2018) Development, modeling and testing of thermal protection systems in small, slow-burning solid rocket motors. Doctoral dissertation, Massachusetts Institute of Technology

Staruch R, Chopra R, Hynynen K (2012) Hyperthermia in bone generated with MR imaging–controlled focused ultrasound: control strategies and drug delivery. Radiology 263(1):117–127

Tebbenjohanns J, Höfer C, Bergmann L, Dedroogh M, Gaudin D, von Werder A, Rühmkorf K (2016) Shortening of freezing cycles provides equal outcome to standard ablation procedure using second-generation 28 mm cryoballoon after 15-month follow-up. Europace 18(2):206–210

Tehrani MH, Kashkooli FM, Soltani M (2024) Spatiotemporal modeling of nano-delivered chemotherapeutics for synergistic microwave ablation cancer therapy. Comput Methods Programs Biomed 247:108102

Turcotte S, Averill-Bates DA (2001) Sensitization to the cytotoxicity of melphalan by ethacrynic acid and hyperthermia in drug-sensitive and multidrug-resistant Chinese hamster ovary cells. Radiat Res 156(3):272–282

van Zon-Rabelink IA, Vleugels MP, Merkus HM, de Graaf R (2003) Endometrial ablation by rollerball electrocoagulation compared to uterine balloon thermal ablation: technical and safety aspects. Eur J Obstet Gynecol Reprod Biol 110(2):220–223

Wang L, Zhang M, Tan K, Guo Y, Tong H, Fan X, Fang K, Li R (2014) Preparation of nanobubbles carrying androgen receptor siRNA and their inhibitory effects on androgen-independent prostate cancer when combined with ultrasonic irradiation. PLoS One 9(5):e96586

Wang S, Huang P, Chen X (2016) Stimuli-responsive programmed specific targeting in nanomedicine. ACS Nano 10(3):2991–2994

Winter PM, Neubauer AM, Caruthers SD, Harris TD, Robertson JD, Williams TA, Schmieder AH, Hu G, Allen JS, Lacy EK, Zhang H (2006) Endothelial $\alpha_\nu\beta 3$ integrin–targeted fumagillin nanoparticles inhibit angiogenesis in atherosclerosis. Arterioscler Thromb Vasc Biol 26(9):2103–2109

Yang XH, Li KZ, Bai LT, Zhao ZG, Wang Y (2018a) Thermal ablation behavior of SiC coating for 3D braided carbon fiber reinforced ZrC-SiC composites in different heat fluxes. Vacuum 156:334–344

Yang W, Wang W, Liu B, Zhu B, Li J, Xu D, Ni Y, Bai L, Liu G (2018b) Immunomodulation characteristics by thermal ablation therapy in cancer patients. Asia Pac J Clin Oncol 14(5):e490–e497

You J, Zhang R, Zhang G, Zhong M, Liu Y, Van Pelt CS, Liang D, Wei W, Sood AK, Li C (2012) Photothermal-chemotherapy with doxorubicin-loaded hollow gold nanospheres: a platform for near-infrared light-trigged drug release. J Control Release 158(2):319–328

Zhu XL, Liu SB, Man BY, Xie CQ, Chen DP, Wang DQ, Ye TC, Liu M (2007) Analysis by using X-ray photoelectron spectroscopy for polymethyl methacrylate and polytetrafluoroethylene etched by KrF excimer laser. Appl Surf Sci 253(6):3122–3126

Acoustic Cavitation-Based Drug Delivery

6

Vancha Harish, Aditya Kumar, Malakapogu Ravindra Babu,
Alishia Leo, and Shriyansh Srivastav

Abstract

Acoustic cavitation-based drug delivery systems were gaining more interest by the researchers in recent years and emerged as a transformative approach in modern medicine, offering noninvasive, precise and targeted delivery of drug. By utilising ultrasound waves, these systems enable the efficient transport of therapeutic agents across biological barriers, such as the skin, BBB and cellular membranes, overcoming challenges associated with traditional delivery methods/systems. The present chapter explores the fundamental principles of acoustic-based drug delivery, focusing on mechanisms such as cavitation, acoustic streaming and thermal effects. Additionally, it delves into advancements in microbubble and nanocarrier technologies, which enhance drug encapsulation and release at targeted sites with minimal off-target effects. Applications of these systems were highlighted in various diseases, emphasising their potential to improve therapeutic efficacy and patient outcomes. Despite the promising potential, challenges such as standardisation, scalability and long-term safety remain areas for further exploration. This chapter aims to provide a comprehensive overview of the science, technology and clinical applications of acoustic-based drug delivery, underscoring its role in revolutionising personalised medicine and advancing the frontier of drug delivery systems.

V. Harish (✉) · A. Kumar · A. Leo
School of Pharmaceutical Sciences, Lovely Professional University, Phagwara, Punjab, India
e-mail: harish.24838@lpu.co.in

M. R. Babu · S. Srivastav
Department of Pharmacy, School of Medical and Allied Sciences, Galgotias University, Greater Noida, Uttar Pradesh, India

M. R. Babu et al. (eds.), *Transdermal Applications of Minimally Invasive Drug Delivery Systems*, https://doi.org/10.1007/978-981-96-6275-3_6

Keywords

Acoustic cavitation · Ultrasound waves · Skin · BBB · Microbubble

6.1 Introduction

Cavitation is the term used to describe the oscillation from gas bubbles exposed to compressional waves. Cavitation can be divided into two types of behaviour. At comparatively lower sonic levels, stable cavitation, a recurrent oscillation in bubble diameter without bubble collapse, occurs. The bubble oscillates so much at higher intensities due to the inertia of the inward-moving water that it violently breaks, releasing shock waves, elevated temperatures and free radicals (Fig. 6.1). This type of cavitation is also known as collapse or inertial cavitation (Yasui 2018).

The compressional waves produced by vapour cavities with uninterrupted driving acoustic frequency (f) have subharmonic ($f/2$), ultra-harmonic ($(2n + 1) f/2$, where $n = 1, 2, 3\ldots$) and harmonic (nf, where $n = 1, 2, 3\ldots$). Harmonic and subharmonic frequency observations have been linked to stable cavitation (Hasanzadeh et al. 2011). Alternatively, subharmonic acoustic emission has also been linked to inertial cavitation, which is demonstrated by sonoluminescence, the production of oxidants and the strength of the unwanted backdrop noise. Drug release is caused by

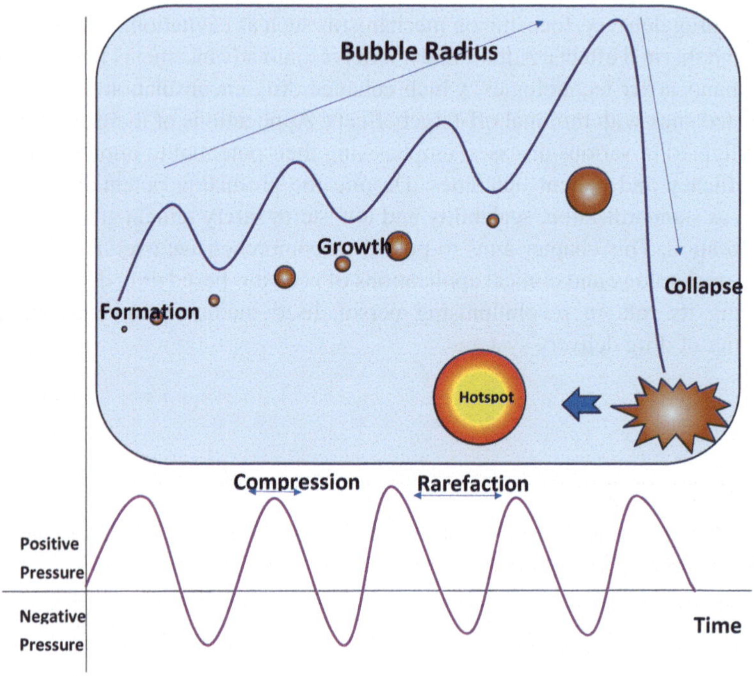

Fig. 6.1 Acoustic pressure cycles

steady (not inertial) cavitation, as evidenced by a favourable association between release of drug and harmonic or subharmonic acoustic emissions when nonharmonic background noise is absent (Cramer and Lauterborn 1982). In the case of collapse cavitation, the broadband emission from the vibrations is responsible for the nonharmonic background. Collapse cavitation may be the cause of medication release if the percentage release matches the degree of change in the acoustic pressure spectrum's baseline (Neppiras 1969).

6.2 Cavitation-Enhanced Extravasation for Drug Delivery

In addition to providing cancer cells with nourishment, the tumour ecosystem and related circulatory system serve as a defensive mechanism. According to recent in vivo research in mice, the therapeutic effectiveness of systemically administered anticancer drugs may eventually be limited by the distinct physiological barriers that tumours offer. Consequently, it is necessary to create new techniques that can alter the tumour environment in order to get over the obstacle that the tumour vasculature presents (Arvanitis et al. 2011).

In the past, a number of tumour environment-targeting techniques have been used to improve therapy procedures and provide effective noninvasive means of delivering anticancer drugs. To better target therapeutic drugs to cancer tumours, many of these strategies rely on taking advantage of specific characteristics of the tumour vasculature. As per research, the extended growth of new blood vessels in tumours and the presence of large gaps (380–780 nm) in certain blood vessels have made it easier for neoplastic drugs to accumulate passively after being delivered intravenously through diffusion or convection (Arvanitis et al. 2011; Jain 2005).

Diffuse agents with a molecular weight of less than 10 nm are unfortunately equally prone to disperse to robust tissues and do not stay in the tumour for an ample period to have therapeutic effects. Additionally, the high interstitial fluid pressure found in tumours limits the ability to attain therapeutically significant levels of macromolecular agents (ranging from 100 to 300 nm) near blood vessels due to weak convective forces. Lastly, the equal distribution of anticancer drugs is hindered by the complex and irregular vascular growth in tumours, which fails to offer consistent and continuous vascular flow through the tumour (Stylianopoulos and Jain 2015).

Elevated dosages of neoplastic drugs require administration through the entire system to raise the concentration throughout the tumour, which has significant unfavourable cytotoxic side effects. In order to lessen systemic cytotoxic effects, it has recently been suggested that encapsulated chemotherapy agents be released via ultrasound (Xie et al. 2018). These agents can be stabilised nanodroplets or thermally sensitive liposomes. The aforementioned physiological barriers nevertheless limit the chemotherapeutic drugs' bioavailability throughout the tumour, even though their bioavailability is higher in the area around the tumour (Maeda 2001). Some tumour cells are situated above 100 μm from blood arteries due to tumour expansion, resulting in hypoxic and necrotic areas. While nearby necrotic zones are situated farther from arteries having dead cells due to a lack of oxygen and nutrition delivery, hypoxic zones are deficient in oxygen concentration and highly active cancer cells. It is still very difficult to extravasate therapeutic medicines to the point where they can reach those distant cancer cells (Allen and Cullis 2013).

Vascular normalisation, a technique that attempts to promote convection and restore tumour blood flow, has been proposed as a pharmaceutical solution to these problems. However, this procedure lessens the enhanced uptake that tumours often demonstrate, raises the potential for inflammation at the tumour's outer region, and only provides a brief window of opportunity for successful treatment (Goel et al. 2011). The search for genomic, dependable and low-impact techniques that could get past the obstacle to medication administration posed by tumour vascular networks and raise the safety margin of powerful antineoplastic elements is still proceeding, even though the majority of the previously mentioned strategies and innovative pharmaceutical compounds show significant potential for future effective and precise treatment (Jain 1990).

Ultrasound, which was once thought of as a diagnostic technique, is quickly becoming a very useful, affordable and completely noninvasive therapeutic tool (Frenkel 2008). Whether used alone or in conjunction with local-level or system-wide administered curative medicines, therapeutic ultrasonography uses either heat or dynamic impacts to cause alterable or permanent tissue transformation (Kennedy 2005). The use of high-intensity focused ultrasound (HIFU) for nonsurgical heat therapy of deeply situated solid masses has made considerable use of thermal effects and is becoming more and more accepted in clinical settings. While mechanical effects have been used to release chemotherapeutic medications from other agents, thermal effects induced by ultrasound have even been used to target the discharge of these drugs by thermosensitive liposomes at a more experimental level (Allen and Cullis 2013).

Acoustic cavitation, which is the regular or irregular movement of gas and vapour pockets influenced by sound waves, often has the most significant effect on ultrasound, along with other factors like acoustic streaming and the force of sound waves on particles. Sonothrombolysis, the reversible breaching of the blood-brain barrier (neurovascular barrier), and the release and transport regarding genes for cardiovascular disorders have all been shown to be significantly impacted by the acoustic cavitation's mechanical effects and bubble interaction with biological environments (Chauhan et al. 2011). Numerous studies have employed stable microbubbles, also referred to as ultrasound contrast agents (UCAs), to deliver abundant delivery of cavitation nuclei in order to significantly enhance the likelihood of cavitation imitation because acoustic cavitation is characterised by randomness in its ecosystem when cavitation sites are absent (Allen and Cullis 2013).

Inertial cavitation, which occurs when a cavity less than the resonant frequency expands rapidly and collapses violently over one or a small number of acoustic cycles under higher tensile pressures, can produce acoustic waves as well as possibly micro-sized liquid jets when it is close to spatial constraints like cell membranes or vessel walls (Deckers and Moonen 2010). Sonoporation, which involves reversibly opening cell membranes to transfer macromolecules and accomplish DNA transfection, has made use of inertial cavitation and its related bioeffects. A recent in vitro research study was the first to directly link the amount of medication being delivered to the acoustic emissions associated with inertial cavitation, further proving that the technique may be used to transport macromolecules into a blocked channel. A number of in vivo investigations also demonstrated that macromolecule extravasation occurs when microbubble destruction is present (Kooiman et al. 2020). It was demonstrated that 200 nm polymer microspheres were distributed in all directions, to an expected extend of 90 μm from UCA-ruptured capillaries using a mouse muscle model. Therefore, regardless of the dimensions of the endothelial gaps within the tumour, sonication-induced cavitation presents the special potential to improve the escape of macromolecular antineoplastic compounds and raise the porosity of microvessels (Sennoga et al. 2017).

Inertial cavitation produces broad-spectrum noise in frequency domain, while stable cavitation is recognised by emissions at harmonics, subharmonics and ultra-harmonics of the main frequency (Chapa et al. 2013). When using therapeutic ultrasound, the most easily affected in situ method for detecting cavitation processes is passive cavitation detection (McLaughlan et al. 2006). This study aims to find the best UCA concentration and ultrasound exposure settings (PRF, duty cycle and pressure amplitude) to enhance ultra-harmonic and broadband acoustic emissions linked to stable and inertial cavitation while minimising cell damage. Stopping the adenovirus from replicating requires achieving a low cavitation activity mortality rate for cancer cells (Shinar and Ilovitsh 2024).

6.3 Exploring the Impact of Cavitation on Low-Frequency Ultrasound-Driven Transdermal Drug Delivery

The increased pressure LFS cell experiment clearly revealed that the main technique by which LFS enhances skin permeability is cavitation induced by ultrasound. By employing a high-viscosity coupling medium to selectively reduce cavitation outside the skin, further evidence suggests that the source of the skin permeability effect (Pitt and Husseini 2008). Under the ultrasonic circumstances analysed, the acoustic measurement technique could not identify intradermal cavitation. Acoustic measures of the two forms of the cavitation activity (temporary vs. steady) indicate that this temporary cavitation is the main source of LFS-triggered skin permeabilisation. By measuring transient cavitation activity at two different points in the LFS system and comparing its relationship with ultrasound intensity to the skin permeabilisation effect, we demonstrated that the key factor in the increased skin permeability from LFS is temporary cavitation occurring on or near the skin membrane (Pitt and Husseini 2008; Sahni and Chaudhuri 2011).

6.4 Trends and Developments in Acoustic Drug Targeting for Solid Tumours: Strategies Surpassing Sonoporation

6.4.1 Drug Encapsulation and Nanomedicine

Over a century ago, gelatin capsules were invented, leading to the development of encapsulation, a protective shell technique for encapsulating medications to prevent leakage and survive harsh environments (Ahmed et al. 2018). By regulating the water solubility, stability or delivery rates of pharmaceuticals, drug carriers progressively evolved into a tool for enhancing their pharmacokinetics. Newer approaches are being researched to enhance medication delivery employing systemically administered nanomedicines, whereas prior efforts mostly focused on drug encapsulation for oral consumption (commonly at the macroscale) (Honari and Sirsi 2023).

One technique that uses aberrant tumour matrix for delivering medications just for tumour tissue is called EPR targeting. The EPR effect demonstrated partial results successfully used in clinics, despite encouraging preclinical results (Jeong et al. 2024). The literature has a wealth of research on the causes of EPR clinical difficulties. It is mostly associated with the diverse character related to the EPR influence and its magnification through preclinical tumour prototypes. Clinical practice has shown that EPR targeting alone is insufficient (Schwartz et al. 2021).

Pharmaceutical scientists are now more aware of cancer biology, thanks to nanomedicine, which helps them spot biological obstacles to drug delivery. Additionally, it has opened the door for the development of drug carrier formulations that contain bioactive substances, offering a platform for cutting-edge delivery and targeting strategies (Haugse 2020). Successful nanocarriers are still being manipulated by scientists to enhance medication delivery, as in the case of COVID-19 vaccinations. Research on classical nanomedicine is still ongoing, with new nanoparticle shapes being studied or altered (Aw et al. 2016).

6.4.1.1 Nanomedicine Beyond the EPR Effect

To discharge their payload locally in tumours, it is possible to create medication carriers responsive towards external energy or inherent factors specific to tumorous tissues. Tumour abnormalities, including decreased pH, hypoxia or aberrant enzyme activity, are used in intrinsic stimuli targeting to cause drug release from carriers. External targeting, on the other hand, causes drug discharge from a stimulus-responsive carrier by locally applying physical triggers like warmth, magnetic effect, voltage or acoustic waves to neoplastic mass. For this aim, heat-sensitive particles are being investigated and have advanced to clinical trials, such as temperature-sensitive liposomes (TSLs) (Haugse 2020; Lyon et al. 2024).

Although thermosensitive drug vehicles have shown encouraging outcomes, they have also encountered a number of drawbacks, such as limited heat resolution, the potential for energy dispersion and challenges in administering hyperthermia in medical environments (Honari 2022). Despite the great potential of this drug delivery method, more effective ways to target deep tissue have been developed via changing the outer stimuli, like ultrasound or high-frequency ablation.

Image-guided drug delivery (IGDD), a more thorough drug delivery method, was created as a result of introduction of targeting actively with stimuli from the outside. "IGDD" refers to the use of an imaging modality to guide drug carriers to the tissue of interest, start the drug discharge and see the position and quantity of drug delivery (ProQuest n.d.-a). By modifying the external energy source, IGDD may provide doctors with a venue to strictly regulate the release of drugs or modify the quantity regarding medication offloading. Because heat-triggered drug release lacks a feedback mechanism, it cannot be considered an IGDD platform. To develop a range of imaging techniques, we examined them in order to begin and measure the release of drugs using a real-time IGDD platform. Ultrasonography is well known among medical imaging modalities for its great penetration depth, cost-effectiveness and relating to space and time accuracy (Awad et al. 2021).

6.4.1.2 Enhancing Delivery Through Drug Uncaging

Investigations revealed that the absence of medication liberation from carriers was one of the main issues in nanomedicine from the beginning. For liposomal doxorubicin and cisplatin liposomes, two therapies that showed just a little increase over their free equivalents, this created difficulties (Lyon et al. 2024). One reason for the failure was the lack of a mechanism in the liposomes to release the medicines for accessibility (Fateh et al. 2021). Researchers looked for creative strategies to cause drug expulsion triggered by a stimulus in order to address this problem and boost success rates, creating a whole new discipline known as "drug uncaging" (Fan et al. 2022). The terms "triggering drug release", "on-demand drug release" and "drug uncaging" all describe the same idea: the release of a drug upon stimulation-induced excitement for drug uncaging. It is common practice to employ external energy like a beam, warmth or ultrasound to disrupt active ingredient transporters. Even with its accuracy in applications, ultrasound's potential has not yet been thoroughly investigated (Purohit et al. 2023).

6.4.2 Ultrasound-Induced Radiation Forces in Drug Delivery

Through positionally regulating and displacing ultrasound-sensitive agents, ultrasound can provide an alternative targeting approach. Low-pressure acoustic waves are known to be able to displace MBs in a liquid medium because the momentum of the sound waves is absorbed. Primary acoustic radiation force is the term for this phenomenon, which is very useful in low-intensity US wave medicine delivery applications (Zhang et al. 2022). Despite its exciting potential, ultrasonic radiation forces are among the lesser-studied ideas on medicine administration. Acoustical data and optical pictures are then used to illustrate the effect of radiation force. Under suitable experimental circumstances, the microsphere transit velocity is calculated (Paliwal and Mitragotri 2006). According to his study, ultrasonic parameters that increase bubble displacement and reduce bubble destruction include a high pulse-repetition frequency (PRF) and a low acoustic pressure. Later, it was shown that using acoustic radiation to show that particles were able to be deposited on a tube was feasible. Via radiation pressures to implant acoustic-responsive particles into compromised tissue remotely is one of the innovative approaches for ultrasound targeting. In a recent investigation, it used ultrasonic exposure to effectively implant PLGA microparticles with many cavities in a foam-cell spheroid model. The damaged tissue might gradually receive the payload from the implanted particle (Kilroy et al. 2012).

6.4.3 Recent Advances in Ultrasound-Based Transdermal Drug Delivery

Although transdermal drug delivery has significantly advanced the field of biomedical applications, it has not yet reached its full potential as a substitute for the widely used oral and intravenous drug administration methods. This modern approach would be a desirable and appealing supplement to the available medication delivery methods. Although oral medicine is frequently used due to its convenience, its pharmacological potential is hindered by absorptive inefficiency. However, there are some disadvantages to this drug delivery method. On the other hand, hypodermic injections allow for the direct administration of a specific amount of medication into the systemic circulation. Up to 30% of people are afraid of needles because of the pain they produce. Additionally, the potential for patients and medical staff to sustain unintentional needle injuries contributes to the undesirable spread of infectious diseases. By applying customised medicated adhesive patches to the skin, transdermal medication administration provides a number of benefits to overcome problems related to oral delivery and hypodermic injections. However, transdermal medication delivery techniques come with a number of drawbacks. The skin's physical barrier primarily limits the effectiveness and application of transdermal medication administration. Keratinocytes are responsible for maintaining the several stratum layers that make up the human epidermis. The mature cells in these cells migrate towards the stratum corneum (SC), the epidermis' topmost layer, as they multiply. This outer barrier is composed of a layer of dead corneocytes laden with keratin that is 15–20 μm thick and appears "brick and mortar" shaped. Numerous state-of-the-art techniques are continuously being used to investigate strong penetration through this natural barrier. Thermal ablation, chemical enhancers, microneedles and ultrasound are examples of third-generation approaches that have been researched (Seah and Teo 2018).

6.4.4 Sonophoretic Effects on the Skin

In order to treat digital polyarthritis, hydrocortisone was initially administered by ultrasound in 1954. This used ultrasound to transfer active components over the skin barrier. There have been several hypothesised processes regarding the effects of sonic cavitation on the skin in the literature that is now available (Rich et al. 2014). The precise mechanism of acoustic cavitation is still unknown; however, it has been shown that the mechanical effects play a significant role in increasing skin permeability (Wu et al. 2023). According to certain ideas, shock waves structurally change the SC's lipids to produce diffusion channels, whereas high-pressure microjets physically shatter the lipid bilayers of the SC. Alternative theories suggest that microstreaming flow close to the skin causes sonophoresis (SP), which in turn causes shear stresses that stretch the SC and open up transdermal distribution channels. It has also been proposed that by increasing the skin permeability and drug diffusion coefficients, the heat effects of ultrasonography can enhance sonophoretic skin contact (Tezel et al. 2002). However, the real SP application might not be strong enough to change the affected medications' and tissue's molecular kinetic energy in a noticeable way. Furthermore, transdermal medications that are thermally unstable may be harmed by heat (Jin et al. 2024).

6.4.5 Noninvasive Therapy by High-Intensity Focused Ultrasound (HIFU)

High-intensity focused ultrasound (HIFU) is used in noninvasive cancer therapy to quickly apply heat to a tiny area that coincides with the focal volume in order to cause thermal necrosis, which kills cells (Coussios et al. 2007). The spatial distribution of acoustic intensity in the beam determines, to the first order, the shape of the zone of thermal necrosis, or lesion, created in tissue by high-intensity focused ultrasound (HIFU) (Jenne 2014). The focal zone, which is an oblate spheroid with a main diameter of roughly 2 mm and a length of roughly 1 cm for a standard HIFU applicator running at 1 MHz, receives the majority of the heating. Solid tumours can therefore be treated noninvasively because they can be found up to 20 cm below the skin's surface. The creation of distinct, dependable and repeatable thermal lesions is therefore ideal, as targeting and real-time monitoring of HIFU therapy present significant obstacles (Marinova et al. 2016). In order to diminish the possibility of unintended damage in the ultrasonic propagation path, it is also crucial to maximise the heat deposition efficiency in the focal volume relative to prefocal heat deposition (Marinova et al. 2016; Yao et al. 2022). We then go on to give a general overview of ultrasonic heat deposition and explain how acoustic cavitation might be used to produce the intended therapeutic effects at far lower intensities than those now used in the majority of clinical systems.

6.5 Making Waves: How Ultrasound-Targeted Drug Delivery Is Transforming Pharmaceutical Approaches

6.5.1 Microbubble Cavitation-Based Drug Delivery

An often utilised method in UTDD is the incorporation of microbubbles, typically comprised of a gas centre surrounded by a stabilising layer composed of lipid, protein or polymer. Because of their special response to sound waves, microbubbles, along with liposomes and micelles, can enhance drug delivery in the body and are commonly employed as UCAs for enhancing image clarity. Microbubbles used for UTDD should ideally be free from toxicity, be administered venously, have a size not exceeding 8 μm in diameter to allow for unhindered passage through the capillary bed, enhance acoustics and be stable enough to last the time taken for administration. Intravenously injected UCA will often reflect ultrasound, elevating contrast by more than 20 dB caused by the acoustic impedance discrepancy in the blood and gas within the UCA (Delaney et al. 2022). When the acoustic wave goes through the bubbles, cavitation causes them to expand and contract. The agent's expansion and contraction, or cavitates, are proportionate to the sound wave's rarefaction and compression, respectively. Microbubbles can be employed in therapeutic settings since they display three distinct behavioural patterns when subjected to ultrasonic waves. The bubbles vibrate at reduced acoustic powers (often $MI < 0.2$) due to the ultrasonic pulses employed for imaging. Minor cell membrane ruptures may result from this, which may improve local but not always regional drug distribution. At MI values between 0.2 and 0.8, changes in vascular permeability take place as the sonic power increases. These changes are caused by increased capillary ruptures that permit blood to escape, pool and enhance local medication absorption. This phenomenon, known as sonoporation, happens when the UCA oscillations start to release shock waves with minimal energy and become nonlinear (also known as stable cavitation) with little microbubble breakdown (Joshi and Joshi 2019). The disruption raises considerably as the acoustic intensity is elevated further (often $MI > 0.8$) when the UCA breaks readily, releasing many intense shock waves that radiate from the microbubbles (so-called inertial cavitation). At decreased ultrasonic frequencies, maintaining the maximum downward pressure enhances cavitation and sonoporation, demonstrating the importance of wave amplitude.

Disrupting and destroying microbubbles (UTMD) may alter vascular structures, including cell junctions and cell membranes, by techniques like radiation force, shock waves, sonoporation and microstreaming. As the bubble collapses against a barrier, fluid will be concentrated and forced through it, forming a liquid jet that may travel over 20 μm in the direction of the boundary at pressures of up to 60 MPa and an average speed of around 80 m/s (Sierra et al. 2017). Ultrasonic treatment will cause microbubbles in the small blood vessels to vibrate and start reactions that improve the nearby blood vessel's permeability, enabling the drugs to exit into the specific tissue. Numerous methods, including rupture or reversible opening of the endothelial cell junctions, fenestration pore modification, endocytosis and sonoporation for intracellular transport and other alterations to the vascular endothelium, might cause these bioeffects. It is crucial to note that, even though UCA is generally given systemically, the ultrasonic pulse can be tailored to produce effects that are unique to the targeted tissue for an anticipated therapeutic application (e.g. opening of the blood-brain barrier (BBB), biofilm destruction, cancer therapies, etc.). During the last couple of decades, various iterations of UCA and microbubbles have been developed with differing degrees of efficacy using various combinations of gases (such as air, oxygen and sulphur hexafluoride) and shell materials (such as proteins, phospholipids, synthetic polymers and surfactants). Optison (GE Healthcare),

Definity (Lantheus Medical Imaging) and Lumason/SonoVue (Bracco) are the only three UCAs presently approved by the FDA for clinical use in the USA. Yet, at different stages of the research process, UTDD can be treated using a range of micro- and nanoscale medications (ProQuest n.d.-b). Three main groups of microbubble types are exploring technologies for drug delivery, as well as simultaneous administration of medication-containing microbubbles: (1) microbubbles containing medicines; (2) drug-filled microbubbles(nanodroplets present in the original location); and (3) focused microbubbles with attached ligands to target receptors on the cell surface. In addition, therapeutic medications have been inserted into UCA, and microbubbles are utilised for drug delivery without the need for administering free systemically. Aiming to optimise the absorption of a substance in the desired area while reducing damage to surrounding tissues which is caused by drug exposure. As mentioned earlier, this concept utilises ultrasonic pressure to induce the therapeutic effect and is aimed at the specific area of interest. Yet, the quantity of medication that can be included into a microbubble shell or membrane is restrained due to restricted space for medication in the process of loading. Additionally, the biological activity of a drug or the acoustic properties of a UCA could potentially be altered, if it is connected to or combined with the UCA microbubble membrane. Lipid-based microbubbles and UCAs presently reign supreme in the field of UTDD literature because of their prominence. Availability for purchase, ease of manufacturing, adaptability and diversity and quick distribution summary. Essentially, the agents are comprised of a gas core surrounded by a lipid monolayer (Chapla et al. 2022).

Selection of gas and lipid can impact the loading and behaviour of the agent. Spherical structures formed by the aggregation of surfactant molecules in a solution likewise, spherical vesicles composed of a single layer of lipids that come together on their own. When placed in a liquid solution, liposomes are sphere-shaped vesicles with two or multiple layers of phospholipids surrounding a liquid centre comprised of water. Numerous research projects are dedicated to producing microbubbles with polymer shells instead of lipid-based agents, which are being considered as an alternative option. Their increased thickness (100–400 nm) may boost drug capacity, which could lead to better results in UTDD (Gallarate et al. 2021). These polymeric microbubbles have a higher tolerance compared to lipid-based microbubbles to ultrasonic pressures; experiments have demonstrated their ability to withstand low pressure levels roughly 0.54 MPa greater than phospholipid agents. Originally, natural compounds such as collagen alginate were utilised, while man-made polymers such as poly(lactic-co-glycolic) acid (PLGA) were employed (Akimoto et al. 2005). PLA has now completely taken the place of these natural ones. Overall, the procedure of producing artificial polymer microbubbles usually involves emulsifying a water phase. By employing a polymeric solution in an organic solvent, one can use either lyophilisation or spray drying or opt for a microfluidic method to produce uniform microbubble populations intended for use by UTDD. In addition to applications, teams are also exploring various types of microbubbles and agents. Protein or bioprotein shells are used in micelles, liposomes and microbubbles, along with lipid- and polymer-based microbubbles (Yang et al. 2020).

6.5.2 Cancer-Related Delivery

The extent of research conducted in this area has been effectively demonstrated in many review papers, which have included studies on microbubbles as a targeted delivery mechanism for a variety of cancer types. A few current initiatives, such as UTDD and sonoporation techniques, will be briefly discussed in this overview. Wu et al. studied how well pluronic polymer micelles, specifically P123/F127, work to produce polymeric micelles transporting curcumin for UTDD purposes (Moradpour et al. 2024).

Jablonowski et al. employed polyethylene glycol (PEG) and tumour necrosis factor-related apoptosis to decorate DOX-loaded PLA microbubbles with the triggering ligand TRAIL, merging the concepts of immune protection and targeting multiple pathways. Following 20 min of ultrasound application. At 5 MHz, 0.94 MPa, these microbubbles demonstrated greater cancer-killing effects on MDA. MB-231 breast cancer cells in culture can reach up to 80%, while MCF7 breast cancer cells can reach up to 60%. They even achieved a reduction in immunological activation of at least 30% (Moradpour et al. 2024; Mulvana et al. 2017). This team recently demonstrated some success in treating MiaPaCa-2 pancreatic cancer cells by encapsulating gemcitabine in a PLA microbubble (Cressey et al. 2019).

The results from experiments performed in a lab setting did not lead to the inhibition of tumours in live mice, demonstrating the constraints of microbubbles in carrying drugs. Teraphongphom et al. showed that embedding quantum dots, magnetic iron oxide nanoparticles and gold nanoparticles into the PLA microbubble's shell is feasible in an attempt to enhance loading and delivery of drugs. This implies that these methods could provide chances for imaging in multiple modes or treatments (Zhang et al. 2022).

Other elements, like DNA or gases, that could enhance cancer treatments can also be transported by microbubbles. For instance, Du et al. conjugated dipalmitoylphosphatidylcholine (DPPC)/1,2-distearoyl-sn-glycero-3-phosphoglycerol (DSPG)/N-(carbonyl-methoxypolyethyleneglycol 2000)-1,2-distearoyl-sn-glycero-3-phosphoethanolamine (DSPE-PEG2000) microbubbles with magnetic mesoporous silica nanoparticles carrying pGCMV/eGFP plasmid DNA (Wang et al. 2024). This method shielded surrounding cells from the vector's nonspecific cytotoxicity while simultaneously preventing the vector from degrading in an ovarian cancer xenograft mice model (Kumar 2019).

Furthermore, they were able to preserve the microbubble's cavitational characteristics for UTDD after 1 min of insonation at 7 MHz, 0.6 MPa and 30% duty cycle. Eisenbrey and colleagues developed ultrasonic-responsive oxygen-carrying microbubbles using insonation (4.2 MHz, 2.5 MPa, 4 s) with Span 60 and water-soluble vitamin E surfactant (SE61 O2) to treat hypoxic breast cancers in mice. By elevating the tumour radiosensitivity, they increased it nearly threefold compared to the hypoxic control increase in tumour oxygen levels by 20 mmHg right before starting radiation treatment. Additionally, this research showed that SE61 O2 microbubbles proved effective in treating brain metastases due to these types of breast cancer, indicating that these agents have a broad range of applications (Sun et al. 2018), to achieve multimodal UTDD against hypoxic cancers like breast and head tumours. In addition to studying neck cancers, this team is also investigating the idea of placing lonidamine in these oxygen-transporting substance microbubbles (Moradpour et al. 2024).

Pancreatic ductal adenocarcinoma (PDAC) stays a fatal cancer with a medianless than half a year survival expectation and an approximately 10% chance of survival within 5 years; considerable attempts have also been made to sonoporate this cancer. Conventional chemotherapy regimens are mainly ineffective at preventing recurrence and eventual death because they have only slightly improved overall survival rates. Logan et al. developed a drug-filled phospholipid compound for a one-step production of a dual-filled UCA, providing a new technique to load multiple drugs into a lipid-based microbubble (Okuda 1996).

Mixing gemcitabine with phospholipase D, calcium chloride, chloroform and DBPC resulted in a gemcitabine-carrying phospholipid complex for the lipid microbubble shell. The final microbubble also contained DSPE-PEG2000 and paclitaxel in a specific weight ratio, encapsulating a perfluorobutane core. Combining this microbubble with ultrasound led to a 40% reduction in tumour growth in a BxPC-3 pancreatic cancer mouse model (Myers et al. 2021).

Dwivedi et al. developed a DOX-loaded magneto-liposome complex with an alternative approach, a microbubble complex containing DOX-loaded iron oxide nanoparticles enclosed within small, multilayered vesicles and later bound chemically to lipid-coated microbubbles (Mittal and El-Serag 2013). DSPC combined with cholesterol and DSPE-PEG2000 maleimide (60 parts per one-sixth). Utilising a perfluorooctane gas core (20 to every one) in a mouse model with pancreatic cancer xenografts, the magneto-liposome-bound microbubble complex improved targeting of pancreatic tumours and had tumouricidal effects, resulting in about an 80% decrease in the size of the tumour. Furthermore, teams are attempting to use ultrasound to improve PDAC tumour permeability and drug uptake (Myers et al. 2021).

Schultz and colleagues employed xenograft mouse models of PDAC in order to investigate the use of PDAC sonoporation with two different methods of ultrasound treatment, consisting of 2.0 MHz pulses of 20 μs duration at both low and high acoustic powers, administered for a duration of 10 min. All four universally clinically approved ultrasound contrast agents (Optison, Definity, Lumason/SonoVue and Asserrectorferm Tema.Sonazoid [manufactured by GE Healthcare]) were used to find the best settings for enhancing treatment effectiveness (Sultan et al. 2021). Across several tests and permeability markers, high-power ultrasound treatment (ISPTA = 200 mW/cm^2) with Sonazoid as the UCA most steadily increased permeabilisation. According to the findings of this and other preclinical research, UCA can produce tiny resealing pores in cells, weaken endothelial cell connections and raise fenestration sizes. These effects can all let chemotherapeutic drugs enter the PDAC tumour more deeply (D'Souza et al. 2019a).

Ten PDAC patients undergoing sonoporation treatment showed a substantial improvement in efficacy when compared to 63 historical controls in a phase I clinical trial designed to enhance the effectiveness of standard-of-care chemotherapy (De Cock et al. 2015; Leow et al. 2015). Half of the patients saw tumour regression, and the number of treatments they could receive increased significantly (8.3 ± 6.0 cycles in controls, 13.8 ± 5.6 cycles after sonoporation treatment; $p = 0.008$). Most significantly, the controls' total survival increased from 8.9 months to 17.6 months ($p = 0.011$). There is presently a larger phase II clinical investigation in progress (D'Souza et al. 2019b).

6.5.3 Infection-Related Delivery

One new use of UTDD is the study of the augmentation of antibiotics activated by ultrasound to heal infections. Recently, Horsley et al. showed that their DSPC/DSPE-PEG/DSPE-PEG-biotin/1,2-dipalmitoyl-sn-glycero-3-phosphoethanolamine-N-(lissamine rhodamine B sulfonyl) increased the intracellular delivery of gentamicin by more than 16 times in the lab using an organoid model of human bladder of urinary tract infection. (79.51/6-em narrow space]: [1/6-em thin space] 10[thin space (1/6-em)]:[thin space (1/6-em)]10[thin space (1/6-em)]:[thin space (1/6-em)]0.5 molar ratio) microbubbles adorned with DSPC/DSPE-PEG liposomes loaded with gentamicin when subjected to ultrasonic treatment (1.1 MHz, 2.5 MPa, 20 s) in contrast to insonated controls (Hu et al. 2019). Additionally, liposomes not conjugated to microbubbles but insonated resulted in double the delivery compared to other methods (Jia et al. 2024). Improving acute transplant rejection is another use of UTDD that has recently been researched; one example of this is the research on rejection of heart transplants that Liu et al. presented. This team created a DSPC/DSPE-PEG2000 (9[thin space (1/6-em)]: [thin space (1/6-em)]1 molar ratio) microbubble loaded with the anti-rejection drug FK506; the rats that were treated with both the infused microbubbles and exposure to HIFU (1 MHz, 2 W/cm^2) showed decreased rejection of cardiac graft (average 4-day prolonged graft survival) and heightened drug distribution (1.64-fold higher) (Guo et al. 2024).

Research has proved both in vitro and in vivo that simple biofilms and catheter-associated biofilms become more vulnerable to vancomycin after 5–10 min of UTMD (80–300 kHz at 0.5–1.0 W/cm^2 and a 50% duty cycle, which is substantially outside the capabilities of clinical scanners) via cavitation. Strong shock waves caused by the bursting of UCA microbubbles cause biofilms to temporarily make the cell membrane permeable, accelerate transportation of antibiotics, increase the permeability of the extracellular matrix of the biofilm and decrease its thickness (Jung et al. 2018). Even though the prevalence of joint infections has been declining, bacterial infections of the synovial fluid will be diagnosed in over 20,000 people annually. Large, proteinaceous, bacterial biofilm aggregates (about 2–4 mm in size) may be the main cause of antibiotic treatment failures for joint infections, which are a major source of morbidity and mortality (ProQuest n.d.-c).

During a preliminary study, *Staphylococcus aureus* septic arthritis was examined in 12 pig femorotibial joints; we evaluated the therapeutic applicability of amikacin injections and Definity microbubbles in conjunction with UTMD application (Bugli et al. 2022). The infection persisted in septic joints despite treatment with amikacin alone or amikacin in combination with ultrasonic insonation (without microbubbles). All five pigs with proven infection demonstrated a decrease in bacterial load to below visible levels ($p = 0.008$) when UTMD was administered concurrently with injection of amikacin (4 s destructive pulses at an MI > 0.6 at 3.5–5.0 MHz). This implies that treating challenging instances of septic arthritis with antibiotics and microbubble cavitation may be a successful therapy strategy (Yan et al. 2024).

6.6 UCAs (Ultrasound Contrast Agents) with Different Cavitation Scales and Technical Challenges

6.6.1 Spontaneous Bubbles

Gas-filled substances can serve as contrast agents to induce cavitation effects due to the contrast in acoustic impedances between tissue and gas (ProQuest n.d.-d). Gas cavities will form as a result of the disruption of the bonds between the molecules of the liquids when the negative ultrasonic pressure is high enough (higher than 4.0 MPa for water). Consequently, the ultrasonic energy focus is in the millimetre range, which determines the size of cavitation for naturally occurring UCAs. Through the use of a fast motion camera, Connor et al. immediately saw that high-pressure ultrasound (28.6 MPa) produced bubble clouds of several millimetres, while increased pressure ultrasound (38.4 MPa) produced enormous gas bubbles ranging from 10 to 100 μm in diameter (Fu et al. 2019). The enormous ultrasonic pressure needed to create spontaneous gas nuclei further compromises the technique's safety, in addition to its comparatively large cavitation scale. One encouraging technological development to address these issues is the use of catheters (diameters ranging from 2.5 to 7.0 mm) to accurately provide intense sound pressure pulses and create bubble clouds (Fernandes 2024).

6.6.2 Microbubble

When UCA concentration and ultrasonic energy are taken into consideration, synthetic UCAs can cause cavitation on a micrometre to millimetre scale. A consistent size distribution of man-made microbubbles containing gas that is fluorinated (C3F8/C4F10) and different shell materials was achieved (1–5 μm) (Fernandes 2024; Hu et al. 2023). Synthetic microbubbles are made with shell materials including phosphatidylcholine lipids (approximately 3 nm thick), polymers/copolymers (100–200 nm thick) and proteins that have been altered in structure (15–20 nm thick). Different types of functional materials, such as liposomes for drug-loading microbubbles, cationic lipids for gene-loading microbubbles and antibodies for tumour-targeting microbubbles, have been incorporated into microbubble shells (Mulvana et al. 2017). Antibodies attached to the shells can actively target and cluster microbubbles in tumours, which is unlike the passive aiming of ultrasound concentration. Green fluorescence indicates that the attached iRGD and/or CCR2 antibodies on the microbubble shells have enhanced the tumour-targeting and tumour-accumulating capacity of the microbubbles (Versluis et al. 2020). However, due to their limited circulation durations (half-life of 3–15 min), the use of microbubbles in few advanced treatment and diagnosis remains difficult. The addition of the PLGA polymer to shell made of lipids surrounding microbubbles enhances their durability in vivo. Additionally, it is demonstrated that the inclusion of hard shell microbubbles with a prolonged cavitation period (about two times) and great circulation stability was created using PLGA (Wang et al. 2024).

6.6.3 Nanodroplet

Another important man-made contrast agent is the nanodroplet, with a size ranging from 100 to 600 nm, containing liquid cores made of perfluoropentane ($C5F12$, BP = 29 °C) having lower boiling point or perfluorohexane ($C6F14$, BP = 56 °C), and using shell materials similar to microbubbles. During the ultrasound rarefaction cycle, the liquid centres of the nanodroplets experienced phase changes and evaporated to produce microbubbles when excited with ultrasound at high pressures (more than 2 MPa) (Zullino et al. 2018). Nanodroplets a few hundred nanometres in size are superior to microbubbles in terms of their ability to penetrate deeper into tumour tissues and blood vessel gaps. Furthermore, nanodroplets had a longer half-life of 2–6 h at 37 °C than microbubbles did. The high level of ultrasonic pressure needed for vaporisation is, however, a barrier to the translational study of nanodroplets in preclinical research. A synergistic vaporisation technology may be developed in the future to vaporise nanodroplets with safer ultrasonic exposure. By combining optical and acoustic energy, the ultrasound pressure required to vaporise 254.8 nm nanodroplets could potentially be reduced to 0.82 MPa (Il Yoon et al. 2017).

6.6.4 Gas-Stabilising Nanoparticles

The recently published synthetic UCAs are gas-stabilising nanoparticles, such as mesoporous silica nanoparticles, nanocups and nanocones. These nanoparticles can confine nanobubbles against gas dissolution because of their structural holes. Throughout the ultrasonic wave's rarefactional pressure phase, the nanobubbles expanded and separated from the cavities. The driving ultrasound's frequency (0.5–2 MHz) was significantly lower than the nanobubble resonance frequency (e.g. 120 MHz for a 100 nm diameter); thus, inertial cavitation was induced mainly, and as the diameter of the nanobubbles increased, so did their inertial cavitation thresholds (2 MPa for a 100 nm diameter). Gas-stabilising nanoparticle-mediated cavitation activity persisted for at least four times as long as microbubbles. Nevertheless, the synthesis of these nanoparticles was far more intricate than that of microbubbles (Yemane et al. 2020).

6.6.5 The Newly Revealed Cavitation Physical Details

Cavitation occurs when UCAs violently collapse or vibrate regularly due to ultrasonic wave stimulation. Cavitation was classified based on the mechanical index (MI) of ultrasound as either inertial cavitation involving microbubble collapse or stable cavitation involving microbubble vibration. The specific form of cavitation was affected by the pressure, viscosity and inertia of the external environment, as well as the materials of the shell and the surface tension of the microbubbles (Tang et al. 2022). According to cavitation imaging and numerical modelling, inertial cavitation causes mechanical effects by microjetting, while stable cavitation causes mechanical effects through microstreaming. Cavitation microstreaming causes the fluid around it to swirl in a cross- or sunflower-shaped pattern. The observed fluid microstreaming velocity ranged widely from 6 mm/s to 525 cm/s, depending on the liquid's viscosity, the microbubble's condition at the limits and the ultrasonic parameters. When microbubbles collapse quickly during cavitation microjetting, high-speed fluid jets and shock waves are produced. Cavitation jetting produced a shock wave with a maximum intensity of 1.8 GPa and a velocity of 5.5–157.74 m/s (Ghiyasiyan-Arani 2024). Lately, researchers have utilised high-speed fluorescence imaging and ultrahigh-speed cameras to showcase the release kinetics of microbubble payloads during cavitation. Lipid release from microbubble shells started occurring when the bubble's relative oscillation amplitude exceeded 30%, as evidenced by Luan et al. Nonspherical bubble oscillations can result in lipid oversaturation in the vicinity before discharging the lipids at a speed of approximately 0.1 m/s. Keep in mind that cavitation's shedding action may sonoprint directly delivering microbubble payloads into tumour spheroids or cells. In sonoprinting, the nanomedicines were applied to the targets as patches rather than via diffusion driven by a concentration gradient. It can be concluded that during sonoprinting, damage was done to the tumour or cell barriers. It is unknown, nevertheless, if biomarkers were also quickly released at the same time as the microbubble cavitation (ProQuest n.d.-e).

6.7 Cellular Responses to Cavitation-Mediated Barrier-Breaking

6.7.1 Membrane Retraction

By employing microscopic imaging with a 10 ms gap between frames, it has been proved that cavitation microstreaming outcomes in the retraction of lipid bilayer membranes. Recent research utilising an ultrahigh-speed camera and position-locating microbeads found that the maximal deformation strain reduced exponentially as the cell-bubble distance decreased, and cavitation jetting caused irregular membrane retraction within 12 μs. Both the membrane and actin cytoskeleton decreased in size within 120 s after cavitation, even though the cell remained undamaged, they also revealed. Significantly, membrane retraction due to cavitation might raise endothelial cell bonding, potentially assisting in delivery of drugs and release of biomarkers (Houthaeve et al. 2022).

6.7.2 Sonoporation

Cavitation leads to the lipid bilayer breaking apart and results in pores forming on the cell membrane, a phenomenon referred to as sonoporation. Transport through the cell membrane that occurs in both directions and is facilitated without intermediaries is possible through the generated holes. By using scanning electron microscopy, it is possible to detect the correlation between ultrasound pressures and pore size. For example, an ultrasound of 0.19 MPa creates pores that range in diameter from 1 to 94.3 nm, while an ultrasound of 0.48 MPa creates pores that range in diameter from 2.03 to 4.31 μm (Schlicher et al. 2006). The high-resolution confocal microscope unveiled that the micrometre-scale pores enlarged before starting to heal. They further demonstrated that pores bigger than 100 μm^2 cannot heal, while those smaller than 30 μm^2 may close in as little as 1 min. Also, they demonstrated that shifting microbubbles produced longer, more substantial disruptions, whereas non-displacing microbubbles produced circular ones. Further research revealed that dislocating microbubbles improved alterable sonoporation and maintained cell viability in comparison to non-displacing microbubbles. These findings highlight that, particularly when employing a lengthy ultrasound pulse duration for excitation, the dislocation of cavitation microbubbles caused by force from ultrasonic radiation should also be considered in sonoporation (Meijering et al. 2009).

6.7.3 Endocytosis/Exocytosis

Nanoscale membrane budding in endocytosis and exocytosis events shields cells from mechanical stress loading. Cellular reactions brought on by cavitation have been documented to exhibit these phenomena. It is hypothesised that cavitation-induced endocytosis facilitated the intracellular transport of macromolecules

(155- and 500-kDa dextrans) (Hu et al. 2023). FITC-dextran (2 MDa) was primarily taken up by cells via endocytosis within 15 min of excitation due to microbubble cavitation caused by low acoustic pressure. In addition to endocytosis, it has been observed that the cavitation-damaged cells repaired their membranes by secreting a significant quantity of lysosomes through exocytosis (Fekri et al. 2016). According to reports, cavitation enhanced the quantity of extracellular vehicles (EVs) bearing CD9 and CD63a by increasing the exocytosis of cancer cells, which in turn raised the quantity of EVs by two and four times, respectively (Wen et al. 2023; De Cock et al. 2015).

6.7.4 Membrane Blebbing

Recent data showed that the microbubble-cell contact caused the cytoskeleton to disassemble and reassemble in addition to causing membrane-level effects. With increasing sonic pressure and bubble-cell distance, the extent of cavitation-induced cytoskeleton damage rose and fell, respectively (Leow et al. 2015). The resealed membrane grew, and a round membrane bulge with a size ranging from 30 to 2000 nm developed the microbubble's diameter when it separated from the injured cytoskeleton (Jia et al. 2021). The irreversible blebs caused cell death, whereas the reversible blebs passed through four phases: nucleation, enlargement, stopping and retraction. They also found a positive correlation between the amount of cell harm caused by cavitation and the lifespan of membrane blebs, which ranged from tens of seconds to minutes (Qin et al. 2018).

6.7.5 Apoptosis

According to reports, irreversible cellular cavitation damage (such as an excessive calcium inflow, generation of reactive oxygen species, reduction in mitochondrial membrane potential and liberation of cytochrome c) may cause cells to undergo apoptosis and then fragment their constituents into apoptotic blebs (Honda et al. 2004). On the cells subjected to cavitation (US + MB), significant membrane blebs linked to apoptosis appeared. Be aware that apoptosis brought on by cavitation was a prolonged cellular occurrence. Twelve hours following cavitation, the quantity of cell apoptotic bodies surged (Cao et al. 2021). Cavitation-mediated drug delivery improved the therapeutic efficacy of anticancer medications by increasing cell death in a linear fashion with ultrasonic acoustic energy. In the case of biomarker release through cavitation effects, cell apoptosis led to the rupture of DNA double strands and the liberation of cytosolic components. The apoptotic bodies were observed to release intracellular DNA fragments from the cells harmed through cavitation (Ashush et al. 2000).

6.7.6 Cavitation-Mediated Drug Delivery

Cavitation facilitated targeted drug distribution in tumours, reducing nonspecific distribution and systemic toxicity by utilising ultrasound and UCAs' ability to target tumours and load drugs. Therapeutics such as neoplastic genes, chemotherapeutic medications, photosensitisers and sonosensitisers have all been introduced into tumours by cavitation. A peak efficiency of approximately 30% was documented for in vitro gene transfection with cavitation using either 3D acoustic field tuning or a specialised microbubble approach (van Blokland 2018). Using plasmid DNA that expressed IFN-β during in vivo transfection led to a 3.4-fold elevation in T cell recruitment and raised IFN-β production from 1.1 to 3.8 pg/mg due to cavitation improvement. In vivo transfection with cavitation-medicated siRNA and shRNA, which work to suppress mRNA associated with cancer, dramatically reduced the depiction of the specific proteins (AKT2 and Nrf2) in tumours (Helfield et al. 2023). Additionally, cavitation improved siRNA (SIK2, anti-miR21, VEGF and survivin) penetration and accumulation in tumour areas during the combination chemo-gene treatment utilising siRNA and paclitaxel, leading to effective tumour growth reduction. When siRNA (HIF1α) and a photosensitiser (porphyrin) were used in synergistic therapy, cavitation-mediated transfection downregulated over 80% of HIF1α mRNA, and the treatment totally stopped tumour development (Bian 2016). Furthermore, cavitation in synergistic therapy with a chemotherapeutic agent (hydroxychloroquine) and a sonosensitiser (chlorin e6) led to increased buildup of drugs (1.88–3.33-fold) in gliomas and extended the average survival period of mice with tumours to 52 days.

Cavitation significantly enhanced the concentration of drugs (paclitaxel, doxorubicin, gemcitabine, etoposide and hydroxychloroquine) within areas of tumour, leading to a four to eight times increase in buildup and ultimately improving the effectiveness of chemotherapy (Bian et al. 2017). The first clinical trial in humans has reported the use of cavitation to improve the delivery of carboplatin chemotherapy to the brain in patients with recurrent glioblastoma. During the research, patients with an intact or slightly disrupted BBB ($n = 8$) had an average survival of 8.64 months, whereas those with significant BBB disruption ($n = 11$) survived for a median of 12.94 months (Pua and Zhong 2009). This approach involves higher doses of drugs (paclitaxel, doxorubicin, gemcitabine, etoposide and hydroxychloroquine) specifically administered within areas of the tumour to enhance the effectiveness of chemotherapy. The first clinical trial in humans has reported on the use of carboplatin chemotherapy through cavitation-enhanced trans-BBB delivery in patients with recurrent glioblastoma (Tang et al. 2002). Patients with little to no disruption in the BBB had a median survival of 8.64 months in the research, in relation to patients with evident BBB disruption who had an average survival of 12.94 months (Paliwal and Mitragotri 2006). Various drug delivery systems developed using cavitation technology were shown in Table 6.1.

Table 6.1 Cavitation-mediated drug delivery

References	Therapeutic agent	Model	UCAs	Ultrasound	Results
Xie (2021)	Plasmid DNA (pCMV-C-mCherry and pCMV-C-EGFP)	YTS cells and HeLa cells	Microbubble: 1.07 ± 0.85 μm Conc.: (1.73 ± 0.61) × 10^8/mL	Freq.: 250 kHz; PNP: 0.1–0.5 MPa; PD: 10–20%; time: 30–60 s	A three-dimensional acoustic field-based conformal transfection (AFCT) system increased in vitro transfection efficiency to ~30%
Ilovitsh (2020)	Plasmid DNA (pGFP and pIFN-β)	Murine breast tumour model (HCC827 cells and NDL cells)	Microbubble: Targeting microbubbles (VisualSonics™) loaded with CD326 antibody, 0.75 μm, Conc.: 2 × 10^7/mL	Freq.: 250 kHz; PNP: 100–500 kPa; PD: 4 ms; PRF: 30 Hz; time: 3 min	The maximum in vitro transfection efficiency (~33%) was achieved at a PNP of 500 kPa. In vivo transfection increased IFN-β secretion from 1.1 pg/mg to 3.8 pg/mg and enhanced T cell recruitment (3.4-fold)
Liu (2021)	Plasmid DNA (pGPU6/GFP) and shRNA (AKT2)	Murine breast tumour model (MCF-7 cells)	Microbubble: dual-targeting microbubbles loaded with iRGD-peptide and CCR2 antibodies, 1.32 ± 0.22 μm Conc.: 1 × 10^5/mL	Freq.: 1 MHz; PRF: 1 kHz; PD: 30%; Intensity: 1.65 W/cm^2; time: 30 s	Targeted microbubble increased in vitro transfection efficiency to ~30%. In vivo transfection decreased the expression of AKT2 protein and inhibited tumour growth
Argenziano (2022)	siRNA (Nrf2)	Murine melanoma model (M14 cells)	Nanodroplet: ~ 100 nm Conc.:1 × 10^{12}/mL	Freq.: 2.5 MHz; time: 0, 5, 10, 15, 30 and 60 s	Nrf2 protein was hardly detected 48 h after ultrasound and nanobubble-mediated delivery. The treatment also sensitised M14
Liu (2020)	siRNA (SIK2 and anti-miR21) nanocomplex and paclitaxel	Murine ovarian cancer model (EOC cells)	Microbubble: SonoVue™ Conc.: 5 mL/kg	Freq.: 1 MHz; PD: 50%; Intensity: 1.8 W/cm^2; time: 10 s	Cells to cisplatin treatment Ultrasound and microbubble (USMB) resulted in a 2.5-fold increase in RNA delivery in EOC xenografts, and USMB + siSIK2 + paclitaxel combination treatment reduced tumour volumes by ~93%

(continued)

Table 6.1 (continued)

References	Therapeutic agent	Model	UCAs	Ultrasound	Results
Han (2020)	siRNA (VEGF) nanoparticles and paclitaxel	B16F10 cells and SCC-7 tumour model	Microbubble: 1.72 ± 0.44 µm concentration data not given	Freq.: 1.5 MHz; PD: 5%; PRF: 1 Hz; I_{SPTP}: 10 W/cm^2; time: 30 s	Both penetration and accumulation of siRNA-nanoparticle were enhanced in tumour tissues. The treatment achieved sufficient VEGF-A inhibition and significantly inhibited tumour growth
Akbaba (2020)	siRNA (survivin) and paclitaxel	Murine lung cancer model (A549 cells)	Nanodroplet: 218.9369.6 nm, Conc.: 12–60 µg/mL	Freq.: 3 MHz; energy: 449 J	Nanobubbles enhanced the delivered siRNA into cancer cells and significantly increased the apoptosis rate of cancer cells to 52.8%
Sun (2018)	siRNA (HIF1α) and porphyrin	Murine breast cancer model (MDA-MB-231 cells)	Microbubble: ~ 1.3 µm Concentration data not given	Freq.: 1.03 MHz; PD: 50%; intensity: 1 W/cm^2; time: 1–3 min	In vitro transfection downregulated nearly 80% of HIF 1α mRNA. Synergistic treatment using ultrasound and laser irradiation completely inhibited the tumour growth
Qu (2020)	Chlorin e6 and hydroxychloroquine	Murine glioma tumour model	Microbubble: Diameter and concentration data not given	Freq.: 1 MHz; PD: 20%; PRF: 1 Hz; intensity: 0.6–1 W/cm^2; time: 60 s	Ultrasound-targeted microbubble destruction caused higher drug accumulation (1.88–3.33-folds) in glioma and increased the median survival durations of tumour-bearing mice to 52 days
Kuo (2019)	Doxorubicin	Murine Osteosarcoma model (MG-63 cells)	Microbubble: 1.1 ± 0.3 µm Conc.: (18.7 ± 5.9) × 10^9/mL	Freq.: 1 MHz; PNP: 0.3 MPa; PD: 10 ms; PRF: 1 Hz; time: 1 m	Microbubbles showed high payload capacity (efficiency of 69.4 ± 9.1%, payload of 1.4 mg/mL) of doxorubicin and provided 3.7-fold suppression of tumour growth

References	Therapeutic agent	Model	UCAs	Ultrasound	Results
Zhang (2021) and Ishida (2021)	Doxorubicin Doxorubicin	Murine breast cancer model (MDA-MB231 cells) Diffuse intrinsic pontine glioma model (SU-DIPG-17 cells)	Microbubbles: 2.2 ± 1.2 μm Conc.: 1 × 10^9/mL Microbubble: Definity™ Conc.: 0.02 mL/kg	Freq.: 3.0 MHz; PD: 0.0087%; PRF: 10 Hz; power: 60%; MI: 0.97; time: 10 min Freq.: 1.78 MHz; PRF: 0.67 Hz; PD: 10 ms; time: 3 min	Ultrasound stimulation released doxorubicin loaded on the cross-linked microbubbles in a rate of six times higher and promoted doxorubicin deposition in tumours. MRI-guided focused ultrasound enhanced doxorubicin delivery by fourfold compared to controls and significantly suppressed the volumetric tumour growth rate (24.4 ± 10.9%)
Ruan (2021)	Gemcitabine	Murine bladder tumour model (RT112 cells)	Microbubble: diameter data not given Conc.: 3 × 10^8/mL	Freq.: 1 MHz; PNP: 1 MPa; PD: 1%; PRF: 0.5 Hz	Ultrasound- and microbubble-mediated gemcitabine delivery delayed the time for tripling tumour volume to 25.6 ± 6.8 days and reduced the acute intestinal toxicity of gemcitabine
Wei (2021)	Etoposide	Murine glioblastoma model (MGPP3 cells)	Microbubble: 1.37 ± 1.02 μm Conc.: 8 × 10^8/mL	Freq.: 1.5 MHz; PNP: 0.7 MPa; PD: 1 ms; PRF: 5 Hz; time: 120 s	Ultrasound and microbubble increased etoposide concentration in tumour tissue by eightfold and decreased tumour growth by 45% as compared to without ultrasound
Wang (2022)	Hydroxychloroquine	Murine breast tumour model (4T1 cells)	Nanodroplet: 266.2 nm Concentration data not given	Freq.: 1 MHz; intensity: 1.5 W/cm^2; time: 1 min	Ultrasound-triggered release selectively and efficiently accumulated hydroxychloroquine-loaded nanodroplets in tumours. The treatment inhibited tumour growth with an inhibition value of 80.02%

(continued)

Table 6.1 (continued)

References	Therapeutic agent	Model	UCAs	Ultrasound	Results
Idbaih (2019)	Carboplatin	Human glioblastoma	Microbubble: SonoVue™ Conc.: 0.1 mL/kg	Freq.: 1.05 MHz; PD: 23.8 ms; PRF: 0.5–1 Hz; time: 50–270 s	And inhibited tumour lung metastasis Patients with clear BBB disruption ($n = 11$) had a median overall survival (OS) of 8.64 months compared with that of 8.64 months with no or poor BBB disruption ($n = 8$)

6.8 Conclusion

Acoustic-based drug delivery systems represent a paradigm shift in the field of therapeutic delivery, offering unparalleled precision, safety and efficiency. By harnessing the unique properties of ultrasound waves, these systems have demonstrated the potential to overcome critical barriers in conventional drug delivery, enabling targeted and controlled release of therapeutics. The integration of microbubble and nanocarrier technologies has further enhanced their efficacy, paving the way of innovative treatments in oncology, neurology and gene therapy.

Despite significant advancements, challenges remain in optimising ultrasound parameters, ensuring long-term safety and achieving clinical scalability. Continued research and interdisciplinary collaboration will be essential to address these limitations and unlock the full potential of this technology. As the field evolves, acoustic-based drug delivery systems are poised to play a transformative role in personalised medicine, offering new hope for effective and minimally invasive treatments across a wide range of diseases.

References

Ahmed SE, Awad N, Paul V, Moussa HG, Husseini GA (2018) Improving the efficacy of anticancer drugs via encapsulation and acoustic release. Curr Top Med Chem 18(10):857–880. https://doi.org/10.2174/1568026618666180608125344

Akbaba H, Erel-Akbaba G, Kotmakçı M, Başpınar Y (2020) Enhanced Cellular Uptake and Gene Silencing Activity of Survivin-siRNA via Ultrasound-Mediated Nanobubbles in Lung Cancer Cells. Pharm Res, 37(8) https://doi.org/10.1007/S11095-020-02885-X

Akimoto T et al (2005) Acute genitourinary toxicity after high-dose-rate (HDR) brachytherapy combined with hypofractionated external-beam radiation therapy for localized prostate cancer: correlation between the urethral dose in HDR brachytherapy and the severity of acute genitourinary toxicity. Int J Radiat Oncol Biol Phys 63(2):463–471. https://doi.org/10.1016/j.ijrobp.2004.11.041

Allen TM, Cullis PR (2013) Liposomal drug delivery systems: from concept to clinical applications. Adv Drug Deliv Rev 65(1):36–48. https://doi.org/10.1016/J.ADDR.2012.09.037

Argenziano M et al (2022) Ultrasound-Responsive Nrf2-Targeting siRNA-Loaded Nanobubbles for Enhancing the Treatment of Melanoma. Pharmaceutics 14(2):341. https://doi.org/10.3390/PHARMACEUTICS14020341

Arvanitis CD, Bazan-Peregrino M, Rifai B, Seymour LW, Coussios CC (2011) Cavitation-enhanced extravasation for drug delivery. Ultrasound Med Biol 37(11):1838–1852. https://doi.org/10.1016/J.ULTRASMEDBIO.2011.08.004

Ashush H, Rozenszajn LA, Blass M, Barda-Saad M, Azimov D, Radnay J, Zipori D, Rosenschein U (2000) Apoptosis induction of human myeloid leukemic cells by ultrasound exposure. Cancer Res 60(4):1014–1020. https://aacrjournals.org/cancerres/article/60/4/1014/507034/Apoptosis-Induction-of-Human-Myeloid-Leukemic. Accessed 27 Nov 2024. [Online]

Aw MS, Paniwnyk L, Losic D (2016) The progressive role of acoustic cavitation for non-invasive therapies, contrast imaging and blood-tumor permeability enhancement. Expert Opin Drug Deliv 13(10):1383–1396. https://doi.org/10.1080/17425247.2016.1192123

Awad NS et al (2021) Ultrasound-responsive nanocarriers in cancer treatment: a review. ACS Pharmacol Transl Sci 4(2):589–612. https://doi.org/10.1021/ACSPTSCI.0C00212/ASSET/IMAGES/MEDIUM/PT0C00212_0010.GIF

Bian S (2016) Real-time monitoring of ultrasound and cavitation mediated drug delivery. University of Oxford, Oxford

Bian S, Seth A, Daly D, Carlisle R, Stride E (2017) A multimodal instrument for real-time in situ study of ultrasound and cavitation mediated drug delivery. Rev Sci Instrum 88(3):034302. https://doi.org/10.1063/1.4978811/1021496

Bugli F et al (2022) Antimicrobial peptides for tackling cystic fibrosis related bacterial infections: a review. Microbiol Res 263:127152. https://doi.org/10.1016/J.MICRES.2022.127152

Cao J et al (2021) Microbubble-mediated cavitation promotes apoptosis and suppresses invasion in AsPC-1 cells. Ultrasound Med Biol 47(2):323–333. https://doi.org/10.1016/J.ULTRASMEDBIO.2020.10.014

Chapa J, Bourgo RJ, Greene GL, Kulkarni S, An G (2013) Examining the pathogenesis of breast cancer using a novel agent-based model of mammary ductal epithelium dynamics. PLoS One 8(5):e64091. https://doi.org/10.1371/JOURNAL.PONE.0064091

Chapla R, Huynh KT, Schutt CE (2022) Microbubble–nanoparticle complexes for ultrasound-enhanced cargo delivery. Pharmaceutics 14(11):2396. https://doi.org/10.3390/PHARMACEUTICS14112396

Chauhan VP, Stylianopoulos T, Boucher Y, Jain RK (2011) Delivery of molecular and nanoscale medicine to tumors: transport barriers and strategies. Annu Rev Chem Biomol Eng 2:281–298. https://doi.org/10.1146/ANNUREV-CHEMBIOENG-061010-114300/CITE/REFWORKS

Coussios CC, Farny CH, ter Haar G, Roy RA (2007) Role of acoustic cavitation in the delivery and monitoring of cancer treatment by high-intensity focused ultrasound (HIFU). Int J Hyperthermia 23(2):105–120. https://doi.org/10.1080/02656730701194131

Cramer E, Lauterborn W (1982) Acoustic cavitation noise spectra. Appl Sci Res 38(1):209–214. https://doi.org/10.1007/BF00385950

Cressey P, Zhang W, Turcanu M, Cochran S, Thanou M (2019) Design of nanoparticles for focused ultrasound drug delivery. In: Nanotechnology characterization tools for tissue engineering and medical therapy, pp 205–239. https://doi.org/10.1007/978-3-662-59596-1_5

D'Souza J, Sultan L, Hunt S, Karmacharya M, Brice A, Wood A, Sehgal C (2019a) Multimodal sonographic assessment of hepatocellular carcinoma response to antivascular therapy |request PDF. https://www.researchgate.net/publication/338203301_Multimodal_Sonographic_Assessment_of_Hepatocellular_Carcinoma_Response_to_Antivascular_Therapy. Accessed 27 Nov 2024. [Online]

D'Souza JC et al (2019b) Multimodal sonographic assessment of hepatocellular carcinoma response to antivascular therapy. In: IEEE International Ultrasonics Symposium, IUS, pp 974–977. https://doi.org/10.1109/ULTSYM.2019.8926209

De Cock I et al (2015) Ultrasound and microbubble mediated drug delivery: acoustic pressure as determinant for uptake via membrane pores or endocytosis. J Control Release 197:20–28. https://doi.org/10.1016/J.JCONREL.2014.10.031

Deckers R, Moonen CTW (2010) Ultrasound triggered, image guided, local drug delivery. J Control Release 148(1):25–33. https://doi.org/10.1016/J.JCONREL.2010.07.117

Delaney LJ, Isguven S, Eisenbrey JR, Hickok NJ, Forsberg F (2022) Making waves: how ultrasound-targeted drug delivery is changing pharmaceutical approaches. Mater Adv 3(7):3023–3040. https://doi.org/10.1039/D1MA01197A

Fan CH, Ho YJ, Lin CW, Wu N, Chiang PH, Yeh CK (2022) State-of-the-art of ultrasound-triggered drug delivery from ultrasound-responsive drug carriers. Expert Opin Drug Deliv 19(8):997–1009. https://doi.org/10.1080/17425247.2022.2110585

Fateh ST, Moradi L, Kohan E, Hamblin MR, Dezfuli AS (2021) Comprehensive review on ultrasound-responsive theranostic nanomaterials: mechanisms, structures and medical applications. Beilstein J Nanotechnol 12(1):808–862. https://doi.org/10.3762/BJNANO.12.64

Fekri F, Delos Santos RC, Karshafian R, Antonescu CN (2016) Ultrasound microbubble treatment enhances Clathrin-mediated endocytosis and fluid-phase uptake through distinct mechanisms. PLoS One 11(6):e0156754. https://doi.org/10.1371/JOURNAL.PONE.0156754

Fernandes DA (2024) Comprehensive review on bubbles: synthesis, modification, characterization and biomedical applications. Bioconjug Chem 35(11):1639–1686. https://doi.org/10.1021/ACS.BIOCONJCHEM.4C00137/ASSET/IMAGES/MEDIUM/BC4C00137_0032.GIF

Frenkel V (2008) Ultrasound mediated delivery of drugs and genes to solid tumors. Adv Drug Deliv Rev 60(10):1193–1208. https://doi.org/10.1016/J.ADDR.2008.03.007

Fu JW et al (2019) Multifunctionalized microscale ultrasound contrast agents for precise theranostics of malignant tumors. Contrast Media Mol Imaging 2019(1):3145647. https://doi.org/10.1155/2019/3145647

Gallarate M, Chirio D, Chindamo G, Peira E, Sapino S (2021) Osteomyelitis: focus on conventional treatments and innovative drug delivery systems. Curr Drug Deliv 18(5):532–545. https://doi.org/10.2174/1567201817666200915093224

Ghiyasiyan-Arani M (2024) Active nanomaterials for Li-ion batteries and advanced nanomaterials for supercapacitors. In: Renewable and clean energy systems based on advanced nanomaterials: basis, preparation, and applications. Elsevier, Amsterdam, pp 125–153. https://doi.org/10.1016/B978-0-443-13950-5.00007-5

Goel S et al (2011) Normalization of the vasculature for treatment of cancer and other diseases. Physiol Rev 91(3):1071–1121. https://doi.org/10.1152/PHYSREV.00038.2010/ASSET/IMAGES/LARGE/Z9J0031125910015.JPEG

Guo J et al (2024) Injectable acoustically responsive multifunctional hydrogel microspheres blasted for the effective treatment of MSRA-infected chronic osteomyelitis. https://doi.org/10.2139/SSRN.4743623

Han H et al (2020) Focused ultrasound-triggered chemo-gene therapy with multifunctional nanocomplex for enhancing therapeutic efficacy. Journal of Controlled Release 322:346–356. https://doi.org/10.1016/J.JCONREL.2020.03.041

Hasanzadeh H, Mokhtari-Dizaji M, Zahra Bathaie S, Hassan ZM, Nilchiani V, Goudarzi H (2011) Enhancement and control of acoustic cavitation yield by low-level dual frequency sonication: a subharmonic analysis. Ultrason Sonochem 18(1):394–400. https://doi.org/10.1016/J.ULTSONCH.2010.07.005

Haugse R (2020) Molecular mechanisms of sonoporation in cancer therapy: optimization of sonoporation parameters and investigations of intracellular signalling. https://bora.uib.no/bora-xmlui/handle/1956/22966. Accessed 27 Nov 2024 [Online]

Helfield B, Sirsi S, Kwan J, Gray M (2023) Cavitation-enhanced drug delivery and immunotherapy. Pharmaceutics 15(9):2207. https://doi.org/10.3390/PHARMACEUTICS15092207

Honari A (2022) Sensitizing conventional drug vehicles to ultrasound for improved delivery (Doctoral dissertation). https://hdl.handle.net/10735.1/10038

Honari A, Sirsi SR (2023) The evolution and recent trends in acoustic targeting of encapsulated drugs to solid tumors: strategies beyond sonoporation. Pharmaceutics 15(6):1705. https://doi.org/10.3390/PHARMACEUTICS15061705

Honda H, Kondo T, Zhao QL, Feril LB, Kitagawa H (2004) Role of intracellular calcium ions and reactive oxygen species in apoptosis induced by ultrasound. Ultrasound Med Biol 30(5):683–692. https://doi.org/10.1016/J.ULTRASMEDBIO.2004.02.008

Houthaeve G, De Smedt SC, Braeckmans K, De Vos WH (2022) The cellular response to plasma membrane disruption for nanomaterial delivery. Nano Converg 9(1):1–17. https://doi.org/10.1186/S40580-022-00298-7/FIGURES/3

Hu Y, Yang M, Huang H, Shen Y, Liu H, Chen X (2019) Controlled ultrasound erosion for transdermal delivery and hepatitis B immunization. Ultrasound Med Biol 45(5):1208–1220. https://doi.org/10.1016/J.ULTRASMEDBIO.2019.01.012

Hu Y, Wei J, Shen Y, Chen S, Chen X (2023) Barrier-breaking effects of ultrasonic cavitation for drug delivery and biomarker release. Ultrason Sonochem 94:106346. https://doi.org/10.1016/J.ULTSONCH.2023.106346

Idbaih A et al (2019) Safety and feasibility of repeated and transient blood-brain barrier disruption by pulsed ultrasound in patients with recurrent glioblastoma. Clinical Cancer Research 25(13):3793–3801. https://doi.org/10.1158/1078-0432.CCR-18-3643/74166/AM/SAFETY-AND-FEASIBILITY-OF-REPEATED-AND-TRANSIENT

Il Yoon Y, Tang W, Chen X (2017) Ultrasound-mediated diagnosis and therapy based on ultrasound contrast agents. Small Methods 1(8):1700173. https://doi.org/10.1002/SMTD.201700173

Ilovitsh T et al (2020) Low-frequency ultrasound-mediated cytokine transfection enhances T cell recruitment at local and distant tumor sites. Proc Natl Acad Sci USA 117(23):12674–12685. https://doi.org/10.1073/PNAS.1914906117/SUPPL_FILE/PNAS.1914906117.SAPP.PDF

Ishida J et al (2021) MRI-guided focused ultrasound enhances drug delivery in experimental diffuse intrinsic pontine glioma. Journal of Controlled Release 330:1034–1045. https://doi.org/10.1016/J.JCONREL.2020.11.010

Jain RK (1990) Vascular and interstitial barriers to delivery of therapeutic agents in tumors. Cancer Metastasis Rev 9(3):253–266. https://doi.org/10.1007/BF00046364/METRICS

Jain RK (2005) Normalization of tumor vasculature: an emerging concept in antiangiogenic therapy. Science 307(5706):58–62. https://doi.org/10.1126/SCIENCE.1104819

Jenne JW (2014) Non-invasive transcranial brain ablation with high-intensity focused ultrasound. Front Neurol Neurosci 36:94–105. https://doi.org/10.1159/000366241

Jeong Y-G, Park J-H, Khang D (2024) Sonodynamic and acoustically responsive nanodrug delivery system: cancer application. Int J Nanomedicine 19:11767–11788. https://doi.org/10.2147/IJN.S496028

Jia C, Shi J, Yao Yao T, Han ACHY, Qin P (2021) Plasma membrane Blebbing dynamics involved in the reversibly perforated cell by ultrasound-driven microbubbles. Ultrasound Med Biol 47(3):733–750. https://doi.org/10.1016/J.ULTRASMEDBIO.2020.11.029

Jia B et al (2024) Emerging polymeric materials for treatment of oral diseases: design strategy towards a unique oral environment. Chem Soc Rev 53(7):3273–3301. https://doi.org/10.1039/D3CS01039B

Jin Y, Zhang T, Yang Y, Chen H, Li Y, Wu H (2024) How the texture density affect dynamics of a single acoustic cavitation bubble near a skin-like wall in Sonophoresis applications. IEEE Access 12:34505–34511. https://doi.org/10.1109/ACCESS.2024.3369182

Joshi B, Joshi A (2019) Ultrasound-based drug delivery systems. In: Bioelectronics and medical devices: from materials to devices-fabrication, applications and reliability, pp 241–260. https://doi.org/10.1016/B978-0-08-102420-1.00014-5

Jung NY, Park CK, Kim M, Lee PH, Sohn YH, Chang JW (2018) The efficacy and limits of magnetic resonance–guided focused ultrasound pallidotomy for Parkinson's disease: a phase I clinical trial. J Neurosurg 130(6):1853–1861. https://doi.org/10.3171/2018.2.JNS172514

Kennedy JE (2005) High-intensity focused ultrasound in the treatment of solid tumours. Nat Rev Cancer 5(4):321–327. https://doi.org/10.1038/nrc1591

Kilroy JP, Klibanov AL, Wamhoff BR, Hossack JA (2012) Intravascular ultrasound catheter to enhance microbubble-based drug delivery via acoustic radiation force. IEEE Trans Ultrason Ferroelectr Freq Control 59(10):2156–2166. https://doi.org/10.1109/TUFFC.2012.2442

Kooiman K et al (2020) Ultrasound-responsive cavitation nuclei for therapy and drug delivery. Ultrasound Med Biol 46(6):1296–1325. https://doi.org/10.1016/J.ULTRASMEDBIO.2020.01.002

Kumar CSSR (2019) Nanotechnology characterization tools for tissue engineering and medical therapy, pp 1–464. https://doi.org/10.1007/978-3-662-59596-1

Kuo TT, Wang CH, Wang JY, Chiou HJ, Fan CH, Yeh CK (2019) Concurrent Osteosarcoma Theranostic Strategy Using Contrast-Enhanced Ultrasound and Drug-Loaded Bubbles. Pharmaceutics 11(5):223. https://doi.org/10.3390/PHARMACEUTICS11050223

Leow RS, Wan JMF, Yu ACH (2015) Membrane blebbing as a recovery manoeuvre in site-specific sonoporation mediated by targeted microbubbles. J R Soc Interface 12(105):20150029. https://doi.org/10.1098/RSIF.2015.0029

Liu Y et al (2020) Ultrasound-Mediated Long-Circulating Nanopolymer Delivery of Therapeutic siRNA and Antisense MicroRNAs Leads to Enhanced Paclitaxel Sensitivity in Epithelial Ovarian Cancer Chemotherapy. ACS Biomater Sci Eng 6(7):4036–4050. https://doi.org/10.1021/ACSBIOMATERIALS.0C00330/SUPPL_FILE/AB0C00330_SI_001.PDF

Liu Y et al (2021) Ultrasound molecular imaging-guided tumor gene therapy through dual-targeted cationic microbubbles. Biomater Sci 9(7):2454–2466. https://doi.org/10.1039/D0BM01857K

Lyon P, Carlisle R, Coussios CC (2024) Triggered temperature-sensitive liposome release by focused ultrasound for localised drug delivery. In: Image-guided focused ultrasound therapy, pp 364–393. https://doi.org/10.1201/9780429162671-25

Maeda H (2001) SMANCS and polymer-conjugated macromolecular drugs: advantages in cancer chemotherapy. Adv Drug Deliv Rev 46(1–3):169–185. https://doi.org/10.1016/S0169-409X(00)00134-4

Marinova M, Rauch M, Schild HH, Strunk HM (2016) Novel non-invasive treatment with high-intensity focused ultrasound (HIFU). Ultraschall Med 37(1):46–55. https://doi.org/10.1055/S-0035-1553318/ID/JR630-19/BIB

McLaughlan J et al (2006) The design and implementation of a passive cavitation detection system for use with ex vivo tissue. AIP Conf Proc 829(1):338–342. https://doi.org/10.1063/1.2205493

Meijering BDM et al (2009) Ultrasound and microbubble-targeted delivery of macromolecules is regulated by induction of endocytosis and pore formation. Circ Res 104(5):679–687. https://doi.org/10.1161/CIRCRESAHA.108.183806/SUPPL_FILE/ZHH183806-SUPPLEMENT.AVI

Mittal S, El-Serag HB (2013) Epidemiology of hepatocellular carcinoma: consider the population. J Clin Gastroenterol 47(Suppl(0)):1105–1115. https://doi.org/10.1097/MCG.0B013E3182872F29

Moradpour M et al (2024) Efficacy of Antivascular ultrasound (AVUS) in hepatocellular carcinoma (HCC). Cancer 16(22):3756. https://doi.org/10.3390/CANCERS16223756

Mulvana H et al (2017) Characterization of contrast agent microbubbles for ultrasound imaging and therapy research. IEEE Trans Ultrason Ferroelectr Freq Control 64(1):232–251. https://doi.org/10.1109/TUFFC.2016.2613991

Myers S et al (2021) NAFLD and MAFLD as emerging causes of HCC: a populational study. JHEP Rep 3(2):100231. https://doi.org/10.1016/J.JHEPR.2021.100231

Neppiras EA (1969) Subharmonic and other low-frequency emission from bubbles in sound-irradiated liquids. J Acoust Soc Am 46:587–601. pubs.aip.org. https://pubs.aip.org/asa/jasa/article-abstract/46/3B/587/752030. Accessed 27 Nov 2024. [Online]

Okuda K (1996) Epidemiology of hepatocellular carcinoma. Jpn J Cancer Chemother 23(9):1105–1115. https://doi.org/10.1097/MCG.0B013E3182872F29

Paliwal S, Mitragotri S (2006) Ultrasound-induced cavitation: applications in drug and gene delivery. Expert Opin Drug Deliv 3(6):713–726. https://doi.org/10.1517/17425247.3.6.713

Pitt WG, Husseini GA (2008) Ultrasound in drug and gene delivery. Adv Drug Deliv Rev 60(10):1095–1096. https://doi.org/10.1016/J.ADDR.2008.03.001

ProQuest (n.d.-a) Translating focused ultrasound combined nanomedicines for treatment of bone infections and canine cancer patients—ProQuest. https://www.proquest.com/openview/884f615f34fdfa50d017a3b19818d8d6/1?pq-origsite=gscholar&cbl=18750&diss=y. Accessed 27 Nov 2024. [Online]

ProQuest (n.d.-b) Docetaxel loaded nanobubbles for ultrasound triggered antivascular therapy and localized drug delivery—ProQuest. https://www.proquest.com/openview/c6bc1f8438db13091b90fcb2ee0f21b8/1?pq-origsite=gscholar&cbl=18750&diss=y. Accessed 27 Nov 2024. [Online]

ProQuest (n.d.-c) The effects of low frequency ultrasound in transdermal drug delivery—ProQuest. https://www.proquest.com/openview/770187d1a7f5e459f023add619e0838b/1?pq-origsite=gscholar&cbl=18750. Accessed 27 Nov 2024. [Online]

ProQuest (n.d.-d) Designing phase-change contrast agents for safe diagnostic ultrasound imaging and enhanced therapeutic outcomes—ProQuest. https://www.proquest.com/openview/2a1c2373bbc3397206f36342dba646a6/1?pq-origsite=gscholar&cbl=18750&diss=y. Accessed 27 Nov 2024. [Online]

ProQuest (n.d.-e) Nanoparticle enabled phase control in solidification processing—ProQuest. https://www.proquest.com/openview/95ef83c3267e80e377f8fe143c440620/1?pq-origsite=gscholar&cbl=18750&diss=y. Accessed 27 Nov 2024. [Online]

Pua EC, Zhong P (2009) Ultrasound-mediated drug delivery: range of ultrasonic interactions and applications in biological systems. IEEE Eng Med Biol Mag 28(1):64–75. https://doi.org/10.1109/MEMB.2008.931017

Purohit MP et al (2023) Acoustomechanically activatable liposomes for ultrasonic drug uncaging. bioRxiv. p 2023.10.23.563690. https://doi.org/10.1101/2023.10.23.563690

Qin D et al (2018) In situ observation of single cell response to acoustic droplet vaporization: membrane deformation, permeabilization, and blebbing. Ultrason Sonochem 47:141–150. https://doi.org/10.1016/J.ULTSONCH.2018.02.004

Qu F et al (2020) Manipulation of Mitophagy by 'All-in-One' nanosensitizer augments sonodynamic glioma therapy. Autophagy 16(8):1413–1435, https://doi.org/10.1080/15548627.2019.1687210

Rich KT, Hoerig CL, Rao MB, Mast TD (2014) Relations between acoustic cavitation and skin resistance during intermediate- and high-frequency sonophoresis. J Control Release 194:266–277. https://doi.org/10.1016/J.JCONREL.2014.08.007

Ruan JL et al (2021) Ultrasound-Mediated Gemcitabine Delivery Reduces the Normal-Tissue Toxicity of Chemoradiation Therapy in a Muscle-Invasive Bladder Cancer Model. International Journal of Radiation Oncology*Biology*Physics 109(5):1472–1482. https://doi.org/10.1016/J.IJROBP.2020.11.046

Sahni E, Chaudhuri B (2011) Experiments and numerical modeling to estimate the coating variability in a pan coater. Int J Pharm 418(2):286–296. https://doi.org/10.1016/J.IJPHARM.2011.05.041

Schlicher RK, Radhakrishna H, Tolentino TP, Apkarian RP, Zarnitsyn V, Prausnitz MR (2006) Mechanism of intracellular delivery by acoustic cavitation. Ultrasound Med Biol 32(6):915–924. https://doi.org/10.1016/J.ULTRASMEDBIO.2006.02.1416

Schwartz MR, Debski AC, Price RJ (2021) Ultrasound-targeted nucleic acid delivery for solid tumor therapy. J Control Release 339:531–546. https://doi.org/10.1016/J.JCONREL.2021.10.010

Seah BCQ, Teo BM (2018) Recent advances in ultrasound-based transdermal drug delivery. Int J Nanomedicine 13:7749–7763. https://doi.org/10.2147/IJN.S174759

Sennoga CA et al (2017) Microbubble-mediated ultrasound drug-delivery and therapeutic monitoring. Expert Opin Drug Deliv 14(9):1031–1043. https://doi.org/10.1080/17425247.2017.1266328

Shinar H, Ilovitsh T (2024) Volumetric passive acoustic mapping and cavitation detection of nanobubbles under low-frequency insonation. ACS Mater Au 5(1):159–169. https://doi.org/10.1063/1.2205493

Sierra C et al (2017) Lipid microbubbles as a vehicle for targeted drug delivery using focused ultrasound-induced blood-brain barrier opening. J Cereb Blood Flow Metab 37(4):1236–1250. https://doi.org/10.1177/0271678X16652630/ASSET/IMAGES/LARGE/10.1177_0271678X16652630-FIG6.JPEG

Stylianopoulos T, Jain RK (2015) Design considerations for nanotherapeutics in oncology. Nanomedicine 11(8):1893–1907. https://doi.org/10.1016/J.NANO.2015.07.015

Sultan LR, Karmacharya MB, Hunt SJ, Wood AKW, Sehgal CM (2021) Subsequent ultrasound vascular targeting therapy of hepatocellular carcinoma improves the treatment efficacy. Biology (Basel) 10(2):1–14. https://doi.org/10.3390/BIOLOGY10020079

Sun Y, Luo J, Chen J, Xu F, Ding X, Huang P (2018) Ultrasound-mediated microbubbles destruction for treatment of rabbit VX2 orthotopic hepatic tumors. Appl Acoust 138:216–225. https://doi.org/10.1016/J.APACOUST.2018.04.007

Tang H, Wang CCJ, Blankschtein D, Langer R (2002) An investigation of the role of cavitation in low-frequency ultrasound-mediated transdermal drug transport. Pharm Res 19(8):1160–1169. https://doi.org/10.1023/A:1019898109793/METRICS

Tang Y, Kang J, Wang YQ (2022) Cavity expansion in nonlinear viscoelastic solids: a non-linear dynamic study. Int J Non Linear Mech 144:104076. https://doi.org/10.1016/J.IJNONLINMEC.2022.104076

Tezel A, Sens A, Mitragotri S (2002) Investigations of the role of cavitation in low-frequency sonophoresis using acoustic spectroscopy. J Pharm Sci 91(2):444–453. https://doi.org/10.1002/JPS.10024

van Blokland A (2018) Instigating and monitoring transdermal drug delivery using ultrasound-mediated cavitation. University of Oxford, Oxford

Versluis M, Stride E, Lajoinie G, Dollet B, Segers T (2020) Ultrasound contrast agent modeling: a review. Ultrasound Med Biol 46(9):2117–2144. https://doi.org/10.1016/J.ULTRASMEDBIO.2020.04.014

Wang X et al (2022) Dual-responsive nanodroplets combined with ultrasound-targeted microbubble destruction suppress tumor growth and metastasis via autophagy blockade. Journal of Controlled Release 343:66–77. https://doi.org/10.1016/J.JCONREL.2022.01.009

Wang J, Wang Y, Zhong L, Yan F, Zheng H (2024) Nanoscale contrast agents: a promising tool for ultrasound imaging and therapy. Adv Drug Deliv Rev 207:115200. https://doi.org/10.1016/J.ADDR.2024.115200

Wei HJ et al (2021) Focused Ultrasound-Mediated Blood-Brain Barrier Opening Increases Delivery and Efficacy of Etoposide for Glioblastoma Treatment. International Journal of Radiation Oncology*Biology*Physics 110(2):539–550. https://doi.org/10.1016/J.IJROBP.2020.12.019

Wen Z et al (2023) Ultrasound meets the cell membrane: for enhanced endocytosis and drug delivery. Nanoscale 15(33):13532–13545. https://doi.org/10.1039/D3NR02562D

Wu H et al (2023) Mechanisms underlying the influence of skin properties on a single cavitation bubble in low-frequency sonophoresis. Ultrason Sonochem 101:106690. https://doi.org/10.1016/J.ULTSONCH.2023.106690

Xie X et al (2018) Challenges and opportunities from basic cancer biology for nanomedicine for targeted drug delivery. Curr Cancer Drug Targets 19(4):257–276. https://doi.org/10.2174/1568009618666180628160211

Xie L, Wang J, Zhao S, Lai ML, Jiang T, Yan F (2021) An acoustic field-based conformal transfection system for improving the gene delivery efficiency. Biomater Sci 9(11):4127–4138. https://doi.org/10.1039/D1BM00251A

Yan K, Yao J, Liu L, Liang W, Cai Y (2024) Effects of low-frequency ultrasound combined with anti-MRSA agents on the mouse model of pulmonary infection. Microbiol Spectr 12(3):e0101623. https://doi.org/10.1128/SPECTRUM.01016-23/SUPPL_FILE/SPECTRUM.01016-23-S0001.DOCX

Yang B, Liang J, Qu ZY, Yang FY, Liao ZY, Gou HF (2020) Transarterial strategies for the treatment of unresectable hepatocellular carcinoma: a systematic review. PLoS One 15(2):e0227475. https://doi.org/10.1371/JOURNAL.PONE.0227475

Yao R, Hu J, Zhao W, Cheng Y, Feng C (2022) A review of high-intensity focused ultrasound as a novel and non-invasive interventional radiology technique. J Interv Med 5(3):127–132. https://doi.org/10.1016/J.JIMED.2022.06.004

Yasui K (2018) Acoustic cavitation. In: Acoustic cavitation and bubble dynamics. Springer, Cham, pp 1–35. https://doi.org/10.1007/978-3-319-68237-2_1

Yemane PT et al (2020) Ultrasound for the delivery of a nanocarrier across biological barriers in tumors: impact of cavitation and acoustic radiation force. Ultrasound Med Biol 45(11):3028–3041. https://doi.org/10.1016/J.ULTRASMEDBIO.2019.07.683

Zhang C, Liu C, He W, Jiao D, Liu Z (2021) Cross-linking-enhanced and ultrasound-mediated drug delivery: From fabrication, mechanisms to translations. Appl Mater Today 22:100897. https://doi.org/10.1016/J.APMT.2020.100897

Zhang L et al (2022) Ultrasound-induced biophysical effects in controlled drug delivery. Sci China Life Sci 65(5):896–908. https://doi.org/10.1007/S11427-021-1971-X/METRICS

Zullino S, Argenziano M, Stura I, Guiot C, Cavalli R (2018) From micro- to Nano-multifunctional Theranostic platform: effective ultrasound imaging is not just a matter of scale. Mol Imaging 17:1536012118778216. https://doi.org/10.1177/1536012118778216/ASSET/IMAGES/LARGE/10.1177_1536012118778216-FIG6.JPEG

Magnetic Hyperthermia Ablation-Based Drug Delivery

7

Vishakha Jaiswal, Narahari Narayan Palei,
Malakapogu Ravindra Babu, and Shriyansh Srivastav

Abstract

Magnetic hyperthermia ablation (MH) is a promising method for targeted drug delivery and cancer therapy. When exposed to an alternating magnetic field, magnetic nanoparticles produce localized heat, which permits tumor ablation and initiates the release of therapeutic medicines. By combining targeted medication release with hyperthermia-induced cell death, this dual-function strategy reduces systemic toxicity and adverse effects while increasing therapy efficacy. By providing better biocompatibility, extended circulation, and increased drug-loading capacity, the combination of lipid nanoparticles and magnetic nanoparticles has advanced the field even further. Overcoming biological barriers to effective delivery, guaranteeing uniform heat dispersion, and optimizing nanoparticle design are among the difficulties. Precision-targeting techniques and patient-specific nanoparticle compositions could transform drug delivery systems in personalized medicine.

Keywords

Magnetic hyperthermia · Magnetic nanoparticles · Tumor ablation · Targeted therapy · Drug delivery

V. Jaiswal · N. N. Palei (✉)
Amity Institute of Pharmacy, Amity University, Lucknow, Uttar Pradesh, India

M. R. Babu · S. Srivastav
Department of Pharmacy, School of Medical and Allied Sciences, Galgotias University, Greater Noida, Uttar Pradesh, India

M. R. Babu et al. (eds.), *Transdermal Applications of Minimally Invasive Drug Delivery Systems*, https://doi.org/10.1007/978-981-96-6275-3_7

7.1 Introduction

Magnetic hyperthermia (MH) is a minimally invasive therapeutic approach that employs magnetic nanoparticles (MNPs) to generate localized heat upon exposure to an alternating magnetic field (AMF) (Zhang et al. 2024).

This heat is harnessed for ablation of cancerous tissues, disrupting the integrity of malignant cells while sparing healthy tissues. The underlying principle of MH lies in the magnetic loss mechanisms of MNPs, such as Néel and Brownian relaxation, which convert electromagnetic energy into thermal energy. The controlled application of heat (typically 42–45 °C) induces apoptosis or necrosis in tumor cells, making MH a promising strategy in oncology. Hyperthermia has emerged as powerful minimally invasive oncological treatment which could be combined with chemotherapy, radiotherapy, phototherapy, and immunotherapy (Liu et al. 2020; Gavilán et al. 2021).

Over the past few decades, MH has gained considerable attention due to its ability to target deep-seated tumors with precision. By modifying the surface of MNPs, their stability, biocompatibility, and functionality can be enhanced, facilitating their selective accumulation in tumors through passive (enhanced permeability and retention effect) or active targeting (ligand-receptor interactions) (Yu et al. 2010; Yao et al. 2020).

7.2 Magnetic Hyperthermia for Drug Delivery

Magnetic hyperthermia is increasingly utilized in clinical applications such as magnetic resonance imaging (MRI) and targeted drug delivery. In drug delivery, this technique employs magnetic nanoparticles (NPs) that are introduced into the body and transported through the circulatory system to the tumor site. Upon exposure to an alternating current (AC) magnetic field, these nanoparticles generate thermal energy, elevating the temperature of the tumor region to 40–45 °C. This localized heating induces apoptosis in cancer cells while minimizing harm to the surrounding healthy tissues (Zhang and Lu 2024; Vassallo et al. 2023).

In this context, MNPs are not only employed for heat generation but also as carriers for therapeutic agents. The dual functionality of MNPs—thermal ablation and drug release—allows for a synergistic approach to cancer therapy. Heat generated during hyperthermia can trigger the release of drugs from thermosensitive carriers or disrupt cellular barriers, enhancing drug penetration into tumor tissues (Ribeiro et al. 2022).

Nanocarriers such as liposomes, hydrogels, and polymeric nanoparticles can be engineered with MNPs to achieve controlled drug release. For instance, thermosensitive liposomes encapsulating MNPs and chemotherapeutic agents release their payload at elevated temperatures induced by MH, thereby minimizing systemic toxicity (Bi et al. 2019).

Additionally, hyperthermia can increase membrane permeability and vasodilation in tumor tissues, further promoting drug uptake. Several magnetic materials can be used for drug delivery, including iron, cobalt, gadolinium, and their compounds. However, superparamagnetic iron oxide nanoparticles stand out as the preferred choice for magnetic hyperthermia treatment. They offer unique benefits: their magnetic response can be controlled through particle size adjustments, and they are versatile in accepting different surface coatings, exhibit strong biocompatibility, have low toxicity, and can be efficiently eliminated from the body (Kacem et al. 2024; Beik et al. 2016).

Targeting specificity is another advantage of MH-based DDS. MNPs functionalized with targeting ligands (e.g., antibodies, peptides, or small molecules) can selectively bind to cancer cell receptors, ensuring that both heat and drugs are delivered directly to the tumor site (Prajapati et al. 2024). This approach not only enhances therapeutic efficacy but also reduces adverse effects associated with nonspecific drug distribution.

Due to its localized nature, magnetic hyperthermia allows for precise targeting of pathological tissues, minimizing damage to surrounding healthy structures. However, this precision comes with the necessity for extra caution, primarily due to the risk of overheating, which can lead to unintended tissue damage or complications. Overheating can occur if the heat generated by the nanoparticles exceeds the tissue's ability to dissipate it, potentially causing burns, necrosis, or inflammation in the surrounding healthy tissue (Bravo et al. 2024).

7.2.1 Mechanisms of Magnetic Hyperthermia

Hyperthermia (HPT) has been confirmed as a minimally invasive thermal therapy that alleviates severe adverse effects to the tumor cells, thereby enhancing the treatment to bring a new hope to cancer patients. HPT could be classified as mild hyperthermia (40–42 °C), moderate hyperthermia (42–45 °C), and thermal ablation (≥45 °C) (Bravo et al. 2024; Rajan et al. 2024).

Cancer cells are particularly sensitive to heat above 41 °C due to their accelerated metabolism and low-oxygen environment, making them more susceptible to thermal damage than healthy cells. By exposing tumor tissue to elevated temperatures for specific durations, therapeutic effects can be achieved that lead to cancer cell death.

Magnetic hyperthermia involves the use of magnetic nanoparticles to stimulate localized heating of tissues for medical intervention. Where magnetic hyperthermia is implemented, the fundamental philosophy consists of delivering or targeting magnetic nanoparticles to the site of the tumor. Such magnetic NPs having high magnetic susceptibility generate heat when exposed to an alternating magnetic field (AMF) (Kacem et al. 2024).

Magnetic hyperthermia occurs when magnetic nanoparticles generate heat in response to an alternating magnetic field (AMF). Two main mechanisms that drive this heating process are hysteresis losses and/or relaxation losses (Néel and Brownian relaxations).

A ferromagnetic material has multi-magnetic domains (Weiss domains) where all the magnetic dipoles are aligned in the same direction. As an alternating magnetic field is applied, the magnetic domains must continuously reorient themselves to match the field's direction until reaching magnetic saturation. This repeated movement of domain boundaries generates heat through energy dissipation during each magnetic cycle. In multi-domain magnetic nanoparticles (MNPs), the primary heating effect stems from hysteresis losses. Multi-domain magnetic nanoparticles typically demonstrate stronger resistance to changes in their magnetization when exposed to an alternating magnetic field. This resistance creates persistent magnetism that requires applying a specific coercive force to overcome. As a result, energy dissipated from the system being directly proportional to the hysteresis loop's area. The amount of heat generated can be quantified by measuring the area enclosed within the hysteresis loop, as this is proportional to the total energy released during the magnetization cycle.

In nanoparticles exhibiting superparamagnetic behavior and magnetic nanoparticles smaller than 20–50 nm, heat production does not occur through hysteresis losses. Instead, heating primarily results from two relaxation processes: Brownian motion (particle rotation) and Néel relaxation (internal magnetic moment reorientation) (Fig. 7.1).

Néel relaxation is a consequence of exposure to an AMF; the magnetic moment within each nanoparticle rapidly switches its direction. This internal reorientation produces heat through magnetic friction. In Brownian relaxation, the whole nanoparticle physically rotates in the surrounding medium to align with the changing magnetic field. This mechanical rotation creates frictional heating with the surrounding fluid (Bravo et al. 2024).

The relative contribution of Néel relaxation depends on nanoparticle size, and Brownian relaxation depends upon hydrodynamic size of nanoparticle and

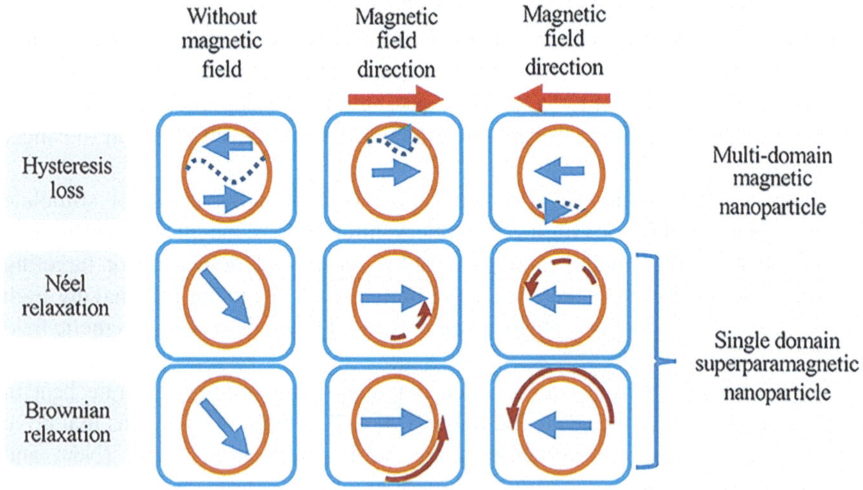

Fig. 7.1 Mechanism of heat generation by MNPs under AMF. (Adapted from Chang et al. (2018))

medium's viscosity. According to both theoretical and experimental data, maghemite nanoparticles of around 15 nm would be ideal diameter to achieve optimal heating efficiency by balancing the contributions from Néel and Brownian relaxation mechanisms.

The amount of heat generated can be calculated using this equation:

$$P = \mu_0 \chi_m f H^2$$

where P is amount of heat generated, μ_0 is the permeability of free space, χ_m is the magnetic susceptibility, f is the frequency of the applied magnetic field, and H^2 is the strength of the applied magnetic field.

The conversion of magnetic energy of NPs in thermal energy exposed to an alternating magnetic field is measured by the specific absorption rate (SAR) or specific loss power (SLP). This value can be determined using the following equation:

$$\text{SAR} = C\left(\frac{dT}{dt}\right)\left(\frac{M_s}{M_m}\right)$$

where C, dT/dt, M_s, and M_m are specific heat capacity of the solvent, initial slope of the time-dependent heating curve, mass of the solvent, and mass of magnetic nanoparticles, respectively.

SAR/SLP is dependent upon size, composition, and magnetism of nanoparticle; surface modifications; magnetic field's frequency and amplitude; and properties of the surrounding environment/fluid (Das et al. 2019).

7.2.2 Magnetic Hyperthermia-Assisted Drug Delivery (MHADD) Mechanisms

Scientists in 1957 pioneered the idea of harnessing magnetic materials' ability to generate heat from dissipated magnetic energy as a method for treating cancer through hyperthermia (Kumar and Mohammad 2011).

Emerging nanotechnology-based magnetic hyperthermia technique may destruct the malignant cells directly, intensify the cytotoxic effect of loaded drugs (thermo-chemosensitization), and in particular cases may trigger drug release, simultaneously (Ribeiro et al. 2022; Beik et al. 2016).

7.2.3 Magnetically Guided Drug Targeting

Recent studies have revealed a major challenge that only 0.7% of injected nonmagnetic nanoparticles successfully reach the tumor site, highlighting an urgent need for better targeting methods. However, there is a promising solution through electromagnetic field (EMF) control; researchers can create sophisticated drug release patterns, enabling either sequential release of different drugs or controlled combination therapy (Day et al. 2021). The magnetic nanoparticles (MNPs) must remain concentrated at the tumor site to achieve effective hyperthermia treatment and drug delivery. Clinicians can use three distinct delivery methods: injecting the particles directly into the tumor site (intratumoral injection), relying on intravenous systemic delivery, or active targeting of surface functionalized NPs (Hervault and Thanh 2014).

Intratumoral injection of MNPs is the simplest way to administer the NPs but only possible in localized and superficial or easily accessible tumor. Systematic delivery is achieved by injecting MNPs intravenously that will accumulate preferentially in tumor tissue due to the enhanced permeability and retention (EPR) effect (Hervault and Thanh 2014; Włodarczyk et al. 2022).

Some drug delivery systems, when injected systemically (passive delivery), take advantage of the accumulation of NPs in certain tumor tissues with leaky vasculatures and impaired lymphatic drainage, known as the enhanced permeability and retention (EPR) effect (Day et al. 2021).

The EPR effect represents a critical advantage in tumor-targeted therapy. Tumors develop abnormally permeable blood vessels to support their rapid growth, allowing them to acquire needed nutrients and oxygen. This increased permeability creates an opportunity: the vessels become "leaky" compared to normal tissue vessels, enabling NPs to pass through vessel walls and accumulate within the tumor mass. This accumulation is further enhanced by the tumor's impaired lymphatic drainage system, which allows the particles to remain in the tumor tissue for extended periods (Day et al. 2021; Hervault and Thanh 2014).

When MNPs are administered intravenously, they can exploit this natural phenomenon, passing through these leaky vessels to concentrate in the tumor tissue. To maximize this effect, clinicians can also opt for direct arterial injection into the blood vessels serving the target organ, thereby increasing particle deposition in the tumor site. However, successful exploitation of the EPR effect requires careful particle design to ensure they maintain sufficient circulation time in the bloodstream, allowing them to reach and penetrate the tumor vessels effectively (Day et al. 2021; Hervault and Thanh 2014).

Cells internalize magnetic nanoparticles (MNPs) through various endocytic pathways, including clathrin-mediated endocytosis, caveolin-mediated endocytosis, macropinocytosis, phagocytosis, and membrane fusion. Once inside the cell, these MNPs either undergo lysosomal degradation or are transported back to the cell membrane through recycling mechanisms (Shirsat et al. 2024).

Magnetic hyperthermia treatment (MHT) using magnetic nanoparticles produces concentrated heat that triggers cancer cell death through multiple pathways. When the local temperature rises, it creates thermal stress that interferes with essential cellular processes and disrupts normal signaling networks. This heat also causes proteins to lose their proper structure and function. While cells attempt to protect themselves by activating heat shock proteins (HSPs), the intense thermal damage overwhelms these defensive mechanisms. The combined effects of thermal stress, protein denaturation, oxidative damage, and disrupted mitochondrial function ultimately result in programmed cell death (apoptosis) of the cancer cells (Rajan et al. 2024) (Fig. 7.2).

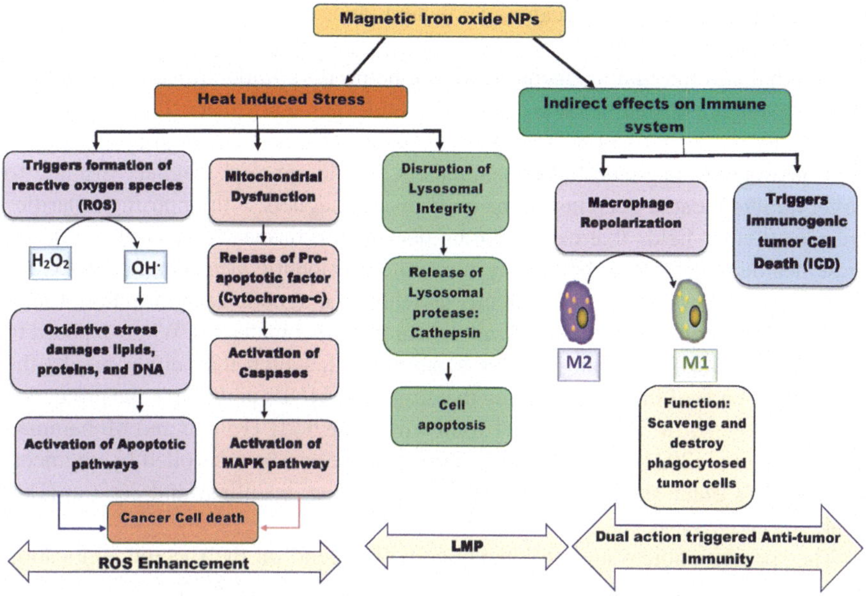

Fig. 7.2 Intracellular MH-induced cancer ablation (*LMP* lysosomal membrane permeabilization)

Nanoparticles must be precisely sized—between 50 and 200 nm—to avoid being filtered out by either the kidneys or liver. Their surface coating is equally important, determining both how well they enter and stay in cancer cells and how long they remain in the bloodstream. Delivering nanoparticles through the EPR effect distributes them more evenly throughout tumors compared to direct injection and works for most cancer types. However, this method struggles to achieve concentrations high enough for effective treatment—increasing the dose is not feasible due to toxicity concerns. To improve tumor uptake, researchers can either guide the particles using external magnetic fields or functionalize with targeting molecules (Day et al. 2021; Hervault and Thanh 2014).

7.2.4 Magnetic Hyperthermia-Assisted Drug Release

Delivery of drug from MNPs under the influence of AMF/EMF can be through breaking of bonds (BB) or release of encapsulated drug in matrix from MNPs.

When exposed to alternating magnetic fields (AMF) for limited periods (60 s), the magnetic iron oxide NP absorbs enough energy leading to rotation, vibration, and development of tiny cracks or crevice, which temporarily releases the drug. This release could be stopped by turning off the magnetic field. However, when the exposure was extended to 5 m, the cracks became larger, resulting in faster drug release, but the damage became permanent and could not be reversed (Day et al. 2021).

External and internal magnetic fields can be used to guide drug carriers to their targets, though external fields are generally preferred. They offer three key advantages: they are easier to use, minimize invasion into the body, and allow precise control over the magnetic field's parameters. While external magnets struggle to target tumors located deep inside the body, using magnets with opposing polarities creates stronger fields that can penetrate deeper into tissues (Day et al. 2021). In these heat-activated drug delivery systems, the therapeutic agents can be delivered in two ways. They can either be directly attached to magnetic nanoparticles through thermosensitive linker or enclosed within heat-sensitive materials. When exposed to an AMF, the magnetic nanoparticles generate heat, which then either breaks the molecular bonds of linker connecting the drugs and MNPs or disrupts the protective coating, allowing the drugs to be released within the body (Kumar and Mohammad 2011; Day et al. 2021; Yu et al. 2014). Drug release can be controlled by engineering the NP's outer shell properties such as diffusion, swelling, and erosion (Day et al. 2021).

The specific choices and optimizations will depend on the specific application and the desired therapeutic outcomes and are depicted in Table 7.1.

7.2.5 Advantages of Combining Hyperthermia with Drug Delivery

Heat-triggered drug delivery systems represent a promising advancement in cancer therapeutics, offering the potential for more targeted, effective treatment approaches while minimizing side effects associated with conventional chemotherapeutics. This synergistic approach leverages the heat generated during hyperthermia to improve drug absorption and effectiveness, particularly in targeting tumor cells. The following sections outline the key benefits of this combination therapy.

7.2.5.1 Enhanced Drug Penetration

Hyperthermia increases tumor perfusion and increases drug penetration, facilitating better drug distribution within the tumor (Chia et al. 2023). Thermoresponsive nanoparticles function as targeted drug delivery systems, transporting therapeutic agents into cells and releasing their contents precisely when exposed to elevated temperatures. This controlled release mechanism enables precise timing and localization of drug delivery through external heat application (Colli et al. 2024).

7.2.5.2 Reduced Side Effects

Utilizing hydrogels and nanomaterials can mitigate the adverse effects commonly associated with chemotherapy, such as toxicity and drug resistance (Jaymand 2024). The combination therapy allows for lower dosages of chemotherapeutic agents, reducing systemic exposure and associated side effects (Patri et al. 2024).

Table 7.1 Comprehensive overview of the key components and considerations for magnetic hyperthermia-assisted drug release

Component	Description	Considerations
Magnetic nanoparticles (MNPs)	Core material: iron oxide (magnetite, maghemite) Size and shape: influence magnetic properties and drug loading Surface coating: determines biocompatibility, targeting, and drug release mechanisms	Size: optimize for efficient heat generation and minimal toxicity Surface coating: choose biocompatible, nontoxic materials that facilitate drug loading and targeting Magnetic properties: ensure sufficient magnetization for effective hyperthermia
Drug	Type: chemotherapeutics, antibiotics, biologics Loading: encapsulation, surface conjugation, or within polymer matrices	Drug properties: consider drug stability, solubility, and release kinetics Loading strategy: optimize drug-loading capacity and prevent premature release
Drug release mechanisms	Temperature-sensitive polymers: polymer chains undergo structural changes at elevated temperatures Porous structures: heat-induced opening of pores in nanoparticles Chemical bonds: heat-labile bonds that break upon heating Phase transitions: nanoparticles undergo phase transitions (e.g., from solid to liquid) at elevated temperatures	Triggering mechanism: select a mechanism that ensures controlled and efficient drug release at the target site
External magnetic field	Frequency and amplitude: determine the intensity of heat generation Application: localized or whole-body application	Field parameters: optimize frequency and amplitude to achieve desired temperature and minimize off-target effects Application method: select appropriate methods for precise field delivery
Targeting strategies	Passive targeting: utilizing the enhanced permeability and retention (EPR) effect Active targeting: conjugating MNPs with ligands that bind to specific receptors on target cells Magnetic guidance: using external magnets to direct MNPs to the target site	Targeting efficiency: optimize targeting strategies to maximize drug delivery to the desired location
Safety and toxicity	Biocompatibility: ensuring minimal toxicity and adverse immune responses Clearance: efficient clearance of MNPs and drug from the body Off-target effects: minimizing potential damage to healthy tissues	Thorough preclinical evaluation: conduct rigorous safety and toxicity studies before clinical translation

7.2.5.3 Drug Targeting

MNPs can be targeted through cancer-specific binding agents making the treatment much more selective and effective. MNPs naturally accumulate in tumors through enhanced permeation of leaky blood vessels or targeted delivery using surface-modified particles that specifically bind to cancer cells (Kumar and Mohammad 2011; Vijayan et al. 2023).

7.2.5.4 Improved Treatment Outcomes

Studies indicate that hyperthermia can enhance the immune response against tumors, potentially leading to better long-term outcomes (Chia et al. 2023). Elevated temperatures trigger specific molecular pathways that activate and enhance the immune system's anticancer response. This thermal regulation of cellular signaling mechanisms serves as a key driver in stimulating targeted immune activity against malignant cells (Jiao et al. 2024). The use of magnetic iron oxide nanogels has shown promise in achieving controlled drug release while simultaneously applying hyperthermia, thus maximizing therapeutic effects (Patri et al. 2024; Yadav et al. 2023).

Conversely, while the combination of hyperthermia and drug delivery shows significant promise, challenges such as the potential for heat-induced tumor cell thermotolerance and the need for precise temperature control remain critical considerations in clinical applications (Colli et al. 2024).

7.2.6 Challenges and Considerations in Hyperthermia-Based Drug Delivery

Traditional cancer treatments often combine hyperthermia with radiotherapy and/or chemotherapy to enhance therapeutic outcomes. This multimodal approach not only destroys more cancer cells directly but also makes treatment-resistant cells more susceptible to other therapeutic interventions. However, conventional hyperthermia methods face several significant limitations:

The first major challenge is the lack of precise targeting, which leads to unintended heating of healthy tissues, causing side effects like burns, blisters, and patient discomfort. Secondly, heat delivery methods such as microwave, laser, and ultrasound have limited tissue penetration depth, reducing their effectiveness for deep-seated tumors. Finally, achieving adequate thermal dosage poses particular difficulties when treating tumors near or behind bony structures like the pelvis or skull, as these bones block heat transmission to deeper tissues. This insufficient heating often results in tumor recurrence (Kumar and Mohammad 2011).

Present research efforts are directed at addressing several key challenges in nanoparticle-based therapies: maintaining colloidal stability, optimizing heat generation, improving targeted delivery, and reducing unwanted toxicity. However, advancing these treatments toward clinical use requires deeper investigation in multiple areas. Scientists need to better understand how nanoparticles interact with cells at a fundamental level and gain more comprehensive knowledge of how these particles move through and distribute within the body (pharmacokinetics and biodistribution). Additionally, thorough evaluation of nanoparticle safety over extended periods is essential through preclinical studies before these treatments can successfully transition to clinical applications (Bravo et al. 2024).

7.3 Considerations

The effectiveness and safety of magnetic hyperthermia therapy (MHT) depends on careful monitoring and precise control of various critical parameters. To optimize treatment outcomes and minimize potential risks, several key factors must be closely regulated throughout the therapeutic process, such as:

1. *Thermal monitoring*: Maintaining the therapeutic temperature range (typically 41–45 °C) is essential for effective treatment preventing overheating and tissue damage. Temperature distribution needs to be uniform across the target area (Colli et al. 2024).
2. *Selection of magnetic carrier*: The thermoresponsive materials must be biocompatible and biodegradable. The material properties should allow for adequate drug-loading capacity (Rajan et al. 2024).
3. *Drug release kinetics*: The use of electromagnetic fields can induce localized heating in superparamagnetic nanoparticles, significantly enhancing drug release rates. The rate and timing of drug release must be carefully controlled. Temperature-sensitive carriers should release drugs at specific thermal triggers (Alarcón-Segovia et al. 2023).
4. *Safety and biocompatibility*: The materials employed should be biocompatible, avoiding any triggering of immune responses or inflammatory reactions. The delivery system must have appropriate clearance mechanisms, ensuring it exits the body within a suitable timeframe without harmful accumulation. Prolonged exposure to magnetic fields can exacerbate thermal effects, so treatment duration must be carefully managed (Bravo et al. 2024; Colli et al. 2024).

Researchers worldwide are developing multifunctional magnetic nanoparticles (MNPs) due to their advantageous magnetic properties: minimal remanence and coercivity, high saturation magnetization, and quick response to external magnetic fields.

7.4 Conclusions

A promising technique for targeted cancer treatment and other biological uses is magnetic hyperthermia ablation. It achieves controlled medication release and accurate thermal ablation of tumor tissues by using magnetic nanoparticles in an alternating magnetic field. This method improves patient outcomes, lowers systemic toxicity, and increases treatment efficacy. There are still issues to be resolved, though, like maximizing heat dispersion, guaranteeing effective medication loading and release, and handling possible off-target effects. Magnetic nanoparticles can enhance ablation-based personalized therapy by enabling targeted, controllable thermal destruction of tumors, improving drug delivery specificity, and allowing imaging-guided treatment, thereby increasing therapeutic efficacy and reducing damage to healthy tissue. Their multifunctional role supports the advancement of personalized cancer treatments tailored to individual tumor characteristics.

References

Alarcón-Segovia LC, Morel MR, Daza-Agudelo JI, Ilardo JC, Rintoul I (2023) Hyperthermic triggers for drug delivery platforms. Nanotechnology 35(3):035704

Beik J, Abed Z, Ghoreishi FS, Hosseini-Nami S, Mehrzadi S, Shakeri-Zadeh A, Kamrava SK (2016) Nanotechnology in hyperthermia cancer therapy: from fundamental principles to advanced applications. J Control Release 235:205–221

Bi H, Xue J, Jiang H, Gao S, Yang D, Fang Y, Shi K (2019) Current developments in drug delivery with thermosensitive liposomes. Asian J Pharm Sci 14(4):365–379

Bravo M, Fortuni B, Mulvaney P, Hofkens J, Uji-i H, Rocha S, Hutchison J (2024) Nanoparticle-mediated thermal cancer therapies: strategies to improve clinical translatability. J Control Release 2024:751–777

Chang D, Lim M, Goos JA, Qiao R, Ng YY, Mansfeld FM, Jackson M, Davis TP, Kavallaris M (2018) Biologically targeted magnetic hyperthermia: potential and limitations. Front Pharmacol 9:831

Chia DK, Demuytere J, Ernst S, Salavati H, Ceelen W (2023) Effects of hyperthermia and hyperthermic intraperitoneal chemoperfusion on the peritoneal and tumor immune contexture. Cancer 15(17):4314

Colli C, Masi I, Jacchetti E, Santoni S, Sponchioni M, Colosimo BM, Rosanò L, Raimondi MT, Mauri E, Moscatelli D (2024) Zwitterionic nanoparticles for thermally activated drug delivery in hyperthermia cancer treatment. Nanoscale 16(26):12635–12649

Das P, Colombo M, Prosperi D (2019) Recent advances in magnetic fluid hyperthermia for cancer therapy. Colloids Surf B Biointerfaces 174:42–55

Day NB, Wixson WC, Shields CW IV (2021) Magnetic systems for cancer immunotherapy. Acta Pharm Sin B 11(8):2172–2196

Gavilán H, Avugadda SK, Fernández-Cabada T, Soni N, Cassani M, Mai BT, Chantrell R, Pellegrino T (2021) Magnetic nanoparticles and clusters for magnetic hyperthermia: optimizing their heat performance and developing combinatorial therapies to tackle cancer. Chem Soc Rev 50(20):11614–11667

Hervault A, Thanh NT (2014) Magnetic nanoparticle-based therapeutic agents for thermo-chemotherapy treatment of cancer. Nanoscale 6(20):11553–11573

Jaymand M (2024) Hydrogel-based drug delivery systems for synergistic chemo/hyperthermia therapy of cancer: a comprehensive review. J Drug Deliv Sci Technol 95(1):105581

Jiao W, Dai L, Yan B, Lyu Y, Fan H, Liu X (2024) Heating up the immune battle: magnetic hyperthermia against cancer. Fundam Res

Kacem M, Essoumhi A, Dib M (2024) Magnetic iron oxide-based materials and their hyperthermia application: a review. Inorg Chem Commun 166:112510

Kumar CS, Mohammad F (2011) Magnetic nanomaterials for hyperthermia-based therapy and controlled drug delivery. Adv Drug Deliv Rev 63(9):789–808

Liu X, Zhang Y, Wang Y, Zhu W, Li G, Ma X, Zhang Y, Chen S, Tiwari S, Shi K, Zhang S (2020) Comprehensive understanding of magnetic hyperthermia for improving antitumor therapeutic efficacy. Theranostics 10(8):3793

Patri S, Thanh NT, Kamaly N (2024) Magnetic iron oxide nanogels for combined hyperthermia and drug delivery. Nanoscale 16(33):15446–15464

Prajapati A, Rangra S, Patil R, Desai N, Jyothi VG, Salave S, Amate P, Benival D, Kommineni N (2024) Receptor-targeted nanomedicine for cancer therapy. Receptor 3(3):323–361

Rajan A, Laha SS, Sahu NK, Thorat N, Shankar B (2024) Recent advancements and clinical aspects of engineered iron oxide nanoplatforms for magnetic hyperthermia-induced cancer therapy. Mater Today Bio 29:101348

Ribeiro TP, Moreira JA, Monteiro FJ, Laranjeira MS (2022) Nanomaterials in cancer: reviewing the combination of hyperthermia and triggered chemotherapy. J Control Release 347:89–103

Shirsat SD, Londhe PV, Gaikwad AP, Rizwan M, Laha SS, Khot VM, Achal V, Tabish TA, Thorat ND (2024) Endosomal escape in magnetic nanostructures: recent advances and future perspectives. Mater Today Adv 22:100484

Vassallo M, Martella D, Barrera G, Celegato F, Coisson M, Ferrero R, Olivetti ES, Troia A, Sozeri H, Parmeggiani C, Wiersma DS (2023) Improvement of hyperthermia properties of iron oxide nanoparticles by surface coating. ACS Omega 8(2):2143–2154

Vijayan V, Sundaram A, Vasukutty A, Bardhan R, Uthaman S, Park IK (2023) Tumor-targeting cell membrane-coated nanorings for magnetic-hyperthermia-induced tumor ablation. Biomater Sci 11(21):7188–7202

Włodarczyk A, Gorgoń S, Radoń A, Bajdak-Rusinek K (2022) Magnetite nanoparticles in magnetic hyperthermia and cancer therapies: challenges and perspectives. Nanomaterials 12(11):1807

Yadav AK, Tripathi H, Bastia A, Singh P, Dubey AK, Anuraag NS, Prasad NK, Rath C (2023) Synergistic effect of CoFe2O4–85S nano bio-glass composites for hyperthermia and controlled drug delivery. Materialia 32:101884

Yao Y, Zhou Y, Liu L, Xu Y, Chen Q, Wang Y, Wu S, Deng Y, Zhang J, Shao A (2020) Nanoparticle-based drug delivery in cancer therapy and its role in overcoming drug resistance. Front Mol Biosci 7:193

Yu BO, Tai HC, Xue W, Lee LJ, Lee RJ (2010) Receptor-targeted nanocarriers for therapeutic delivery to cancer. Mol Membr Biol 27(7):286–298

Yu L, Liu J, Wu K, Klein T, Jiang Y, Wang JP (2014) Evaluation of hyperthermia of magnetic nanoparticles by dehydrating DNA. Sci Rep 4(1):7216

Zhang YF, Lu M (2024) Advances in magnetic induction hyperthermia. Front Bioeng Biotechnol 12:1432189

Zhang L, Li Q, Liu J, Deng Z, Zhang X, Alifu N, Zhang X, Yu Z, Liu Y, Lan Z, Wen T (2024) Recent advances in functionalized ferrite nanoparticles: from fundamentals to magnetic hyperthermia cancer therapy. Colloids Surf B Biointerfaces 234:113754

Advancements in Minimally Invasive Probes for Drug Delivery

8

Saravanakumar Kasimedu, Ashok Thulluru, Nawaz Mohammed, Reshma Thadipatri, Nagaveni Pommala, and Mallikarjuna Gandla

Abstract

Minimally invasive probes are revolutionizing drug delivery by enhancing precision and reducing patient discomfort. These probes, including needle-based devices, microneedles, nanoparticles, liposomes, and hydrogels, have been developed to target specific sites while minimizing side effects. Their use in cancer treatment, diabetes management, and pain relief highlights their broad clinical applications. Polymeric and biodegradable materials play a key role in constructing these biocompatible devices, while technologies like micro-electromechanical systems and nano-electromechanical systems further improve their functionality. Despite their advantages, challenges remain, such as biocompatibility and safety concerns. Current research is focused on advancing nanotechnology, implantable systems, and artificial intelligence to push the boundaries of minimally invasive drug delivery. With ongoing developments, these technologies promise to enhance patient outcomes and redefine therapeutic strategies in the future.

S. Kasimedu · M. Gandla
Seven Hills College of Pharmacy (Autonomous), Venkatramapuram, Tirupati, Andhra Pradesh, India

A. Thulluru (✉)
Department of Pharmaceutics, Sandip University School of Pharmaceutical Sciences, Nashik, Maharashtra, India

N. Mohammed · R. Thadipatri
Raghavendra Institute of Pharmaceutical Education and Research (Autonomous), Anantapur, Andhra Pradesh, India

N. Pommala
S.V.U. College of Pharmaceutical Sciences, Sri Venkateswara University, Tirupati, Andhra Pradesh, India

M. R. Babu et al. (eds.), *Transdermal Applications of Minimally Invasive Drug Delivery Systems*, https://doi.org/10.1007/978-981-96-6275-3_8

Keywords
Minimally invasive probes · Drug delivery · Microneedles · Nanoparticles · Biocompatibility · Nanotechnology

8.1 Definition and Importance

Minimally invasive probes (MIPs) have become a vital technology in modern medicine, particularly drug delivery systems. As the demand for more precise, patient-friendly, and efficient therapeutic interventions grows, MIPs offer a unique solution that addresses some of the limits of traditional medication delivery techniques. To fully appreciate their role, it is crucial to explore what minimally invasive probes are, why they are important, and how they fit into the evolving healthcare landscape.

Definition of Minimally Invasive Probes: Minimally invasive probes are medical devices specifically engineered to enter the body through tiny incisions or natural orifices, providing access to internal organs or tissues with minimal disruption to surrounding structures. These probes allow for the administration of drugs directly to targeted areas of the body, often using precision-based delivery techniques. The primary advantage of MIPs lies in their ability to access deep tissue layers or specific disease sites without requiring major surgery, thereby minimizing the risk of collateral damage to healthy tissues and reducing overall recovery time for patients (Sant Anna et al. 2021).

In the context of drug delivery, minimally invasive probes can take many forms, from microneedles to catheter-based systems and even nano-sized probes designed to travel through the bloodstream. These devices are able to deliver a large range of therapeutic agents, encompassing chemotherapy drugs, antibiotics, pain medications, and even biologics like proteins or gene therapies. The probes can be made to release their drug payload in a regulated way, guaranteeing that the therapeutic agent reaches the intended site in the required dose over a specified period of time (Li and Kwon 2020).

Importance in Drug Delivery: The importance of minimally invasive probes in drug delivery cannot be overstated. Conventional medication delivery techniques, such as oral administration/intravenous injection, often suffer from several key limitations. For example, drugs administered orally must pass through the digestive system, where they may be metabolized before reaching their target, leading to reduced efficacy. Systemic injections, on the other hand, deliver drugs throughout the entire body, which can lead to unwanted side effects as the drug affects healthy tissues in addition to the targeted disease site.

Minimally invasive probes overcome these limitations by providing a direct route to the target tissue, allowing for localized drug delivery. This approach significantly enhances the efficacy of the drug while minimizing systemic side effects. For instance, in cancer treatment, delivering chemotherapeutic agents directly to a tumor site can raise the drug's concentration at the target while reducing the exposure of healthy tissues to the toxic effects of the chemotherapy. This not only improves treatment outcomes but also reduces the likelihood of complications, such as organ damage or immune suppression, which are common in systemic chemotherapy.

Additionally, minimally invasive probes can be designed to incorporate advanced drug release mechanisms that are responsive to specific stimuli. These stimuli can be physiological, such as changes in pH or temperature at the disease site, or they can be externally triggered, such as by light, magnetic fields, or ultrasound. Stimuli-responsive probes allow for precise control over when and how much of a drug is released, further enhancing the efficiency of the drug delivery system. This type of targeted, controlled-release delivery is particularly beneficial in treating chronic diseases, where maintaining a stable drug concentration at the disease site over a long period is critical for achieving optimal therapeutic outcomes.

Moreover, MIPs improve patient comfort and reduce the overall invasiveness of medical procedures. Patients undergoing traditional surgery for drug delivery purposes may face long recovery periods, pain, and a higher risk of complications, such as infection or scarring. Minimally invasive probes, on the other hand, reduce the need for large incisions and can often be performed on an outpatient basis. This not only shortens recovery times but also decreases healthcare costs associated with longer hospital stays and more complex postoperative care.

In recent years, the importance of minimally invasive probes has also been recognized in the context of customized healthcare. The goal of personalized medicine is to adjust medical care to each patient's unique needs, and minimally invasive probes offer a way to deliver treatments that are specifically designed for a patient's unique physiological profile. For example, in patients with genetic mutations that cause certain types of cancer, a minimally invasive probe could be used to deliver a targeted therapy that directly addresses the mutation at the molecular level. This kind of precision treatment is becoming increasingly important as our understanding of the genetic and molecular underpinnings of diseases continues to expand (Liang et al. 2021).

Applications in Disease Treatment: The applications of MIPs extend across a wide range of medical conditions. In oncology, minimally invasive probes are often utilized for delivery of chemotherapy drugs directly to a tumor, reducing the overall exposure of the body to toxic agents. This approach not only improves the effectiveness of the treatment but also reduces the likelihood of common side effects, such as nausea, fatigue, and hair loss. Similarly, MIPs have shown promise in delivering targeted therapies to treat localized infections, where antibiotics can be delivered straight to the illness site, limiting damage to healthy tissues and lowering the possibility of antibiotic resistance.

Another important application of MIPs is in the treatment of cardiovascular diseases. For instance, drug-eluting stents, a type of minimally invasive probe, are commonly used in patients with coronary artery disease. These stents release medication that prevents the re-narrowing of arteries after angioplasty, improving long-term outcomes for patients while reducing the need for repeat procedures. Similarly, catheter-based probes are used to deliver medications directly into blood vessels or organs, offering a targeted and efficient approach to treating vascular diseases.

In neurological conditions, minimally invasive probes are being explored for the delivery of drugs across the blood-brain barrier (BBB), which is a significant difficulty in treating illnesses such as Parkinson's and Alzheimer's diseases and brain tumors. MIPs can bypass the BBB by delivering therapeutic agents directly to the brain tissue, providing new hope for patients with these hard-to-treat conditions. For example, ongoing research into nanoparticle-based probes suggests that these devices may offer a way to deliver drugs directly to specific regions of the brain, improving the treatment of neurodegenerative diseases (Parodi et al. 2020).

Economic and Practical Impact: The economic and practical impact of MIPs is another critical factor in their growing importance. Healthcare systems are under growing pressure to deliver care that is more economical and efficient. Minimally invasive procedures reduce the need for extended hospital stays and lower the risk of complications, which translates into significant cost savings. These savings are not only beneficial for healthcare providers but also for patients, who face reduced out-of-pocket expenses and shorter recovery times.

MIPs also provide practical advantages in terms of patient compliance. For chronic conditions that require long-term treatment, minimally invasive drug delivery systems offer a more manageable solution. Instead of repeated invasive surgeries or systemic drug administrations, MIPs can be used to deliver drugs continuously over an extended period, improving the consistency of treatment and reducing the burden on patients.

Finally, the ability of MIPs to integrate with diagnostic tools, such as imaging technologies, allows for real-time monitoring of drug delivery. This integration improves treatment precision and enables healthcare providers to make immediate adjustments to treatment plans based on how well a patient is responding to therapy. For instance, probes equipped with sensors can monitor drug concentrations at the target site, providing feedback that ensures the correct dosage is delivered without under- or overtreatment (Kim et al. 2021).

8.2 Historical Overview

Early Developments in Minimally Invasive Procedures: The history of minimally invasive procedures dates back to the early twentieth century, with early efforts aimed at reducing the size of incisions required for surgical interventions. However, the major leap forward occurred in the late twentieth century with the advent of laparoscopic surgery, which enabled doctors to use tiny incisions to carry out intricate surgeries using endoscopic cameras and specialized instruments. Laparoscopic techniques were initially used in gastrointestinal and gynecological surgeries, but their success paved the way for broader applications in other fields, including urology, cardiology, and orthopedics.

The development of MIPs for the distribution of drugs started in earnest in the late 1990s and early 2000s, as advances in nanotechnology, biomaterials, and microfabrication enabled the creation of more sophisticated and precise devices. Early drug delivery probes were often simple catheter-based systems, which could be threaded through blood vessels to deliver medication directly to specific sites, such as in coronary angioplasty. However, these early systems were limited in their capacity to transport medications in a controlled or sustained manner, and their use was largely restricted to specific procedures (Needleman and Coburn 2021).

The Nanotechnology Revolution: The early twenty-first century saw a revolution in MIP development with the rise of nanotechnology. Nanoparticles, nanorobots, and other nanoscale devices offered new possibilities for drug delivery, allowing for the creation of probes that could travel through the bloodstream, navigate the body's complex anatomy, and release drugs in response to specific triggers. These developments opened up new avenues for treating diseases that were previously difficult to address, such as cancer, where targeting specific cells with high precision is critical.

Today, minimally invasive probes are more sophisticated than ever, with capabilities that include real-time imaging, diagnostic feedback, and multimodal drug delivery. The integration of robotics and artificial intelligence (AI) into MIP systems has further enhanced their precision and functionality, allowing for highly targeted and personalized drug delivery that is guided by real-time data from sensors embedded in the probes. This combination of technologies represents the future of medicine, offering the potential for more effective and less invasive remedies for a variety of illnesses (Hartman and McMullen 2020).

8.3 Types of Minimally Invasive Probes

Minimally invasive probes (MIPs) for drug delivery represent a class of technologies aimed at delivering therapeutic agents with minimal disruption to normal tissues. These probes enhance precision in drug administration, minimize side effects, and promote faster recovery times. The types of MIPs include needle-based probes, microneedles, nanoparticles, liposomes, and hydrogels. Each category of these systems holds distinct advantages for improving the efficiency and precision of drug delivery.

8.3.1 Needle-Based Probes

Needle-based probes are one of the traditional yet evolving methods for drug delivery. These devices use thin, hollow needles to access deep tissue or specific organs, directly delivering drugs to the target site. While the basic concept has remained constant, modern advancements have improved their precision and functionality.

Mechanism and Design: Needle-based probes are typically designed to penetrate skin barriers and access specific tissue types, enabling direct drug administration. Traditional needles primarily focus on subcutaneous or intramuscular injections, but more advanced designs allow for reaching deeper tissues, including targeting tumors or even accessing cerebrospinal fluid. Imaging techniques, such as ultrasound or MRI, can guide these probes to the desired location.

Advancements: Technological innovations have transformed needle-based probes into sophisticated tools capable of precise drug delivery. For instance, retractable needles and automated pumps have been developed to enhance precision. In addition, "smart" needles that integrate biosensors offer real-time feedback on physiological changes, optimizing drug dosage based on immediate biological responses. Coatings on needles that react to environmental stimuli, including pH or temperature changes, by releasing medications represent a significant step forward in localized drug delivery (Sant Anna et al. 2021).

Applications: Needle-based probes are utilized in chemotherapy, where they deliver drugs directly to tumors to minimize systemic toxicity. They are also employed in neurology, allowing for the precise medication distribution to the brain while bypassing the BBB. These probes are additionally used in pain management, providing localized delivery of pain relief agents to specific regions such as the spine, offering longer-lasting effects than systemic treatments (Li and Kwon 2020).

8.3.2 Microneedles

Microneedles are a relatively new form of minimally invasive probes that have gained attention for their ability to painlessly deliver drugs transdermally. These are extremely small needles, typically less than 1 mm in length, designed to penetrate only the outermost layers of the skin, minimizing discomfort while still delivering drugs effectively. Recent advances in microneedle delivery are depicted in Fig. 8.1 (Goudanavar et al. 2024).

Mechanism and Design: Microneedles are made from a range of substances, such as polymers, silicon, and metals. They are typically designed as arrays of tiny projections that painlessly pierce the skin's stratum corneum, the outermost skin layer, and deliver therapeutic agents directly into the dermal layer. These microneedles can either be solid or hollow, depending on whether the drug is coated on their surface or delivered through the hollow structure (Kim et al.

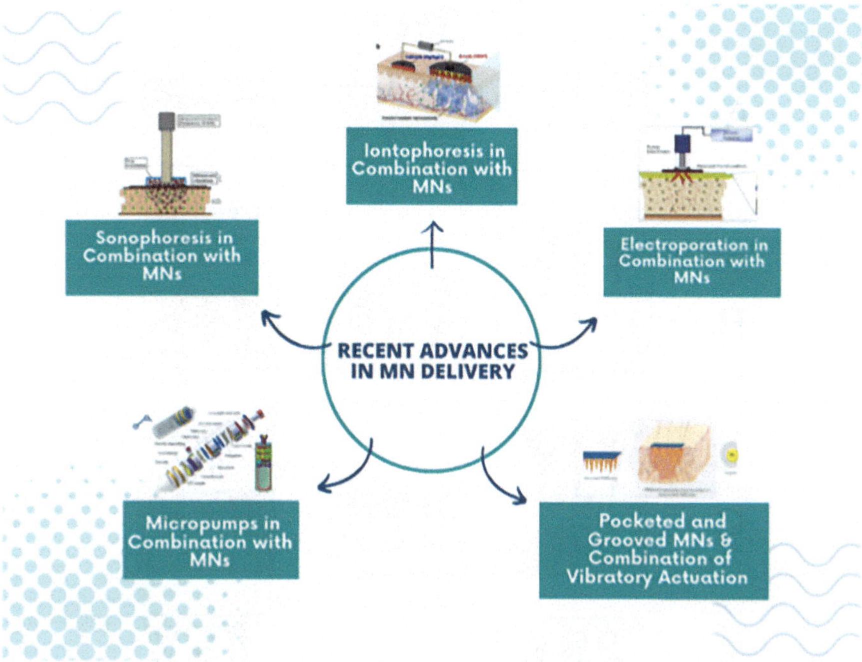

Fig. 8.1 Recent advances in microneedle delivery. (Reprinted with permission from Goudanavar et al. (2024). Copyright 2024 Elsevier)

2021). Types of microneedles and delivery methods are shown schematically in Fig. 8.2 (Hamdan 2024).

Advancements: The development of dissolvable microneedles is a significant advancement in this field. These microneedles are made from biocompatible materials that dissolve after penetrating into the skin, leaving no residue. This eliminates the need for disposal and minimizes the risk of infections. Another breakthrough is stimuli-responsive microneedles, which dissolve upon exposure to certain physiological triggers, such as moisture or body heat, enabling precise, controlled drug release (Zhang and Chan 2020).

8.3.2.1 Applications

Microneedles possess several applications, particularly in vaccine delivery, where they offer painless, efficient administration without the need for syringes. They are also gaining traction for insulin delivery, providing a noninvasive solution for diabetes management. Research has demonstrated the potential of microneedles in delivering anticancer drugs, peptides, and even gene therapies (Yeo et al. 2022). Diagrammatic illustration of the microneedles' mode of action in various diseases is depicted in Fig. 8.3 (Priya and Singhvi 2022).

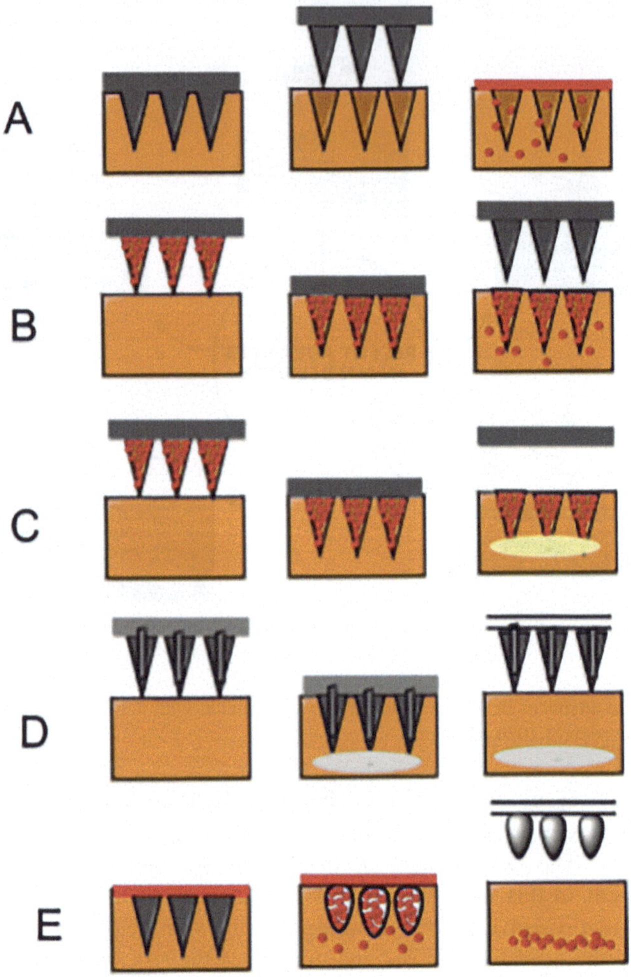

Fig. 8.2 Schematic representation of microneedle types and delivery approaches. (**a**) Solid non-coated. (**b**) Solid coated. (**c**) Dissolving. (**d**) Hollow. (**e**) Swellable. (Reprinted with permission from Hamdan (2024). Copyright 2024 Elsevier)

8.3.3 Nanoparticles

Nanoparticles (NPs) are at the forefront of nanomedicine due to their size, which enables them to penetrate biological barriers, target specific tissues, and deliver drugs with high precision. These particles are typically between 1 and 100 nm in size and possess a large surface area, allowing for functionalization with various drugs, targeting ligands, or imaging agents.

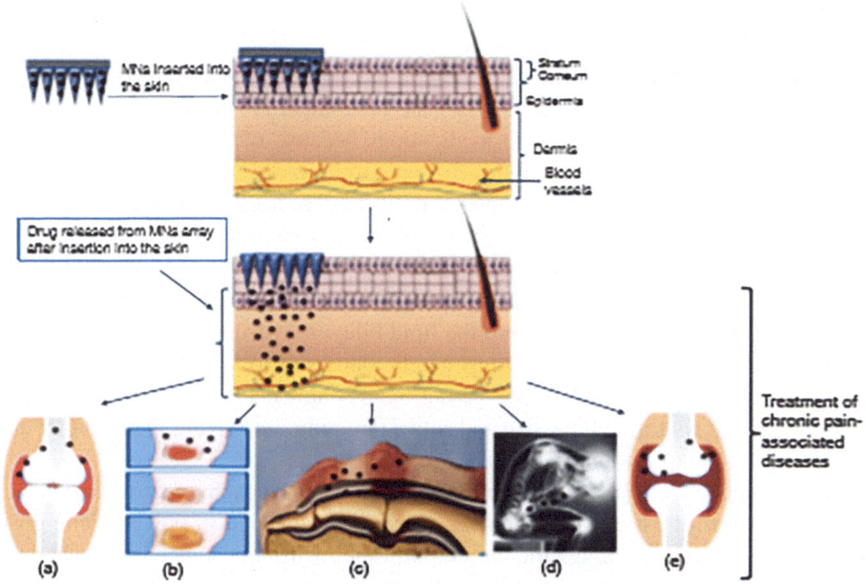

Fig. 8.3 Schematic representation of mechanism of microneedles in (**a**) osteoarthritis, (**b**) atopic dermatitis, (**c**) psoriatic arthritis, (**d**) neuropathic pain, and (**e**) rheumatoid arthritis. (Reprinted with permission from Priya and Singhvi (2022). Copyright 2022 Elsevier)

Mechanism and Design: Metals, polymers, and lipids are just a few of the components that can be used to create nanoparticles. The drugs can be encapsulated within the nanoparticles or attached to their surface. Targeting molecules, such as peptides or antibodies, can be coupled to the nanoparticles' surface to direct them to specific cells or tissues. Once the nanoparticles reach their target, they can release their drug payload in reaction to environmental cues, like variations in temperature or pH (Liang et al. 2021).

Advancements: "Theranostic" nanoparticles have been created as a result of recent developments in nanoparticle technology, which combine therapeutic and diagnostic capabilities. These multifunctional particles enable real-time imaging and drug delivery simultaneously, providing valuable feedback during treatment (Needleman and Coburn 2021). Additionally, biodegradable nanoparticles are being explored, which safely degrade after delivering their drug cargo, minimizing potential long-term side effects.

Applications: Nanoparticles are extensively used in cancer therapy to deliver chemotherapeutic medications straight to tumor cells, lowering the toxicity across the body typically associated with chemotherapy. Their ability to cross the blood-brain barrier also makes them promising candidates for treating neurological disorders, such as Alzheimer's disease and Parkinson's disease (Gao and Wang 2021). Furthermore, nanoparticles have been explored for gene therapy, delivering nucleic acids to specific cells to correct genetic mutations or silence harmful genes (Zhang et al. 2021a).

8.3.4 Liposomes

Drugs can be encapsulated within the core of liposomes, which are spherical vesicles made of phospholipid bilayers. They are biocompatible, perfect for delivering both hydrophilic and hydrophobic medications because of their structure, which resembles biological cell membranes. It is possible to design liposomes to target particular tissues, increase medication solubility, and boost bioavailability.

Mechanism and Design: Liposomes are composed of phospholipids that self-assemble into bilayered vesicles when exposed to water. Whereas hydrophobic medications are incorporated into the lipid bilayer, hydrophilic medications are encased in the aqueous core. Drugs can be delivered to illness locations selectively via liposomes that have targeting ligands attached to their surface. Their structure allows for both passive and active targeting, improving therapeutic outcomes (Bao and Mitragotri 2021).

Advancements: One of the significant advancements in liposome technology is the development of "stealth" liposomes, which are embedded in polyethylene glycol. This coating prolongs the circulation time of the liposomes by preventing their recognition and clearance by the immune system. Additionally, researchers have developed triggered-release liposomes that release their contents in response to specific stimuli, such as heat, light, or ultrasound (Lim et al. 2020).

Applications: Liposomes are widely used in cancer therapy to deliver chemotherapeutic agents, such as doxorubicin, directly to tumors, reducing the drug's systemic toxicity. They have also been explored for delivering vaccines, enhancing immune responses by presenting antigens to immune cells more effectively. In addition to cancer treatment, liposomes are utilized in gene therapy and the treatment of various autoimmune diseases (Parodi et al. 2020).

8.3.5 Hydrogels

Three-dimensional networks of polymers called hydrogels have the ability to swell and absorb a lot of water without losing their structural integrity. Their biocompatibility, high water content, and tunable properties make them a versatile tool for drug delivery. Hydrogels can be designed to release medications over a long length of time under strict monitoring.

Mechanism and Design: Hydrogels are formed through the cross-linking of natural or synthetic polymers, which create a hydrophilic network capable of absorbing water. The drugs can be physically entrapped within the hydrogel matrix or chemically conjugated to the polymer chains. Once inside the body, the drug is released slowly as the hydrogel degrades or through diffusion. Additionally, hydrogels can be made to react to particular environmental cues, including pH or temperature changes, allowing for controlled and prolonged medication release (Hartman and McMullen 2020).

Advancements: Smart hydrogels that respond to physiological changes are among the most significant advancements in hydrogel technology. These materials can swell, shrink, or degrade in response to stimuli such as pH, temperature, or enzymes, releasing drugs in a controlled manner (Liu et al. 2020). Injectable hydrogels are also gaining attention for their ease of administration and ability to form a gel at the target site after injection, providing localized drug delivery.

Applications: Hydrogels are frequently used in wound healing because they release growth factors or antibiotics to stop infection and encourage healing while creating a wet environment that supports tissue regeneration. In order to minimize systemic side effects, they are also utilized in cancer therapy to administer chemotherapeutic drugs locally, straight to the tumor site. Hydrogels are being investigated further for use in regenerative medicine and tissue engineering (Yadav and Gupta 2021).

8.4 Advantages and Challenges of Minimally Invasive Drug Delivery

They are gaining widespread attention due to their potential to revolutionize the way drugs are administered. The ability to deliver therapeutic agents with minimal trauma, while maintaining high precision, significantly enhances treatment efficacy and patient satisfaction. Despite these advantages, several challenges remain. In this chapter, we will explore the key advantages and challenges of MIDDs, focusing on enhanced patient compliance, reduced side effects, precision targeting, and biocompatibility and safety concerns.

8.4.1 Enhanced Patient Compliance

Definition and Significance: Patient compliance, sometimes referred to as adherence, is the degree to which a patient takes their medications as directed by their doctor at the correct times and dosages. Noncompliance is a common issue in conventional drug therapies, particularly in chronic diseases where patients may need to take medication regularly over long periods. MIDDs are designed to simplify the treatment process, facilitating patients' adherence to their prescribed regimen (Needleman and Coburn 2021).

Mechanisms of Improved Compliance: One of the primary mechanisms by which MIDDs enhance patient compliance is through reduced invasiveness. Traditional drug delivery methods, such as intravenous injections or oral medications, often lead to discomfort or inconvenience, which can cause patients to avoid or forget their treatments. MIDDs, including microneedles, nanoparticle-based delivery systems, and transdermal patches, minimize pain and discomfort, thus improving the likelihood of consistent use. Microneedles, for example, penetrate only the outer layers of the skin without causing pain, which is particularly advantageous for patients who are needle-averse (Sant Anna et al. 2021).

Furthermore, MIDDs can be designed for sustained or controlled release of therapeutic agents. This allows for less frequent dosing, eliminating the need for patients to remember multiple doses throughout the day. For instance, injectable hydrogels have been developed to provide continuous drug release over weeks or months, reducing the frequency of administration and improving long-term adherence. This approach is especially beneficial for patients with chronic diseases like diabetes, where long-term compliance is crucial for effective disease management.

Patient-Centric Designs: Patient-friendly designs play a significant role in improving compliance. Wearable drug delivery systems, such as insulin pumps and transdermal patches, allow patients to continue their daily activities without the need for constant monitoring or interruption. These devices provide a seamless integration of treatment into everyday life, reducing the burden on patients. Self-administration is another key factor, as many MIDDs are designed to be easily administered by the patient without the need for healthcare professionals, empowering patients to manage their own treatments (Gao and Wang 2021).

Evidence of Improved Compliance: Numerous clinical studies have demonstrated improved patient compliance with MIDDs. For example, a study on insulin delivery using microneedles showed that patients preferred this method over traditional injections due to the reduced pain and ease of use, leading to higher adherence rates. Similar findings have been reported for cancer patients receiving chemotherapy through targeted nanoparticle systems, where the reduced side effects and convenience of treatment improved compliance (Gao and Wang 2021).

8.4.2 Reduced Side Effects

Traditional Drug Delivery Challenges: This non-targeted approach means that healthy tissues are exposed to the therapeutic agent, which can lead to adverse reactions. For example, chemotherapy drugs are notorious for causing damage to non-cancerous cells, resulting in severe side effects like nausea, fatigue, and hair loss (FDA 2020).

Minimally Invasive Delivery for Localized Treatment: MIDDs offer the ability to deliver drugs directly to the target site, thereby minimizing exposure to healthy tissues. This localized delivery reduces the likelihood of systemic side effects, as the therapeutic agent is concentrated in the area of interest. For example, needle-based probes can deliver chemotherapeutic drugs directly to a tumor, sparing surrounding healthy tissue from exposure. Similarly, transdermal patches and microneedles can deliver drugs locally, avoiding the gastrointestinal system and reducing gastrointestinal side effects commonly associated with oral medications (Hartman and McMullen 2020).

Controlled Release and Side Effect Reduction: MIDDs also allow for controlled or sustained release of drugs, which further mitigates side effects. By maintaining consistent drug levels within the therapeutic window, these systems prevent the peaks and troughs in drug concentration that are typically associated with side effects in conventional therapies. For instance, controlled-release nanoparticles are designed to release their drug payload gradually over time, avoiding the high concentrations that can lead to toxicity (Ratner et al. 2013).

Clinical Evidence of Reduced Side Effects: Clinical trials have provided substantial evidence that MIDDs can reduce side effects compared to traditional drug delivery methods. A study on the use of liposomes for chemotherapy delivery found that patients experienced significantly fewer side effects, such as gastrointestinal distress and hair loss, compared to those receiving traditional chemotherapy. Similarly, nanoparticle-based delivery systems for rheumatoid arthritis have shown reduced inflammation in off-target tissues, improving the overall safety profile of the treatment (Parodi et al. 2020).

8.4.3 Precision Targeting

Importance of Targeted Drug Delivery: One of the greatest challenges in drug therapy is ensuring that the therapeutic agent reaches the desired site of action at an effective concentration while minimizing its distribution to non-target tissues. Precision targeting is a key advantage of MIDDs, enabling the delivery of drugs directly to specific cells or tissues. This targeted approach not only improves the efficacy of treatment but also reduces the risk of side effects, as non-target tissues are spared from unnecessary exposure.

8.4.3.1 Mechanisms of Precision Targeting

MIDDs utilize various strategies to achieve precision targeting. One approach is through the use of nanoparticles that are functionalized with targeting ligands, such as antibodies or peptides, which bind to specific receptors on the surface of target cells. This allows the nanoparticles to home in on the diseased tissue, such as a tumor, and deliver the drug payload directly to the site (Liang et al. 2021). Similarly, liposomes can be engineered to target specific tissues by incorporating targeting molecules on their surface. These delivery systems can also be designed to respond to specific stimuli, such as changes in pH, temperature, or enzyme activity, ensuring that the drug is released only at the target site.

Enhanced Efficacy Through Precision Targeting: MIDDs can greatly increase the effectiveness of treatment by delivering medications straight to the intended location. Through the enhanced permeability and retention (EPR) effect, for instance, targeted nanoparticles can accumulate in the tumor during cancer therapy, providing larger quantities of the chemotherapeutic agent to the tumor while minimizing exposure to healthy tissues. In addition to increasing therapeutic efficacy, this targeted administration lowers the risk of adverse effects by enabling lower drug dosages (Gao and Wang 2021).

Challenges in Precision Targeting: While precision targeting offers significant advantages, it also presents certain challenges. One of the primary obstacles is the complexity of developing delivery systems that can effectively navigate the body's biological barriers, such as the blood-brain barrier or dense tumor stroma. Additionally, the heterogeneity of diseases like cancer, where different cells within a tumor may express different receptors, can complicate the targeting process. Despite these challenges, ongoing research is focused on improving the specificity and efficiency of targeted delivery systems (Yadav and Gupta 2021).

8.4.4 Biocompatibility and Safety Concerns

Importance of Biocompatibility: The capacity of a material to carry out its intended function without causing a negative biological reaction is known as biocompatibility. To guarantee that the delivery mechanism does not result in toxicity, immunological responses, or other undesirable outcomes, biocompatibility is essential in the context of MIDDs. Ensuring the safety of these systems is paramount, especially when they are designed for long-term use or repeated administration (Ratner et al. 2013).

Biocompatibility in Various MIDDs: Different MIDDs face distinct biocompatibility challenges depending on their materials and design. For example, microneedles, which are typically made from polymers, metals, or silicon, must be designed to avoid causing irritation or damage to the skin. Dissolvable microneedles, which degrade after drug delivery, are often made from biocompatible polymers, such as polyvinyl alcohol (PVA) or hyaluronic acid, which are safe for the body to absorb. Nanoparticles, on the other hand, require careful design to avoid triggering an immune response or causing toxicity to non-target organs. Liposomes, due to their phospholipid composition, are generally considered biocompatible, but modifications such as PEGylation can improve their safety profile by prolonging circulation time and reducing clearance by the immune system (Allen et al. 2016).

Challenges in Ensuring Safety: Despite the advancements in material design, ensuring the long-term safety of MIDDs remains a challenge. One concern is the accumulation of materials in the body, particularly in the case of nanoparticles, which may not be easily cleared by the kidneys or liver. This can lead to toxicity if the particles accumulate in vital organs. Additionally, repeated administration of MIDDs, such as in the case of chronic diseases, raises concerns about potential immunogenicity or the development of resistance to the drug or the delivery system (Gadde 2017).

Regulatory and Ethical Considerations: Regulatory agencies, such as the FDA and EMA, have established guidelines for the safety and efficacy of drug delivery systems. MIDDs must undergo rigorous testing to ensure that they meet safety standards before they can be approved for clinical use. Ethical considerations also come into play, particularly in the development of delivery systems for vulnerable populations, such as pediatric or geriatric patients, where safety concerns may be heightened. Ongoing research is focused on addressing these challenges by developing safer, more biocompatible materials and improving the long-term safety of MIDDs (FDA 2020).

8.5 Materials and Technologies in Minimally Invasive Probes

Minimally invasive probes are widely used in modern medical diagnostics and treatments. The advancement of materials and technologies used in the development of these probes has significantly improved their effectiveness, biocompatibility, and functionality. This chapter focuses on three key aspects: polymeric materials, biodegradable materials, and the integration of micro-electromechanical system and nano-electromechanical system technologies. These materials and technologies enable probes to meet the rigorous demands of modern healthcare by minimizing patient discomfort, reducing recovery times, and offering high precision.

8.5.1 Polymeric Materials

Polymeric materials are essential to the development of minimally invasive probes due to their versatile properties. These materials can be designed with varying degrees of flexibility, chemical resistance, and biocompatibility, making them ideal for use in medical applications where patient safety and comfort are paramount.

8.5.1.1 Properties and Advantages

Biocompatibility: Polymers like polyethylene (PE), polyurethane (PU), polydimethylsiloxane (PDMS), and polyimides are frequently used in the construction of medical devices due to their biocompatibility. This ensures that the material does not trigger adverse immune responses when in contact with bodily tissues or fluids (Karp and Langer 2021).

Flexibility: Many polymeric materials are highly flexible, which allows the probes to be maneuvered through the body's natural pathways with minimal damage to surrounding tissues. The flexibility also aids in navigating complex anatomical structures without causing discomfort or trauma to the patient.

Chemical Resistance: Polymers used in minimally invasive probes often demonstrate resistance to various chemicals and bodily fluids. This is crucial for maintaining the integrity of the probe over long durations, especially in procedures that involve the delivery of drugs or monitoring internal conditions over time.

8.5.1.2 Key Types of Polymeric Materials

Polydimethylsiloxane (PDMS): PDMS is an organic polymer based on silicon that is renowned for its flexibility, oxygen permeability, and biocompatibility. It is widely used in soft biomedical devices, such as catheters and wearable sensors, where high flexibility and patient comfort are essential (Zhao et al. 2023a). PDMS also exhibits low toxicity and chemical inertness, making it suitable for long-term implants.

Polyimides: Polyimides are another commonly used polymer in minimally invasive probes. They offer excellent thermal stability and mechanical strength and are compatible with microfabrication processes. Polyimide-based probes are frequently used in neural interfaces due to their ability to withstand harsh environments while maintaining high conductivity for signal transmission (Rodgers et al. 2020).

Polyethylene (PE) and Polyurethane (PU): These polymers are often used for coatings or as part of the structural components in medical devices. Polyethylene is valued for its chemical resistance and flexibility, while polyurethane is often chosen for its elastomeric properties, which provide enhanced durability in dynamic environments (Basu and Kumar 2023).

8.5.1.3 Applications of Polymeric Materials in Minimally Invasive Probes

Neural Probes: Polymeric materials are extensively used in neural interfaces where the probes must interact with delicate brain tissue. Polymers such as polyimide and PDMS are utilized to create flexible, thin-film neural probes that can record or stimulate brain activity with minimal tissue damage (Altun and Inan 2022).

Catheters and Endoscopes: Polymers like PU and PE are used to manufacture flexible catheters and endoscopes. These devices must navigate through blood vessels or other internal organs, requiring a combination of flexibility and strength to ensure patient safety during diagnostic and therapeutic procedures.

8.5.2 Biodegradable Materials

The development of biodegradable materials in minimally invasive probes has been a breakthrough in reducing the long-term risks associated with medical devices. Biodegradable probes can safely dissolve or be eliminated surgically by being absorbed by the body after serving their role and reducing potential complications.

8.5.2.1 Advantages of Biodegradable Materials

Reduced Risk of Complications: Traditional probes that require surgical removal after use pose risks such as infections, tissue damage, and additional recovery time. Biodegradable materials, however, disintegrate within the body after a predetermined time, minimizing the need for secondary surgeries (Wang et al. 2022a).

Improved Patient Outcomes: Since biodegradable probes dissolve naturally, patients experience shorter recovery times and reduced discomfort, making these materials particularly useful in temporary therapeutic or diagnostic applications.

Environmental Impact: Beyond their medical benefits, biodegradable materials also offer a lower environmental impact compared to conventional polymers, which may persist in biological environments for extended periods.

8.5.2.2 Key Biodegradable Materials in Minimally Invasive Probes

Polyglycolic Acid (PGA) and Polylactic Acid (PLA): PGA and PLA are biodegradable polymers that have been extensively used in sutures, implants, and drug delivery systems. These materials degrade by hydrolysis into non-toxic byproducts like glycolic acid and lactic acid, which are naturally metabolized by the body (Sun et al. 2021).

Polycaprolactone (PCL): PCL is another biodegradable polymer used in medical applications due to its slow degradation rate. Its biocompatibility and ability to degrade in physiological conditions make it ideal for long-term implants and drug delivery systems (Zhao et al. 2023b). PCL-based scaffolds are also used in tissue engineering applications, where the material degrades as the tissue regenerates.

8.5.2.3 Applications of Biodegradable Materials in Minimally Invasive Probes

Drug Delivery Systems: Biodegradable polymers are commonly used in drug-eluting stents and implants, where the material degrades as it releases a therapeutic agent. This allows for a controlled release of medication directly at the target site without the need for additional interventions (Khan et al. 2021).

Temporary Neural Interfaces: Temporary neural interfaces made from biodegradable materials are being developed to provide short-term monitoring or stimulation of neural activity. After completing their function, these probes degrade, avoiding the risks associated with long-term implantation (Seo et al. 2020).

Tissue Engineering: Biodegradable materials are used in the development of scaffolds that support tissue regeneration. These scaffolds provide temporary support for cells to grow and regenerate tissues, gradually degrading as the new tissue forms, thus eliminating the need for scaffold removal (Langer and Peppas 2021).

8.5.3 MEMS and NEMS Technologies

The integration of MEMS and NEMS technologies in minimally invasive probes has opened new frontiers in precision medicine, enabling the development of smaller, more efficient, and more functional devices. These technologies allow for the fabrication of highly miniaturized sensors, actuators, and electronic components, which are essential for advanced diagnostic and therapeutic applications.

8.5.3.1 Overview of MEMS and NEMS

MEMS: This refers to the integration of mechanical and electrical components at the microscale, typically ranging from 1 to 100 μm. MEMS devices often include sensors, actuators, and microfluidic components, which are fabricated using micro-fabrication techniques similar to those used in semiconductor manufacturing (Wu et al. 2022).

NEMS: This operates at the nanometer scale (1–100 nm) and incorporates the same principles as MEMS but at a much smaller scale. NEMS devices are capable of ultra-sensitive measurements, such as detecting single molecules or vibrations at the atomic level (Garcia et al. 2020).

8.5.3.2 Advantages of MEMS and NEMS in Minimally Invasive Probes

Miniaturization: MEMS and NEMS technologies allow for the development of ultra-miniaturized probes that can be inserted into small, hard-to-reach areas of the body. This reduces tissue damage during insertion and enhances the precision of diagnostic or therapeutic procedures (Shen et al. 2021).

High Sensitivity: The integration of MEMS and NEMS sensors into minimally invasive probes enables real-time monitoring of physiological parameters with high sensitivity. These sensors can detect minute changes in temperature, pressure, or chemical composition, providing valuable data for early diagnosis and targeted treatments.

Functional Integration: MEMS and NEMS devices can combine multiple functions, such as sensing, actuation, and drug delivery, into a single compact platform. This multifunctionality enhances the capability of minimally invasive probes by enabling them to perform multiple tasks simultaneously (Lee et al. 2020).

8.5.3.3 Applications of MEMS and NEMS Technologies in Minimally Invasive Probes.

Smart Catheters: MEMS-based sensors are integrated into smart catheters that can monitor blood pressure, flow rates, and oxygen levels in real time. These catheters provide critical information during cardiovascular procedures and can assist in adjusting therapies on the fly (Toral et al. 2023).

Neural Interfaces: NEMS technology is being used to develop ultra-sensitive neural probes capable of recording electrical signals from individual neurons. These probes enable researchers to study brain activity at an unprecedented level of detail, opening new possibilities for brain-machine interfaces and neuroprosthetics (Cheng et al. 2022).

Drug Delivery Systems: MEMS-based drug delivery systems allow for precise control over the release of therapeutic agents. These systems can be programmed to release medication at specific times or in response to environmental triggers, such as pH changes or temperature fluctuations (Yu et al. 2021a).

8.6 Applications of Minimally Invasive Probes in Drug Delivery

Minimally invasive probes have revolutionized the delivery of drugs, allowing for precise, controlled administration of therapeutic agents with reduced discomfort and faster recovery times for patients. These devices offer an effective alternative to traditional drug delivery methods by minimizing the need for invasive surgeries, enhancing drug targeting, and improving patient compliance. The following sections explore the applications of minimally invasive probes in three major therapeutic areas: cancer treatment, diabetes management, and pain management.

8.6.1 Cancer Treatment

Cancer treatment has greatly benefited from the advancements in minimally invasive drug delivery systems. The need for precision in targeting cancerous tissues while sparing healthy ones makes these probes a crucial component in modern oncological treatments. Minimally invasive probes allow for site-specific drug administration, reducing the systemic toxicity associated with traditional chemotherapy methods.

8.6.1.1 Advantages of Minimally Invasive Probes in Cancer Treatment

Targeted Drug Delivery: Therapeutic drugs can be given straight to the tumor site via minimally invasive probes, increasing the drug's concentration in the cancerous tissue while minimizing exposure to healthy tissues. Chemotherapy side effects like nausea, exhaustion, and hair loss are reduced with this focused strategy (Jackson and Chester 2021).

Controlled-Release Mechanisms: Probes can be designed to release drugs in a controlled manner, either continuously or in response to specific triggers such as pH or temperature changes within the tumor microenvironment. This allows for sustained drug delivery, increasing the therapeutic efficacy while reducing the frequency of dosing (Wang et al. 2023).

Enhanced Imaging and Monitoring: Minimally invasive probes often incorporate imaging modalities, such as optical sensors or MRI-compatible materials, to provide real-time feedback on drug delivery and tumor response. This integration enhances the accuracy of treatment and allows for dynamic adjustments during therapy (Li et al. 2022).

8.6.1.2 Key Technologies in Cancer Treatment

Drug-Eluting Probes: Drug-eluting probes are equipped with reservoirs or coatings that release anticancer drugs in a controlled manner. These devices have been used in treating brain tumors, where direct drug administration via a probe bypasses the blood-brain barrier, increasing the effectiveness of the treatment (Prasad et al. 2021). An example includes the use of biodegradable polymeric probes loaded with chemotherapy agents such as paclitaxel or doxorubicin for localized delivery.

Nanoparticle-Loaded Probes: Nanotechnology plays a significant role in enhancing drug delivery through minimally invasive probes. Probes can be loaded with drug-containing nanoparticles that target specific cancer cells based on their unique surface markers. Once delivered, these nanoparticles release their payload, ensuring that the drug concentration remains high in the tumor while minimizing systemic toxicity (Zhao et al. 2022).

Photothermal Therapy (PTT): Minimally invasive probes are also being utilized in photothermal therapy, where they deliver nanoparticles that can absorb light and convert it into heat to destroy cancer cells. This approach has been effective in treating cancers like melanoma and prostate cancer, where localized heating of the tumor can result in selective destruction of cancerous cells (Chen et al. 2020).

8.6.1.3 Applications

Glioblastoma Treatment: Glioblastoma, a highly aggressive brain tumor, has been treated with drug-eluting probes designed to deliver chemotherapy agents directly to the tumor site, overcoming the limitations posed by the blood-brain barrier (Reardon et al. 2022). This targeted delivery method enhances drug penetration and reduces the toxicity associated with systemic chemotherapy.

Liver Cancer: Minimally invasive probes are also used in liver cancer treatment, where they facilitate the localized administration of chemotherapy drugs or radiofrequency ablation in combination with drug delivery. These probes minimize damage to surrounding liver tissue while maximizing the treatment's impact on the tumor (Lee et al. 2023).

8.6.2 Diabetes Management

Minimally invasive probes have become an essential tool in the therapy of diabetes because they allow for the regulated administration of insulin and other therapeutic medications. By integrating sensors with medication delivery systems, it is possible to monitor blood glucose levels in real time and modify drug release accordingly, improving patient outcomes and lessening the need for daily insulin injections.

8.6.2.1 Advantages of Minimally Invasive Probes in Diabetes Management

Real-Time Glucose Monitoring: Many minimally invasive probes used in diabetes management integrate glucose sensors that continuously monitor blood sugar levels. Based on real-time data, these probes can automatically adjust the release of insulin, ensuring optimal glycemic control and reducing the risk of hypoglycemia (Homma et al. 2021).

Reduced Invasiveness: Traditional insulin delivery methods, such as subcutaneous injections, are often uncomfortable for patients and lead to poor compliance. Minimally invasive probes can deliver insulin or other antidiabetic agents with minimal discomfort, improving patient adherence to therapy (Zisser et al. 2021).

Long-Term Drug Delivery: Advances in biodegradable materials have led to the development of insulin-releasing probes that can be implanted for long-term drug delivery. These devices provide a steady release of insulin over several weeks or months, reducing the frequency of administration and improving patient convenience (Kim et al. 2022).

8.6.2.2 Key Technologies in Diabetes Management

Insulin Infusion Probes: Insulin infusion systems are one of the most common applications of minimally invasive probes in diabetes management. These devices can deliver insulin continuously or on-demand in response to blood glucose levels, closely mimicking the body's natural insulin regulation (Yu et al. 2021b).

Closed-Loop Systems: Minimally invasive probes are used by closed-loop insulin delivery systems to autonomously modify insulin supply in response to glucose levels. These systems have significantly improved glycemic control in diabetic patients by reducing both hyperglycemia and hypoglycemia episodes (Boughton et al. 2022).

Microneedle Patches: Microneedle patches are a novel form of minimally invasive drug delivery that allows for the painless administration of insulin. These patches contain microscale needles that penetrate the skin and deliver insulin directly into the bloodstream without the need for hypodermic needles (Yu et al. 2021c).

8.6.2.3 Applications

Type 1 Diabetes: For patients with type 1 diabetes, where insulin dependence is critical, minimally invasive probes have transformed disease management. Continuous glucose monitoring combined with insulin delivery systems has allowed for more precise control over blood sugar levels, leading to better long-term outcomes and fewer complications (Shulman et al. 2022).

Type 2 Diabetes: In patients with type 2 diabetes, insulin resistance often complicates treatment. Minimally invasive probes are used not only to deliver insulin but also to administer GLP-1 receptor agonists or other antidiabetic agents that improve insulin sensitivity and promote better glycemic control (Min et al. 2023).

8.6.3 Pain Management

Minimally invasive probes have also had a significant impact on pain management, providing patients with targeted, controlled relief from chronic or acute pain. These probes offer an alternative to systemic pain medications, reducing the side effects associated with oral or intravenous analgesics and improving patient comfort.

8.6.3.1 Advantages of Minimally Invasive Probes in Pain Management

Targeted Drug Delivery: One of the most significant benefits of minimally invasive probes in pain management is their ability to deliver analgesics directly to the site of pain. This targeted approach reduces the need for high systemic doses, minimizing side effects such as sedation, nausea, or addiction risks associated with opioids (Duan et al. 2021).

Controlled Release: Probes can be designed to release pain relief medications in a controlled fashion, either continuously over time or in response to specific triggers, such as the patient's movement or body temperature. This ensures that patients receive a steady supply of analgesics, maintaining consistent pain relief without the need for frequent doses (Rathmell et al. 2023).

Minimization of Invasive Procedures: Traditional pain management strategies, such as nerve blocks or spinal cord stimulators, often require invasive procedures that carry risks of infection or complications. Minimally invasive probes reduce the need for such procedures by providing localized pain relief with minimal trauma to the surrounding tissue (Deer et al. 2021).

8.6.3.2 Key Technologies in Pain Management

Drug Infusion Probes: These devices are used to deliver analgesics, such as local anesthetics or opioids, directly to the affected area. Drug infusion probes can be used in a variety of pain conditions, including postoperative pain, cancer pain, and chronic back pain, providing targeted relief with fewer systemic effects (Abdo et al. 2022).

Spinal Cord Stimulation (SCS) Probes: These devices are used to treat chronic pain problems like complicated regional pain syndrome or failed back surgery syndrome, which are not complete without minimally invasive probes. These probes modulate pain signals and provide long-term comfort without the use of opioids by sending electrical impulses to the spinal cord (Sayed et al. 2020).

Microneedle Arrays for Pain Relief: Microneedle arrays are being developed for transdermal delivery of analgesics, offering a painless and effective method for managing localized pain. These devices use tiny needles to penetrate the skin and release medications such as lidocaine or nonsteroidal anti-inflammatory drugs (NSAIDs) (Xu et al. 2021).

8.6.3.3 Applications

Chronic Pain Conditions: Minimally invasive probes have been widely used in addressing persistent pain syndromes such neuropathy, arthritis, and back pain. By delivering analgesics directly to the affected area, these probes provide sustained relief and improve patients' quality of life (Apkarian et al. 2021).

Postoperative Pain: After surgical procedures, minimally invasive drug delivery probes are used to administer local anesthetics or opioids at the surgical site, reducing the need for systemic pain medications and speeding up recovery times (Kehlet et al. 2021).

8.7 Current Research and Future Directions

The landscape of drug delivery systems has undergone significant advancements in recent years, largely driven by the need for precision medicine and improved patient outcomes. Researchers are exploring innovative materials and technologies that are revolutionizing the way drugs are administered, monitored, and controlled. In this chapter, we will discuss three key areas at the forefront of current research: nanotechnology, implantable drug delivery systems, and the integration of artificial intelligence (AI) into drug delivery.

8.7.1 Nanotechnology in Drug Delivery

Nanotechnology has emerged as a game-changer in drug delivery, offering unparalleled precision and efficacy. The ability to engineer materials at the nanoscale allows for the development of highly specialized drug delivery systems that can target specific tissues, improve bioavailability, and reduce toxicity. The use of nanoparticles, nanocarriers, and nanocomposites enables drugs to be delivered more efficiently and with fewer side effects compared to conventional methods.

8.7.1.1 Advantages

Targeted Delivery: The capacity of nanotechnology to target particular cells or tissues is one of its main advantages. Nanoparticles can identify and attach to certain receptors on diseased cells, including cancer cells, by functionalizing their surface with ligands, antibodies, or peptides. This guarantees that the therapeutic drug is delivered precisely where it is required (Wang et al. 2022b).

Enhanced Bioavailability: Nanoparticles can improve the solubility and stability of drugs, which is particularly beneficial for hydrophobic drugs that have poor bioavailability. By encapsulating the drug in a nanocarrier, it can be protected from degradation, allowing for more of the drug to reach its target (Mitragotri and Anselmo 2021).

Controlled Release: Nanotechnology enables the design of drug delivery systems that release their therapeutic payload in a controlled manner. This can be achieved through stimuli-responsive systems that release the drug in response to changes in pH, temperature, or enzyme levels within the body (Zhang et al. 2023).

8.7.1.2 Key Technologies in Nanotechnology for Drug Delivery

Lipid-Based Nanoparticles: Because of their biocompatibility and capacity to encapsulate both hydrophilic and hydrophobic medicines, liposomes and solid lipid nanoparticles (SLNs) are frequently utilized for drug administration. These nanoparticles have been successfully employed in cancer therapy and vaccine delivery, offering controlled and targeted release of drugs (Kulkarni and Shaw 2021).

Polymeric Nanoparticles: Polymeric nanoparticles, such as those made from poly(lactic-*co*-glycolic acid) (PLGA), are biodegradable and can be engineered to deliver drugs over an extended period. These particles are often used in the delivery of chemotherapy agents, where long-term release is critical for maintaining therapeutic drug levels (Kumar et al. 2021).

Dendrimers: Dendrimers are highly branched, tree-like macromolecules that provide multiple attachment points for drugs, targeting ligands, and imaging agents. Their unique structure allows for the simultaneous delivery of therapeutic agents and diagnostic tools, making them ideal for theranostic applications (Lee et al. 2022).

8.7.1.3 Current Applications of Nanotechnology in Drug Delivery

Cancer Therapy: Nanoparticle-based drug delivery systems have shown remarkable success in cancer treatment, particularly in delivering chemotherapy drugs directly to tumors. By targeting the tumor microenvironment, nanoparticles reduce off-target effects and improve the therapeutic index of anticancer agents like paclitaxel and doxorubicin (Lim et al. 2021).

Gene Therapy: Nanotechnology is being used to improve the delivery of nucleic acids, such as siRNA and CRISPR-Cas9, for gene therapy applications. Nanocarriers can protect the genetic material from degradation and deliver it to specific cells, allowing for the correction of genetic disorders at the molecular level (Gonçalves et al. 2020).

Vaccines: The COVID-19 pandemic highlighted the potential of nanotechnology in vaccine delivery. mRNA vaccines, such as those developed by Pfizer-BioNTech and Moderna, utilize lipid nanoparticles to deliver mRNA into cells, where it triggers an immune response. This approach has paved the way for future vaccine development using nanotechnology (Hou et al. 2021).

8.7.2 Implantable Drug Delivery Systems

They represent a promising approach for the long-term, controlled release of therapeutic agents. These systems are designed to be placed inside the body, where they can deliver drugs over extended periods, ranging from weeks to months, without the need for repeated administration. This technology is particularly useful in chronic conditions that require continuous medication or when patient adherence is a challenge.

8.7.2.1 Advantages of Implantable Drug Delivery Systems

Long-Term Release: One of the primary benefits of implantable systems is their ability to deliver drugs continuously over long periods, reducing the need for frequent dosing. This is especially beneficial for patients with chronic diseases, such as diabetes or cardiovascular conditions, where consistent drug levels are required for optimal management (Gratieri et al. 2021).

Improved Patient Compliance: By eliminating the need for daily or weekly dosing, implantable systems improve patient compliance, which is a critical factor in the success of long-term therapies. Patients can receive treatment without the need for frequent clinic visits or the burden of remembering to take their medications (Santos et al. 2021).

Localized Delivery: Implantable systems can deliver drugs directly to the site of action, minimizing systemic exposure and reducing side effects. This is particularly important in cancer therapy, where localized drug delivery can target the tumor while sparing healthy tissues (Choi et al. 2022).

8.7.2.2 Key Technologies in Implantable Drug Delivery Systems

Biodegradable Implants: Biodegradable polymer-based implants, such as those made from PLGA or polycaprolactone (PCL), are designed to degrade over time, releasing the drug as the polymer matrix breaks down. These systems are often used in the treatment of cancer, where sustained drug release is critical for maintaining therapeutic efficacy (Mohammed et al. 2022).

Reservoir-Based Systems: A drug reservoir encased in a semipermeable membrane that regulates the pace of drug diffusion makes up reservoir-based implants. These systems can be designed to release medications in response to external stimuli like ultrasonic or electric fields, providing for fine control over drug delivery (Wang et al. 2021).

Osmotic Pumps: Osmotic pumps are implantable devices that use osmotic pressure to drive the release of a drug from a reservoir. These pumps are particularly useful in delivering small molecules and peptides over extended periods, and they have been used in pain management and hormone replacement therapy (Gao et al. 2021).

8.7.2.3 Current Applications of Implantable Drug Delivery Systems

Diabetes Management: Implantable insulin pumps have been developed to provide continuous insulin delivery for patients with type 1 diabetes. These pumps can be programmed to release insulin based on real-time glucose readings, offering a more precise and automated method of glycemic control compared to traditional insulin injections (Robertson et al. 2021).

Cancer Therapy: Biodegradable implants loaded with chemotherapy drugs are used to treat various types of cancer, including brain tumors and prostate cancer. These implants allow for localized delivery of the drug, reducing the side effects associated with systemic chemotherapy (Dudek et al. 2023).

Chronic Pain Management: Implantable drug pumps are used to deliver analgesics for the treatment of chronic pain disorders including cancer-related pain or failed back surgery syndrome, straight to the spinal cord or other target locations. These pumps provide long-term pain relief while minimizing the risk of systemic opioid exposure (Deer et al. 2020).

8.7.3 Artificial Intelligence in Drug Delivery

Artificial intelligence (AI) is transforming the field of drug delivery by enabling more personalized and optimized treatment strategies. AI technologies, such as machine learning and predictive analytics, are being integrated into drug delivery systems to enhance drug formulation, optimize dosing regimens, and improve patient outcomes. By analyzing large datasets, AI can help predict how different patients will respond to a particular drug, leading to more effective and individualized treatments.

8.7.3.1 Advantages of Artificial Intelligence in Drug Delivery

Personalized Medicine: AI can analyze patient-specific data, such as genetic profiles, disease history, and lifestyle factors, to predict the optimal drug and dose for each individual. This approach allows for more personalized treatments, which can improve therapeutic outcomes and reduce adverse effects (Topol 2022).

Optimization of Dosing Regimens: AI can be used to develop algorithms that optimize drug dosing regimens based on real-time data. For example, AI algorithms can analyze blood glucose levels in diabetic patients to adjust insulin delivery in real time, ensuring more precise control over blood sugar levels (Samuel et al. 2022).

Drug Discovery and Formulation: AI is playing a significant role in accelerating drug discovery by predicting the properties of potential drug candidates and optimizing drug formulations. This helps to identify promising drugs more quickly and efficiently, reducing the time and cost associated with bringing new therapies to market (Segler et al. 2021).

8.7.3.2 Key Technologies in AI-Driven Drug Delivery

Machine Learning Algorithms: Machine learning (ML) algorithms are being used to analyze large datasets to identify patterns and correlations that would be difficult to detect through traditional methods. These algorithms can predict how a patient will respond to a drug, allowing for more personalized and effective treatments (Khan et al. 2022).

AI-Driven Drug Delivery Devices: AI is being integrated into drug delivery devices, such as insulin pumps and implantable drug delivery systems, to optimize drug release in real time. These devices use data from sensors to monitor physiological parameters and adjust drug delivery accordingly (Chen et al. 2023).

AI in Drug Formulation: By using AI to forecast the chemical and physical characteristics of medication formulations, more reliable and efficient drug delivery systems can be created. Additionally, AI algorithms can tailor nanoparticles' surface characteristics, size, and shape for targeted medication delivery (Zhang et al. 2021b).

8.7.3.3 Current Applications of AI in Drug Delivery

Diabetes Management: AI-powered insulin delivery systems, such as closed-loop insulin pumps, use machine learning algorithms to analyze real-time glucose levels and adjust insulin delivery accordingly. These systems have significantly improved glycemic control in diabetic patients by reducing both hyperglycemia and hypoglycemia episodes (Buckingham et al. 2021).

Oncology: AI is being used to predict how individual cancer patients will respond to different chemotherapy regimens, allowing for more personalized treatment plans. AI algorithms analyze genetic data, tumor characteristics, and patient history to optimize drug delivery and reduce toxicity (Chen et al. 2022).

Neurodegenerative Diseases: AI is being employed in the delivery of drugs for neurodegenerative diseases, such as Parkinson's and Alzheimer's, to optimize the timing and dosage of medications. By analyzing patient data, AI can help maintain therapeutic drug levels and reduce fluctuations that can worsen symptoms (Korber et al. 2023).

8.8 Conclusion

The advancements in minimally invasive probes for drug delivery have revolutionized the medical field by offering precision, efficiency, and patient comfort. These innovations allow for targeted and controlled drug administration, minimizing systemic exposure and reducing side effects. Nanotechnology, implantable systems, and AI-driven methods have further expanded the scope of possibilities, enabling personalized medicine and improved therapeutic outcomes. As research continues to evolve, minimally invasive probes are expected to become even more integral in managing chronic diseases, cancer, and other conditions that require long-term or highly specialized drug delivery. The future of drug delivery lies in these technologies, which not only improve patient outcomes but also streamline the overall treatment process. These advancements mark a significant step toward more effective, efficient, and patient-friendly healthcare solutions, transforming the way treatments are administered and managed.

References

Abdo T, Bojanic S, Alhajri N et al (2022) Intrathecal drug infusion systems in pain management. Pain Med 23(7):1243–1252

Allen TM et al (2016) Liposomes as drug delivery systems. Drug Deliv Syst 20(2):123–145

Altun T, Inan Y (2022) Development of flexible polymer-based neural probes. J Neural Eng 19(4):046009

Apkarian AV, Baliki MN, Geha PY (2021) Advances in brain-targeted minimally invasive therapies for chronic pain. Nat Neurosci 24(3):365–372

Bao G, Mitragotri S (2021) Delivering the future: emerging technologies in drug delivery. Sci Transl Med 13(596)

Basu B, Kumar S (2023) Advances in polymeric materials for medical devices. Adv Mater Interfaces 10(2):2102020

Boughton CK, Daly A, Choudhary P (2022) The artificial pancreas in Type 1 diabetes: current status. Diabetologia 65(4):610–619

Buckingham BA, Forlenza GP, Pinsker JE et al (2021) AI-powered closed-loop insulin delivery systems: clinical advances. Diabetes Technol Ther 23(8):609–616

Chen G, Roy I, Yang C, Prasad PN (2020) Nanotechnology approaches for targeted photothermal therapy. Nanoscale 12(3):2536–2546

Chen H, Qiu Y, Zou Z et al (2022) AI in oncology drug delivery: current status and future potential. Nat Rev Drug Discov 21(4):307–325

Chen R, Cui X, Ren Y et al (2023) AI-enhanced drug delivery devices for precision medicine. Adv Drug Deliv Rev 194:114730

Cheng MY, Chiou JC, Tsai ZT et al (2022) NEMS-based ultra-sensitive neural probes for in vivo monitoring of neuronal activity. J Neural Eng 19(5):056012

Choi S, Jeong J, Lee JH, Lee KB (2022) Implantable drug delivery systems for cancer therapy. Adv Healthc Mater 11(1):2101313

Deer TR, Pope JE, Hayek SM et al (2020) Advances in implantable drug pumps for chronic pain management. Pain Med 21(4):671–684

Deer TR, Pope JE, Hayek SM et al (2021) Minimally invasive neuromodulation techniques for pain management. J Pain Res 14:2989–3003

Duan X, Li Y, Ma X et al (2021) Minimally invasive drug delivery systems for chronic pain management. Pain Manag 11(2):187–199

Dudek AZ, Zhao Y, Day JD (2023) Biodegradable implants for sustained chemotherapy: a review of recent advances. Cancer Drug Targets 22(5):369–378

FDA (2020) Guidelines for drug delivery systems. U.S. Food and Drug Administration, Silver Spring

Gadde S (2017) Nanoparticles for drug delivery: safety concerns. J Control Release 242:265–277

Gao W, Wang J (2021) The rise of nanomachines. Nat Nanotechnol 16(6):653–660

Gao Y, Yang W, Wu Y (2021) Osmotic pumps in drug delivery: mechanisms and advances. J Control Release 336:125–139

Garcia R, Knoll AW, Riedo E (2020) Advanced nanoelectromechanical systems for biological applications. Nat Nanotechnol 15(1):245–255

Gonçalves GA, Paiva RM, Lopes-Junior AG (2020) Advances in nanotechnology for gene therapy and its implications in cancer treatment. J Gene Med 22(5)

Goudanavar PS, Koteswara Rao GSN, Kumar MP, Kumar GBK, Ramesh B, Naveen NR (2024) Microneedle technology—an insight into advancements, latest scholarly and patent data. Intelligent Pharmacy 2:114–121

Gratieri T, Thomazini JA, Mercuri LP (2021) Implantable drug delivery systems: current developments and future directions. Expert Opin Drug Deliv 18(12):1735–1744

Hamdan I (2024) Microneedle and drug delivery across the skin: an overview. Pharmacia 71:1–12

Hartman RL, McMullen JP (2020) High-throughput flow chemistry: a new paradigm in chemical synthesis. Science 369(6499):774–778

Homma T, Hirose S, Nagata N (2021) Continuous glucose monitoring and insulin delivery probes for diabetes management. Diabetes Care 44(5):1176–1185

Hou X, Zaks T, Langer R, Dong Y (2021) Lipid nanoparticles for mRNA delivery. Nat Rev Mater 6(12):1078–1094

Jackson SE, Chester JD (2021) Targeted therapies in cancer treatment. J Cancer Pharmacol 42(3):135–145

Karp JM, Langer R (2021) Materials for minimally invasive medical devices. Nat Rev Mater 6(7):520–535

Kehlet H, Jensen TS, Woolf CJ (2021) Minimally invasive strategies for postoperative pain relief. Lancet 397(10279):2187–2195

Khan AM, Asadujjaman M, Rahman MS et al (2021) Biodegradable materials for controlled drug delivery systems. J Control Release 331:161–182

Khan Z, Kolašinac U, Aliakbarzadeh M (2022) Machine learning in personalized medicine: drug delivery systems. J Control Release 351:529–540

Kim BH, Rho JY, Kim J (2021) Nanomedicine for brain disorders: overcoming the blood-brain barrier. Adv Mater 33(7)

Kim YS, Kim HJ, Lee JY et al (2022) Long-term biodegradable insulin-releasing probes for diabetes treatment. Biosens Bioelectron 198:113837

Korber B, Zeilinger M, Streitberger K et al (2023) AI-driven drug delivery systems in neurodegenerative disease management. J Neural Eng 20(1):015003

Kulkarni CV, Shaw CP (2021) Lipid nanoparticles for drug delivery: advances and applications. Nanomedicine 16(5):419–433

Kumar A, Ghosh D, Misra SK (2021) Polymeric nanoparticles for cancer therapy: design, synthesis, and biomedical applications. Polymers 13(19):3285

Langer R, Peppas NA (2021) Advances in biomaterials for tissue engineering. Nat Rev Mater 6(8):765–784

Lee W, Jang J, Kim DH (2020) MEMS and NEMS technologies for bio-integrated electronics. Adv Mater 32(36)

Lee H, Kim D, Kim WJ (2022) Dendrimers in nanomedicine: a review of applications in cancer drug delivery. Biomaterials 288:121669

Lee YH, Chung WJ, Kim HB et al (2023) Advances in minimally invasive drug delivery for liver cancer therapy. Adv Healthc Mater 12(4)

Li X, Kwon YJ (2020) Nanotechnology in targeted drug delivery and cancer therapy. Nanomedicine 15(5):693–704

Li Y, Huang Z, Shen X et al (2022) Real-time imaging technologies in targeted cancer drug delivery systems. Biosens Bioelectron 194:113623

Liang JF, Huh YM, Lee D (2021) Precision nanomedicine and the emerging role of theranostics. Adv Drug Deliv Rev 177:113949

Lim EK, Kim T, Paik S, Haam S, Huh YM, Lee K (2020) Nanomaterials for theranostics: recent advances and future challenges. Chem Rev 120(15):9982–10029

Lim EK, Jang E, Lee K (2021) Recent advances in nanotechnology for cancer therapy. J Nanobiotechnol 19(1):249

Liu G, Wang W, Yang X, Li C, Fu J (2020) pH-responsive nanoprobes for tumor-specific drug delivery. Adv Drug Deliv Rev 164:11–21

Min S, Kim D, Lee Y et al (2023) Emerging technologies in the management of Type 2 diabetes. Nat Rev Endocrinol 19(2):125–136

Mitragotri S, Anselmo AC (2021) Nanoparticle-mediated drug delivery: challenges and future directions. Adv Drug Deliv Rev 168:25–31

Mohammed EA, Abdullah A, Reddy SM et al (2022) Biodegradable implants in cancer therapy: advances and challenges. J Control Release 345:360–376

Needleman BW, Coburn NG (2021) The evolution of minimally invasive surgery: past, present, and future. Surg Oncol Clin N Am 30(2):279–290

Parodi A, Molinaro R, Sushnitha M (2020) Nanotechnology and vascular disease: current trends. J Control Release 320:277–285

Prasad A, Morris B, Rao R (2021) Drug-eluting probes for cancer therapy: challenges and future perspectives. Cancer Drug Targets 21(4):255–264

Priya S, Singhvi G (2022) Microneedles-based drug delivery strategies: a breakthrough approach for the management of pain. Biomed Pharmacother 155:113717

Rathmell JP, Viswanath O, Connolly AT et al (2023) Advances in targeted and controlled-release analgesics for pain management. Pain Res Manag 2023:2345678

Ratner BD et al (2013) Biomaterials science: an introduction to materials in medicine, 3rd edn. Academic Press, New York

Reardon DA, Wen PY, Mellinghoff IK et al (2022) Drug delivery systems for glioblastoma. Nat Rev Clin Oncol 19(6):336–351

Robertson PR, Hovorka R, Wilinska ME (2021) Continuous subcutaneous insulin infusion and implantable systems for diabetes management. Lancet Diabetes Endocrinol 9(9):592–600

Rodgers SD, Joyce AD, Humphrey JD (2020) Polyimide materials for implantable neural interfaces. Acta Biomater 111:30–43

Samuel O, Richard K, Henry D (2022) AI-driven algorithms for drug dosing optimization in diabetes. Diabetes Technol Ther 24(1):27–35

Sant Anna LB, Van Horne AE, Burns WH (2021) Minimally invasive drug delivery: evolving technologies and trends. Front Pharmacol 12:631728

Santos JL, Pandey A, Wan S et al (2021) Implantable biomaterials for long-term drug delivery. Adv Mater 33(10)

Sayed D, Kapural L, Standiford H et al (2020) Spinal cord stimulation in the management of chronic pain: innovations and future directions. Expert Rev Med Devices 17(10):989–1001

Segler MH, Preuss M, Waller MP (2021) AI in drug discovery and formulation design. Adv Drug Deliv Rev 175:113763

Seo D, Neely RM, Shen K et al (2020) Fully integrated, foldable, and biodegradable neural interfaces. Adv Mater 32(26)

Shen W, Li L, Wang Y et al (2021) Recent progress on MEMS-based biosensors for health applications. Micromachines 12(8):931

Shulman GI, Tamborlane WV, Sherwin RS (2022) Minimally invasive technologies for Type 1 diabetes management. J Clin Invest 132(12)
Sun H, Mei L, Song C, Cui X, Wang P (2021) The in vivo degradation, absorption and excretion of PCL-based implants. Biomaterials 280:121290
Topol EJ (2022) AI in medicine: transforming drug delivery and therapeutics. Nat Med 28(2):261–273
Toral M, Pérez-Sánchez M, Dominguez-Viqueira W et al (2023) Smart catheter devices and MEMS technology in minimally invasive surgery. Biosens Bioelectron 202:114054
Wang B, Zhang X, Zhang Q et al (2021) Reservoir-based drug delivery systems: a review of applications. Biomaterials 276:120935
Wang Y, Wei Q, Huang X (2022a) Biodegradable materials for medical implants and scaffolds. Adv Healthc Mater 11(6)
Wang AZ, Langer R, Farokhzad OC (2022b) Nanoparticle delivery of cancer drugs. Annu Rev Med 73:365–377
Wang Y, Zhang H, Li H (2023) Controlled drug delivery systems in cancer therapy. Adv Drug Deliv Rev 150:46–58
Wu H, Liu J, Li Y (2022) MEMS and NEMS technologies in medical applications. Biosens Bioelectron 199:113866
Xu J, Ju Z, Tang S et al (2021) Microneedle arrays for transdermal pain management. Adv Mater Technol 6(2):2000836
Yadav PS, Gupta AK (2021) Applications of nanotechnology in drug delivery systems for treating infections. Expert Opin Drug Deliv 18(5):597–611
Yeo SJ, Kwak SE, Lee JH (2022) Recent trends in minimally invasive drug delivery systems. Biomater Sci 10(14):3886–3901
Yu X, Zeng Y, Zhu H et al (2021a) MEMS-based controlled drug delivery systems: from design to applications. Micromachines 12(11):1360
Yu T, Thabit H, Hovorka R (2021b) Closed-loop insulin delivery systems: the next generation. Lancet Diabetes Endocrinol 9(6):473–485
Yu J, Qian C, Zhang Y et al (2021c) Microneedle patches as a minimally invasive strategy for insulin delivery. Adv Drug Deliv Rev 173:231–251
Zhang L, Chan JM (2020) Stimuli-responsive delivery systems for minimally invasive drug administration. Adv Drug Deliv Rev 158:95–108
Zhang Y, Yu J, Kahkoska AR (2021a) Advances in injectable smart hydrogels for diabetes management. Adv Mater 33(7)
Zhang Y, Zhao X, He Z et al (2021b) AI-driven nanoparticle design for drug delivery. Nat Commun 12(1):7157
Zhang P, Wu H, Zhang H (2023) Stimuli-responsive nanocarriers for controlled drug release. J Control Release 355:31–43
Zhao Y, Li X, Yang X et al (2022) Nanoparticle-loaded minimally invasive probes for cancer therapy. Nano Today 41:101312
Zhao W, Xu X, Fu Y et al (2023a) Polydimethylsiloxane-based biomedical materials and applications. Mater Sci Eng C 145:113664
Zhao Y, Zhang W, Liao J et al (2023b) Recent advancements in biodegradable and biocompatible polymers for biomedical applications. Mater Today 56:90–106
Zisser H, Robinson D, Bevier W et al (2021) Minimally invasive technologies in diabetes: the future of insulin delivery. J Diabetes Sci Technol 15(1):101–108

Recent Developments in Microneedle-Based Drug Delivery

9

Prathap Madeswara Guptha, Raghavendra Kumar Gunda, Vijayaraj Surendran, Brito Raj S, and Monika Kaushik

Abstract

To get around the serious problems with oral medication distribution, the transdermal route is being adopted more and more in clinical settings. Using transdermal microneedle arrays to administer drugs topically has a lengthy history. It has been shown that microneedles are highly effective and multipurpose tools. Microneedle technology has been presented as a method to penetrate the outer layer of the skin, known as the stratum corneum, in a way that is less invasive. This allows molecules to travel through the skin by establishing temporary channels. The remarkable qualities of this technique including its relative safety, low cost, high medicinal efficiency, and painless penetration have stimulated the interest of both the industrial and scientific communities. Microneedles are incredibly useful in many biological applications because of their exceptional physicochemical and ionic characteristics, which allow them to transport

P. M. Guptha (✉)
Department of Pharmaceutical Sciences, School of Biotechnology and Pharmaceutical Sciences, Vignan's Foundation for Science, Technology and Research (Deemed to be University), Guntur, Andhra Pradesh, India

R. K. Gunda
Department of Pharmaceutics, Narasaraopeta Institute of Pharmaceutical Sciences (Autonomous), Narasaraopeta, Andhra Pradesh, India

V. Surendran
Department of Pharmaceutical Chemistry and Pharmaceutical Analysis, Dr. Kalam College of Pharmacy, Thanjavur, Tamil Nadu, India

B. R. S
Department of Pharmaceutics, School of Pharmacy, Dhanalakshmi Srinivasan University, Trichy, India

M. Kaushik
Amity Institute of Pharmacy, Amity University Madhya Pradesh (AUMP), Gwalior, India

M. R. Babu et al. (eds.), *Transdermal Applications of Minimally Invasive Drug Delivery Systems*, https://doi.org/10.1007/978-981-96-6275-3_9

extremely big molecules. This chapter discusses microneedle technology's evolution and summarizes studies to date. Also it investigates the component materials and production processes that enhance the administration of crucial therapeutic substances via the skin.

Keywords

Microneedle technology · Transdermal microneedle arrays · Physicochemical and ionic characteristics

9.1 Introduction

The pediatric population is incredibly diverse, spanning from newborns to adolescents; therefore, developing new drugs for this group has always been difficult. The maturation pattern of the skin, the ratio of body water to fat, the metabolic capability, and the degree of protein binding also fluctuate with age (Duarah et al. 2019). Moreover, compared to older infants or teenagers, the content and regulation of enzymes vary dramatically, especially in the first few months of life (Kim et al. 2012). As a result, different age groups and medical conditions have varying requirements for acceptable excipients and dose formulations. As a result, designing and developing a single medication formulation for the whole range of developmental phases is almost impossible. In both normal and emergency cases, parenteral administration has consistently been the primary line of treatment for newborn.

A hypodermic needle is used to inject the majority of biotherapeutics and vaccinations. Delivering practically any kind of molecule directly, quickly, and affordably into the body is made possible via injection. Hypodermic needles, however, are difficult for patients to use on their own; as a result, patients who have had specific instruction on safe needle disposal, proper injection technique, and other topics typically use them at home or in clinics. Hypodermic needle use does have certain drawbacks, though, such as the mental anguish and discomfort of the injections, which can negatively impact patient compliance. Additionally, pediatric patients find frequent clinical visits and the possibility of illness transfer to be bothersome and inconvenient (Nguyen and Nguyen 2023). Reusing needles to spread bloodborne diseases is another serious concern, particularly in developing nations. Moreover, hypodermic needles generate waste that is biohazardous and sharp, which can be deadly (Kim et al. 2012; Jeong et al. 2017; Khare and Shende 2021). Although oral administration mostly solves these issues, it is not always possible to provide many medications this way because of poor absorption and drug breakdown in the liver and gastrointestinal tract. Despite the examination of several other procedures, none demonstrate the same extensive effectiveness as direct needle injection.

Transdermal medication administration has several advantages, including the circumvention of hepatic and gastrointestinal metabolism and mitigating risks related to hepatotoxicity, palatability, and disease transmission. Additionally, it aids in lowering total dosages and enhancing bioavailability. As a result, the transdermal route becomes a viable substitute for the parenteral and oral routes now in use (Akhtar et al. 2020; Pires et al. 2019; Pathak and Aggarwal 2014; Brogden et al. 2012). Transdermal medication delivery also lessens the risk of needle stick injuries and psychological anguish, which improves patient compliance. These benefits, which provide a noninvasive substitute for oral and intravenous administration, are especially alluring (Alkilani et al. 2015). Regretfully, many medications cannot be administered transdermally due to the skin's strong barrier qualities. Using chemical enhancers, ultrasonic, electric fields, and thermal approaches has not always resulted in adequate efficient penetration of some active substances, especially those with high molecular weights (Alkilani et al. 2015).

A microneedle must be sufficiently diminutive to mitigate discomfort, anxiety, and the necessity for specialized training for administration while being adequately sized to effectively deliver virtually any medicine or tiny particle composition. Furthermore, a microneedle enables accurate tissue localization of delivery, including the cell nucleus, the suprachoroidal area of the eye, and the skin. Robert Chambers coined the name "microneedle" in 1921 when he employed MN to microdissect echinoderm eggs by inserting the needle into the egg's nucleus. A novel and inventive technique known as microneedle(s) was initially presented in 1976 by Martin S. Gerstel and Virgil A. with the intention of resolving the shortcomings of traditional transdermal systems (Halder et al. 2021). Both solid and hollow MNs for local or systemic drug delivery were mentioned in their patent. Pistor Michel Louis Paul disclosed the first drug-coated MN gadget. Since the field started with technologies from the microelectronics sector, silicon was used to produce the first microneedles. Active research into the use of microneedles for drug delivery began in the mid-1990s, driven primarily by three parallel initiatives taking place at the Georgia Institute of Technology, Becton Dickinson (BD), and Alza Corporation (Zahn et al. 2000).

The application of several micron-scale needles fastened to a supporting membrane constitutes the novel and exciting approach known as microneedle (MN) technology, which guarantees painless and efficient drug penetration into the skin. Over the last 20 years, there has been a notable progression in the creation of medication delivery systems based on Minnesota. By puncturing the stratum corneum, the MNs can produce microchannels without upsetting the blood vessels and nerve endings in the dermis and epidermis, the inner layers of the skin (Zhang et al. 2022; Aung et al. 2019; Adepu and Ramakrishna 2021). As a result, more drug kinds can be administered in a less invasive way and with much greater efficiency (Liang et al. 2020). MNs may provide special advantages for the pediatric population, including painless and minimally invasive children immunization. In addition to their painlessness, MNs have other benefits such as a relatively simple application process, which increases their appeal for long-term use (Liang et al. 2020). It also results in easier disposal and a lower chance of cross-contamination and needle stick injuries. In addition, children's lesser dosage requirements make it easier to reach therapeutic concentrations.

Over the past two decades, substantial progress has been made in the domain of MN-mediated drug delivery. Advancements in micro-fabrication technology have facilitated the production of some MN-based devices. The viability of MN technology for pediatric applications is influenced by the positive attitudes and perceptions of children, physicians, and the general public. The effective distribution, improved patient compliance, and convenient storage of MNs have augmented their market demand and potential.

9.2 Ideal Characteristics of Microneedle Transdermal Patch

Microneedles must range in length or height from 50 to 900 μm and in diameter from 1 μm. The designed microneedles must enter the skin thoroughly without fracturing. MNs must be at their optimal size. Developed microneedles must have mechanical stability and withstand an insertion force of around 10 N. These are also intended to administer the medications in a regulated manner (Halder et al. 2021). These should provide a rapid initiation of action and effective drug delivery, akin to a hypodermic needle. These are intended to provide personalized medicine at diverse dose levels and enhance self-medication. The durability of such products must be improved, and they should be impermeable when delivering liquid dosage forms (transdermal/percutaneous drug absorption). These patches must adhere effectively, comparable to standard transdermal patches (Waghule et al. 2019).

9.3 Material Used for Fabrication of Microneedle-Based Drug Delivery

Microneedles come in a range of heights from 25 to 2000 m and are made of different materials and geometries. A wide range of materials have been utilized to create therapeutic microneedles, including silicon, glass, metals, and non-biodegradable and biodegradable polymers (Aldawood et al. 2021). The principal elements influencing the strength, adaptability, and permeability of microneedles are their materials, which must be meticulously chosen according to the desired use (Nguyen and Nguyen 2023). The selection of materials used for the fabrication of microneedles in drug delivery is based on their variety of options including pros and cons.

The potential of microneedle-based drug delivery devices to improve medication administration, particularly for transdermal and less invasive applications, has drawn a lot of interest. The material used to fabricate microneedles has a significant impact on the use, safety, efficacy, and applicability of the devices. Depending on the microneedles' intended usage, distribution system, and design, different materials are employed. The following are essential elements for the production of microneedles:

Table 9.1 Type and materials used for fabrication of microneedle-based drug delivery

S. no.	Material	Application	Type of microneedle	References
1	Silicon	Drug delivery, vaccine delivery	Solid, hollow, coated	Park et al. (2005)
2	Metal (e.g., stainless steel)	Transdermal drug delivery, insulin administration	Solid, hollow, coated	Roxhed et al. (2008)
3	Polymer (e.g., PLGA, PVP)	Vaccine delivery, cosmetic applications	Dissolving, coated	Ita (2015)
4	Hyaluronic acid	Cosmetic applications (anti-aging, skin hydration)	Dissolving	Kim et al. (2012)
5	Carboxymethyl cellulose (CMC)	Drug delivery, especially for sustained release	Dissolving	Luo et al. (2023)
6	Glass	Vaccine and drug delivery	Solid, hollow	Sabri et al. (2022)
7	Sugar (e.g., maltose)	Drug delivery (e.g., insulin, vaccines)	Dissolving	Lee et al. (2008)
8	Ceramics (e.g., alumina)	Drug delivery, biosensing	Solid	Ito et al. (2011)
9	Hydrogels	Transdermal drug delivery, diagnostics	Dissolving, coated	Zhang et al. (2019)

9.3.1 Silicon

One of the first materials utilized to make microneedles was silicon. It has exceptional mechanical strength and can be made accurately with micro-fabrication methods. For the administration of drugs and vaccines, silicon microneedles are often used in solid, hollow, and coated forms (Table 9.1). However, concerns over skin fragmentation and biocompatibility have limited its widespread clinical use. In spite of this, it is nevertheless a valuable material for preliminary studies and niche applications (Nguyen and Nguyen 2023).

9.3.2 Metals

Metals (such as titanium and stainless steel) are frequently employed because of their simplicity of manufacture and toughness. Microneedles made of titanium and stainless steel are robust and biocompatible and easily pierce the skin to deliver medications. They can be layered with medication formulations and are frequently utilized in both solid and hollowed forms. Insulin injection and other transdermal medication delivery applications make extensive use of metal microneedles (Bao et al. 2022).

9.3.3 Polymers

For the creation of microneedles, biodegradable polymers like poly(lactic-*co*-glycolic acid) (PLGA) and polyvinylpyrrolidone (PVP) are becoming more and more popular. After the medicine is administered, these components disintegrate in the skin, so there is no need to remove them. Polymer microneedles are therefore perfect for applications involving sustained-release medication and vaccine administration. They are frequently created as coated or dissolving microneedles (Gera and Burra 2022).

9.3.4 Hyaluronic Acid

A naturally occurring biopolymer with exceptional hydrophilic and biocompatibility qualities is hyaluronic acid. It is widely utilized in the cosmetic sector for microneedles used for transdermal administration of cosmetic chemicals, anti-aging therapies, and skin hydration. Usually dissolving, hyaluronic acid microneedles give the active substance while the skin absorbs it (Salih et al. 2024).

9.3.5 Carboxymethyl Cellulose (CMC)

Another biocompatible polymer that is utilized to dissolve microneedles for medication administration is carboxymethyl cellulose. For sustained-release applications, where the gradual disintegration of the microneedle permits regulated drug release over time, CMC-based microneedles are very helpful.

9.3.6 Glass

Glass microneedles are employed for precision medicine and vaccine delivery applications, despite being less prevalent than other materials. They are perfect for hollow microneedles because of their strength and transparency, which makes it simple to see how drugs are delivered. But in contrast to alternatives made of polymer and metal, they are more brittle.

9.3.7 Sugar

Drug delivery using sugar-based microneedles, such those derived from maltose, is an appealing sustainable alternative. Upon entry, they disintegrate quickly, delivering the medication into the skin and leaving no residue behind. Proteins, peptides, and vaccinations are frequently administered using these microneedles (Nguyen and Nguyen 2023).

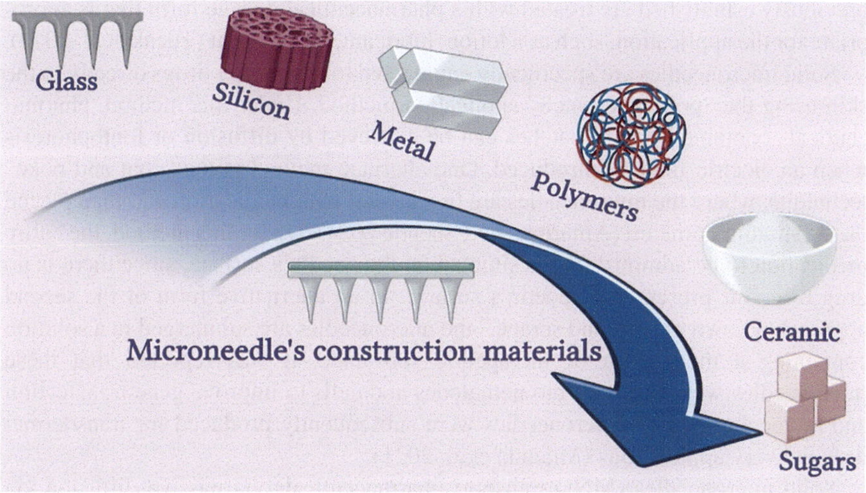

Fig. 9.1 Various materials utilized in microneedle manufacturing

9.3.8 Ceramics

In particular, alumina-based ceramic microneedles deliver remarkable mechanical durability and toughness (Fig. 9.1). They are perfect for medication delivery systems that need materials that are strong and long-lasting. Certain special qualities of ceramics make them suitable for biosensing applications as well.

9.3.9 Hydrogels

The capacity of hydrogel-based microneedles to absorb moisture and expand when in contact with the skin has led to their increasing popularity. It is possible to employ these microneedles for both diagnostic and medication administrations. Their versatility and biocompatibility render them appropriate for a broad array of uses, such as transdermal delivery systems (Shriky et al. 2023).

9.4 Recent Development in Microneedle-Based Transdermal Drug Delivery Systems

9.4.1 Solid Microneedles

The delivery of drugs in solid microneedles often requires a two-step process. In summary, solid MNs are initially inserted into the epidermis and subsequently extracted to form temporary microchannels. Then, the microchannels that were previously established are treated with a pharmaceutical dosage form that is appropriate for the application, such as a lotion, lubricant, or ointment (Tucak et al. 2020).

Solid microneedles are specifically engineered to administer drugs directly to the skin using the "poke-and-patch" application method. Under this method, pharmaceuticals contained within patches can be conveyed by diffusion or iontophoresis when an electric field is introduced. One alternate method is the "coat and poke" technique, where the microneedles are first coated with medication and then placed on the stratum corneum (Amarnani and Shende 2021). Under this method, the entire medication to be administered is situated on the needle's surface, since there is no drug reservoir present on the skin's surface. In an alternative form of the second approach, known as "dip and scrape," the microneedles are submerged in a solution containing a medication or therapeutic substance. It was reported that these microneedles were inserted into nematodes and cells to improve gene transfection and molecular uptake. Microneedles were subsequently produced for transdermal drug delivery applications (Mdanda et al. 2021).

Solid microneedles (MNs) transport pharmaceuticals via passive diffusion via the microchannels formed in the skin. Hence, the drug penetration from solid MNs used for skin pretreatment would be influenced by the length and density design. Furthermore, the pharmacological characteristics also impact the effectiveness of medication administration. Instead of the conventional transdermal administration, the microchannels created by the processing of solid MNs will enhance the infiltration of hydrophilic substances. Various types of solid microneedles are currently being developed. The dimensions of solid silicon microneedles that devised for optimized transdermal drug delivery are as follows: an average height of 158 μm, a base width of 110.5 μm, an aspect ratio of 1.43, a tip angle of 19.4 μm, and a tip diameter of 0.40 μm (Srinivasan and Singaravelu 2017). Microneedle arrays (MA) have been intensively explored for transdermal drug administration because of their painlessness, low skin damage, and low infection risk, according to Jiyu Li et al. However, fabrication issues and fragility have hampered porous MA research. Improved mass manufacturing of titanium porous microneedle arrays (TPMA) is possible with modified metal injection molding (MIM). TPMA showed 30.1% porosity, biocompatibility, and micropore drug delivery, suggesting a solution for transdermal drug administration (Li et al. 2017). The work by Aydin Sadeqi et al. effectively led to the development of a new category of polymeric porous microneedles, which provide a dependable and efficient approach for transdermal drug administration. Upon histological examination, it was shown that the microneedles had strong mechanical characteristics and exceptional skin penetrating capabilities (Sadeqi et al. 2022). The drug release profiles of lidocaine and ibuprofen have verified that the microneedles have the capacity to provide effective local pain relief with few changes in molecular composition during the manufacturing process. The results emphasize the adaptability of microneedles in a diverse array of solid drug formulations, therefore opening up possibilities for future use in flexible and elastic drug delivery systems.

Some intrinsic limitations exist with solid microneedles. An administration method consisting of pretreatment with a MN array followed by the application of medicinal formulations is deemed cumbersome and maybe leads to inaccurate dosing. Given the adverse effects on patient adherence, medication delivery methods utilizing alternative MN have gained increased popularity (Zhang et al. 2022).

9.4.2 Hollow Microneedles

The hollow MN consisted of ceramics, metal, silicon, and glass, with an internal cavity within each needle and an aperture at the needle tip, as previously stated (Cárcamo-Martínez et al. 2021). Consequently, micro-volumes of medication solutions can be injected into the skin. This form of MN has a greater drug delivery capability compared to solid, coated, and dissolving MN arrays (Larrañeta et al. 2016), which can include chemicals, proteins, or vaccines (Larrañeta et al. 2016). Hollow microneedles feature pores at their tips and internal cavities that can accommodate a medicinal solution. The drug is released immediately in the epidermis or upper cutaneous layer as a result of tissue injection. Hollow microneedles may also be employed to deliver molecular entities with a significant molecular weight, such as oligonucleotides, proteins, and vaccines (Mdanda et al. 2021). The production of hollow microneedles is far more intricate, and those with a high aspect ratio do not have an internal support system like a solid needle, which might lead to potential failure if put wrongly. Inadequate management of the patch assembly or unit, both during the process of inserting and removing it, can cause stress that may manifest as fracturing and subsequent failure of the needles. An inherent benefit of hollow microneedles is their ability to administer a substantial and precise amount of medications directly into the skin. The mechanical strength of an optimized microneedle design should be sufficient to prevent needle breakage during skin insertion and reduce the likelihood of bore obstruction, which is a significant problem with hollow microneedles (Avcil and Çelik 2021). Researchers have developed hollow microneedles with asymmetrical bores on the periphery of the tips to avoid obstruction of the bore and allow the medication to reach the adjacent skin tissue.

Tianwei Wu and team conducted an experiment on the integration of hollow microneedles with a paper-based glucose sensor, which presents a highly promising method for noninvasive glucose monitoring (Wu et al. 2022). Through capillary action, the microneedles efficiently permeate the skin to capture interstitial fluid (ISF), which is subsequently analyzed using the paper sensor. This gadget exhibits user-friendliness, cost efficiency, and quick identification, rendering it appropriate for pre-diabetes screening and ongoing glucose monitoring. The prospective applications of this technology encompass a wide range of biosensors, providing a foundation for more extensive diagnostic endeavors. Potential future enhancements may prioritize the augmentation of performance and flexibility for various biomarkers. Significantly, Xenikakis et al. employed 3D printing and liquid crystal display techniques to create two distinct designs of hollow microneedles. The researchers employed scanning electron microscopy to quantify the diameters of the needle, microfocus computed tomography to determine the volumetric characteristics of microneedles and microchannels, and finite element analysis simulation to assess the mechanical properties and skin penetration efficiency (O'Mahony et al. 2023). In vitro, the artificially created hollow microneedles enabled the successful administration of insulin through human skin.

9.4.3 Dissolvable and Swellable Microneedles

Dissolving microneedles, a novel design, have garnered significant interest from both academic and industrial sectors. The structural durability of microneedles is the primary factor in their dissolution, and it is inversely correlated with the quantity of substance present (O'Mahony et al. 2023). Moreover, the physicochemical characteristics of the materials and the design parameters of the microneedle significantly influence the mechanical strength and mechanism of drug release of the needles. Recent studies on dissolving microneedles have focused on the optimization of microneedle design and morphologies (Fig. 9.2). Numerous innovative microneedle designs have been suggested and assessed for efficient penetration of the skin and administration of drugs. Water-soluble and biodegradable polymers suitable for dissolving microneedles are carboxymethyl cellulose, maltose, chitosan, polyvinyl alcohol (PVA), hyaluronic acid, and polyvinylpyrrolidone (Bauleth-Ramos et al. 2023). Moreover, optimizing the stability and safeguarding the bioactivity of biopharmaceutical medications is best achieved by using moderate production settings.

Dissolvable microneedles have been shown to effectively transport insulin into the epidermis, leading to a reduction in blood glucose levels in diabetic rats, mice, and canines, according to research findings. More precisely, there was considerable

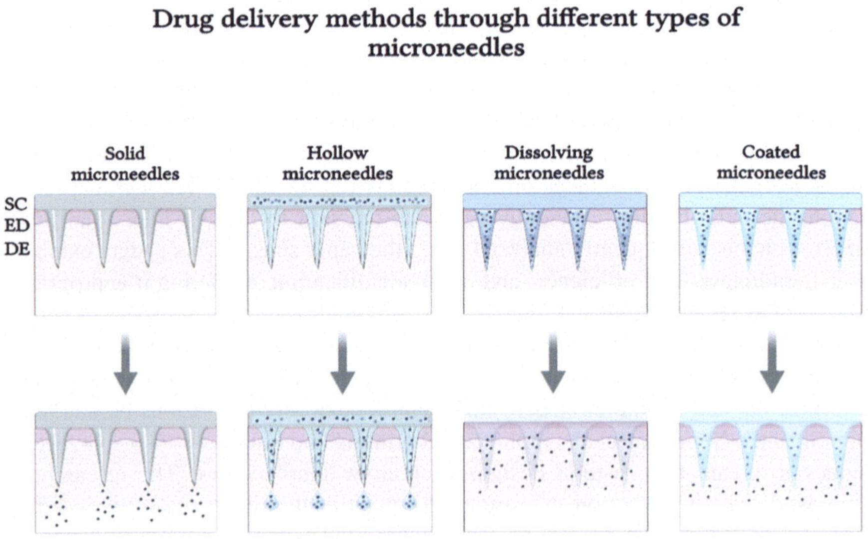

Fig. 9.2 Drug delivery methods through different types of microneedles

interest drawn to a glucose-responsive microneedle-mediated transdermal administration of insulin (Nguyen and Nguyen 2023). Specifically, the pharmacokinetic characteristics of insulin were similar either by microneedle therapy or subcutaneous administration. Furthermore, the medication stability, bioactivity, and bioavailability were enhanced by encapsulating insulin into dissolving microneedles. Once insulin is enclosed within starch and gelatin microneedles, it can maintain greater than 90% of its bioavailability even after being stored for 1 month at temperatures of 25 or 37 °C. In their study, Irina Iachina et al. talk about using dissolving microneedles (MNs) to deliver drugs through the skin. They focus on treating psoriasis with calcipotriol and betamethasone 17,21-dipropionate (Iachina et al. 2023). The study showed that these microneedles could successfully go through the skin of both humans and pigs, making a drug store that could release the drug slowly over time. The biodegradable plastics that make up the microneedles showed promise in increasing drug bioavailability and lowering the number of times a patient had to take their medicine, which could lead to better patient compliance. Using imaging methods like computed tomography (CT) and coherent anti-Stokes Raman scattering (CARS) microscopes showed that the needles went through the skin and stayed there for up to 8 days. The microneedles also had the same therapeutic effects as topical treatments that are used every day. This suggests that the MNs could be a good option for delivering drugs for a longer time. Overall, the results show that dissolving microneedles have a lot of promise for treating long-term skin conditions with less need for regular use (Avcil and Çelik 2021).

Swellable microneedles (SMNs) are presently regarded as one of the most promising varieties of microneedles owing to their simplicity of application, less infection risk, and minimal polymer residue remaining in the epidermis (Omidian and Dey Chowdhury 2024). Moreover, the initial stiffness of the SMNs enhances their ability to penetrate the stratum corneum, in accordance with other MN types. The glass transition temperature (Tg) of the MN polymer decreases due to the diffusion of interstitial fluid from the epidermis into the MNs (Lyu et al. 2023). This activity results in the expansion of the polymer network structure. The hydrophilic properties of a hydrogel efficiently draw moisture from adjacent tissue, leading to the expansion of the microneedle core and the creation of holes that facilitate the release of the therapeutic agent. Swellable microneedle patches are predicated on this design. The principal benefit of this method is the use of the baseplate as a drug reservoir capable of facilitating the flow of the underlying microcirculation via the expanded microneedle design (Mdanda et al. 2021). This is a highly inventive application of the swelling mechanism; they possess a particularly innovative capacity to eliminate interstitial fluid.

9.4.4 Coated Microneedles

Sharp, micrometer-scale needle shafts are affixed to a base substrate in coated microneedle arrays. The needle shafts are coated with a pharmacological substance and water-soluble inactive excipients on their surfaces. Coated microneedles are currently being assessed for the transdermal delivery of a variety of active compounds, such as peptides, small molecules, viruses, and microparticles (Ingrole and Gill 2019). When employing coated microneedles for medication administration, it is imperative to evaluate the reliability of the coating process, the efficacy of drug delivery, and the characteristics of the coatings. In order to ensure that the entire microneedle array and the base substrate were uniformly coated in a single experiment, the microneedle patches were immersed in a coating solution. Another study demonstrated that the use of specific coating processes that restrict coatings to microneedle shafts resulted in increased delivery efficiency and reduced medication waste. A coated microneedle is made up of a solid microneedle with a pointed tip that is insoluble. The microneedle is enclosed in water-soluble inactive excipients and active compounds. When the microneedle is introduced into the epidermis and exposed to interstitial fluids, a hydrophobic covering material is used to facilitate its detachment. The excipients in the microneedle coating dissolve when they come into contact with aqueous interstitial fluids, which leads to the detachment process from the microneedle surface. It is contingent upon the interstitial solubility of the coated excipients that the rate of drug release in the epidermis is determined. The microneedles must be removed from the epidermis after the coating detachment process has been completed from their surfaces.

Bo Zhi Chen et al. devised a microneedle (MN) system that involved the initial mounting of dissolvable drug carriers with precise concentrations onto solid microneedles. Subsequently, the MNs were subjected to a high moisture environment, resulting in uniformly coated MNs with precise drug loading (He et al. 2019). The study achieved precise concentrations ranging from 4.8 to 13.5 ng per needle with minimal standard deviation by utilizing microneedle molds with variable volumes (4.71–12.56 nL per cavity) to control drug loading. The coated microneedles were found to be sufficiently durable for reliable skin insertion, as evidenced by mechanical tests. The hypoglycemic effect of insulin-coated microneedles on diabetic rodents was demonstrated in in vivo trials, which were comparable to subcutaneous administrations (Chen et al. 2022). The clinical translation of microneedle technology is facilitated by the potential of this micro-molding-based fabrication process to mass-produce coated microneedles with homogeneous and controllable drug loading.

9.5 Marketed Microneedle-Based Transdermal Products

Transdermal goods employing microneedles are a novel way to medicine delivery and skincare. These products use small needles to penetrate the skin's outer layer without causing pain or irritation. By enhancing the absorption of active substances, these treatments aim to overcome the epidermal barrier, which frequently restricts the efficacy of conventional topical or oral drugs. Pharmaceutical and cosmetic companies offer marketed solutions that include vaccines (like Vaxxas' vaccine patches), skincare products (like Raphas' acne patches) (Table 9.2), and migraine

Table 9.2 Status of clinical trials based on MN technique (https://clinicaltrials.gov/)

NCT no.	Clinical trial	Condition and diseases	Drug and device	Phase	Location	Status
NCT04732195	Pilocarpine microneedles for inducing sweating	Cystic fibrosis	Pilocarpine microneedle patch	NA	Emory University, USA	Completed
NCT03207763	Microneedle patch study	Vaccination Skin absorption	Microneedle patch	Phase I	Emory University	Unknown
NCT00837512	Insulin administration by microneedles in type 1 diabetes	Type 1 diabetes mellitus	Tool: a hollow microneedle (1 mm) and insulin tube for the subcutaneous (SC) area	I, II	Emory University	Completed
NCT02682056	Using microneedle patches to measure glucose	Diabetes	Intravenous (IV) catheter	I	Emory University	Recruiting
NCT01518478	Pilot study of the flu vaccine from the Atopic Dermatitis Research Network (ADRN)	Atopic dermatitis	Fluzone® intradermal	I	National Institute of Allergy and Infectious Diseases (USA)	Completed
NCT04064411	A transdermal system for treating osteoporosis in women who have gone through menopause: an evaluation of the efficacy and safety of abaloparatide	Postmenopausal osteoporosis	Abaloparatide formulation Microstructured transdermal system; abaloparatide-SC	III	Radius Health, Inc.	Completed
NCT03203174	The combination of microneedles and botulinum toxin applied to the skin to treat palmar hyperhidrosis	Hyperhidrosis	Solid (sham microneedle) Botulinum toxin type A	I	University of California, Davis	Completed
NCT02459938	Comparing the efficacy and safety of ZP-glucagon to injectable glucagon for the treatment of hypoglycemia	Hypoglycemia	Solid/metal (drug-coated titanium microneedles) Zosano microneedle patch	I	Nucleus network (Australia)	Completed

(continued)

Table 9.2 (continued)

NCT no.	Clinical trial	Condition and diseases	Drug and device	Phase	Location	Status
NCT05508191	Comparing the effectiveness of microneedle lasers and fractional CO_2 lasers for treating skin aging with stem cell secretome in Indonesian adult women	Adipose-derived mesenchymal stem cell (ADMSC) secretome to alleviate skin aging and skin rejuvenation	Fractional CO_2 laser; microneedle	III	Fakultas Kedokteran Universitas Indonesia	Completed
NCT06378983	Microneedle radiofrequency in conjunction with oral isotretinoin for moderate to severe acne	The acne vulgaris syndrome, moderate to severe	Oral isotretinoin combined with microneedle radiofrequency therapy	I	Sun Yat-Sen Memorial Hospital of Sun Yat-Sen University	Completed
NCT03607903	Adalimumab microneedles in healthy volunteers	Autoimmune/ auto-inflammatory diseases including juvenile idiopathic arthritis (JIA)	Pain, injection site	I	Centre for Human Drug Research, Netherlands	Active
NCT02594644	Actinic keratoses (AKs) are effectively treated with photodynamic therapy (PDT), which is also cosmetically acceptable	Keratosis, actinic	Aminolevulinic acid	NA	University of California, Davis	Completed
NCT03097315	Suprachoroidal injection of CLS-TA in patients with non-infectious uveitis (AZALEA)	Uveitis, posterior uveitis, anterior uveitis, intermediate panuveitis	4 mg CLS-TA suprachoriodal injection	III	Clearside Biomedical, Inc.	Completed

patches (like Zosano's Qtrypta). These products provide more effective, painless, and frequently self-administered alternatives to traditional treatments.

Numerous transdermal products utilizing microneedles have been created and distributed, mostly in the healthcare and cosmetic sectors. Here are a few noteworthy examples:

9.5.1 Qtrypta

A titanium microneedle patch with a coated formulation of zolmitriptan for the immediate treatment of migraines was created by Zosano Pharma. It is among the first pharmaceutical microneedle items to be filed for FDA clearance. The medication uses microneedles to apply zolmitriptan topically. The medicine can enter the circulation more rapidly because of the microneedles' fast dissolution.

9.5.2 Microneedle Vaccines

A range of microneedle patches designed to provide vaccines against illnesses including COVID-19 and influenza (Table 9.3) are either being developed or are already on the market. These goods promise to offer a painless and efficient immunization procedure. Medi-Ject is one such product, it applies microneedle technology painlessly, and this device improves patient compliance by delivering vaccinations and other therapeutic substances. MicronJet is a microneedle-based platform developed by NanoPass Technologies for painless delivery of vaccines and insulin. Vaxxas is a high-density microarray-based vaccine patch (HD-MAP) designed to administer vaccines. Vaxxas is used especially in the delivery of COVID-19 vaccines. It is useful for worldwide distribution since vaccines may be maintained without refrigeration.

9.5.3 Hollow Microneedle Patch

Hollow microneedle patches developed by 3M are used to deliver more viscous liquids. These hollow needles, in contrast to dissolving microneedles, enable the direct administration of liquid medications through the skin. Given that stability in liquid form is maintained, a broad variety of medication quantities (up to 2 mL) and molecule types can be used with the Microstructured Transdermal System. These hollow microneedles are included in injectable drugs, vaccinations, and pain relief medications.

Table 9.3 Microneedle-based drug delivery with its applications

S. no.	Technology	Type	Applications	References
1	Biodegradable polymer microneedles	Biodegradable	Effective for insulin delivery	Pere et al. (2018)
2	Influenza subunit vaccine-coated microneedles	Vaccine	Comparable immune response to intramuscular injection	Zheng et al. (2018), Kommareddy et al. (2013)
3	Rapidly dissolving microneedle for therapeutics	Dissolving	Fast and painless drug delivery	Zheng et al. (2023)
4	Microneedle system for ocular drug delivery	Ocular delivery	Targeted delivery to the eye	Glover et al. (2023)
5	Hollow microneedle with bevel opening	Hollow	Enhanced fluid delivery	Wang et al. (2023)
6	Zein microneedles for localized chemotherapy	Chemotherapy	Effective for treating breast cancer	Bhatnagar et al. (2018)
7	Analgesic microneedle patch for neuropathic pain	Pain management	Significant pain relief in tested subjects	Xie et al. (2017)
8	Microneedle-mediated treatment of bacterial biofilms	Antibacterial	Effective in biofilm disruption	Xu et al. (2019)
9	3D polymer microneedle array fabrication	Fabrication	Advanced fabrication techniques discussed	Pere et al. (2018)

9.5.4 Dermaroller and Related Devices

Although mostly intended for aesthetic applications, gadgets such as the Dermaroller employ microneedles to improve the application of topical therapies for skin ailments. These dermarollers are used to enhance the cosmetic products to penetrate to the skin effectively.

9.5.5 Cosmetic Patches

Cosmetic manufacturers like L'Oreal and Raphas are producing hyaluronic acid and retinol-filled dissolving microneedles to cure acne, to enhance the look of the skin, and to treat aging.

Microneedle-based products are more patient-friendly because they are often painless or less invasive. A lot of microneedle patches make self-administration simple, which lessens the requirement for medical professionals. By avoiding the outer skin barrier, they increase the potency of the active substances. These devices are providing a more efficient, pleasant, and effective alternative to conventional techniques, which is transforming the realms of cosmetics, vaccinations, and medicine administration.

With continuous research and development aimed at developing novel formulations and applications for both medicinal and cosmetic uses, the market for goods based on microneedles is found to be rapidly growing.

9.6 Applications of Microneedle Array

Recent developments in microneedle array technology have significantly broadened its use in the diagnosis management of a number of illnesses, including vitiligo, actinic keratosis, alopecia, cancer, diabetes, hyperhidrosis, melasma, acne, skin inflammatory diseases, and osteoporosis.

9.6.1 Cancer Diagnosis and Treatment

MN arrays are currently being investigated as potential devices in cancer detection and therapy, notably owing to their ability to penetrate the skin, hence obviating the necessity for invasive biopsies and intravenous medication delivery. Delivering anticancer drugs directly to the location of the action, MNs avoid first-pass metabolism by the liver and subsequently reduce the side effects usually related to systemic drug administration. They provide a relatively noninvasive means for diagnosis, vaccination, and targeted drug delivery, especially in breast cancer. A personalized treatment approach can be presented by designing MN arrays to administer anticancer agents based on parameters such as body weight, tumor size, and drug circulation time (Alimardani et al. 2021).

Conventional diagnostic methods have the drawback of requiring invasive sampling at the time of diagnosis, while frequent samplings are used to track therapy response and can be quite physically and psychologically demanding on patients. The primary advantages of MNs counteract this by providing a noninvasive means of detecting cancer markers during the course of treatment, so enabling routine assessment of the disease's progression. MNs are exceptionally effective at treating malignant illnesses, particularly breast cancer, because of their ability to distribute drugs locally, despite the challenges associated with repeatability and complexity in manufacturing (Khan et al. 2021).

Numerous researches have been conducted on the possible applications of MNs for the treatment of different cancer types, including skin cancer, prostate cancer, and breast cancer. These applications include drug delivery, vaccine administration, and the use of hormone and gene therapy agents, among other therapeutic drugs. Currently, MNs lead the way and show great potential as a breast cancer treatment system since they enable noninvasive breast cancer screening and treatment. An overview of several MN formulations used in breast cancer detection and treatment, as well as their use in other malignancies, is provided by the review. The utilization of sensor technology in MNs; MN applications in the treatment of skin cancer, prostate cancer, and breast cancer; and MNs manufactured with additive manufacturing methods like 3D printing are among the main areas of focus. The review addresses recent advancements in clinical studies pertaining to the use of MNs in cancer treatment, as well as its limitations and safety concerns (Desai et al. 2023).

9.6.2 Polymeric MNs for Controlled Drug Delivery

Research has been done on the use of polymeric MNs for controlled drug delivery applications, such as encasing drugs like calcein and bovine serum albumin for the treatment of lung cancer. The delivery of anticancer drug in the form of both microparticles and nanoparticles with increased therapeutic efficacy was made possible by MNs. MNs can identify indicators unique to cancer, including nitrogen monoxide. This will allow it to assess the presence of cancer, the size of the tumor, and its growth—all extremely important data for developing a personalized treatment plan.

9.6.2.1 3D-Printed Microneedles

Using 3DP is one of the newest and most advanced methods for fabricating MNs since it can create devices with intricate geometries layer by layer with extreme precision. Complex MN designs that may be required for the implementation of precise drug delivery might be created using appropriate methods. MN-based drug delivery systems appear to have promise for the treatment of skin, breast, and prostate cancer. Further research is necessary to determine their applicability to deeper-lying tumors that are inaccessible to MN patches applied to the skin's surface.

9.6.2.2 MNs as Vectors for Prostate Cancer Therapy

Men are most likely to get prostate cancer, which is also the second most common disease in developed countries and has a very high death rate. Anticancer vaccines employing DNA vaccines like DNA vaccines against PC have been used in targeting cells expressing TAAs. However, recent results from clinical trials including that using PROSTVAC raise doubts about DNA vaccines as potential therapies on their own. Such vaccinations can be made more effective by the inclusion of MN-based vaccine delivery systems, which offer a focused and highly effective delivery of therapeutic agents. The major advantages of the application of MNs in cancer treatment are increased patient compliance due to reduced pain from recurrent injections and sampling. MNs also avoid under- or overdosing of drugs, which is a problem with oral administration. Besides, MNs can deliver small and large molecules including nanoparticles of the anticancer drugs, thus improving the outcomes of the treatment. However, there are several challenges in the development and application of MNs for cancer therapy. For example, hydrogel-forming and hollow MNs are suitable for higher doses but are usually restricted for repeated use for safety reasons. Hydrogel-forming and hollow MNs have a great potential for high dose delivery, but the safety issue, especially after repeated applications, restricts their use. Research needs to be continued in this direction to overcome these drawbacks and search for new methodologies with regard to optimizing the drug delivery schedule and sequential release. MNs offer a much safer, more localized, and controlled approach to anticancer agent delivery by reducing systemic exposure and side effects. Due to the unfolding research, it is expected that MNs will become one of the traditional treatment modalities in many cancers due to the less invasive and targeted mode of therapy compared to conventional modalities (Seetharam et al. 2020).

9.6.3 Diabetes Mellitus

Diabetes mellitus is a metabolic disorder and is relatively common; it is characterized by hyperglycemia, resulting from a deficiency of insulin or abnormalities in the usage of insulin. Management is effectively done by monitoring glucose and delivering insulin, where, in recent times, the approach has been focused on using microneedle technology. Microneedle arrays have emerged as a promising tool for continuous glucose monitoring through the sampling of interstitial fluid, hence offering a less painful alternative to traditional blood glucose monitoring methods. Microneedles can also be used for insulin delivery. It offers a painless alternative route, which is transdermal instead of by subcutaneous injection. This method makes patients more likely to follow through, especially kids and teens, but it slows down the health benefits of better glucose control after a meal. Two trials have validated the effectiveness of microneedle devices in administering insulin to children and adolescents with type 1 diabetes. In the preliminary study, five subjects were given insulin intradermally with a single hollow microneedle and insulin subcutaneously with a normal tube. This study has surmised that intradermal microneedles allow for improved postprandial glucose control and better patient compliance, which may be more beneficial in younger patients for the management of diabetes and its complications. Other research has focused on glucose measurement, using a microneedle patch, for improved epidermal antigen-specific immunotherapy for type 1 diabetes. Clinical trials like NCT02837094 and NCT00837512 have also been performed on the use of microneedles in insulin delivery for type 1 diabetes, indicating different aspects in which these devices can be helpful in managing diabetes (McAlister et al. 2021).

Recent advances in microneedle technology have been focused on improving insulin permeability, stability, and bioavailability via alternative route. For instance, an open-loop microneedle (MN) delivery system was designed featuring a microcavity-like structure, in which insulin powder is cast into microcavities formed by carboxymethyl cellulose (CMC), helping to prevent insulin degradation and dilution of the active dose. Their research presented that 10% CMC has proper penetration and is highly capable of insulin loading. The modified MNs improved insulin stability by approximately 5% relative to traditional dissolving MNs throughout an 8-week duration at 25 °C. Insulin distribution was delayed in both in vitro and in vivo investigations (Nazary Abrbekoh et al. 2022).

Recent research has also explored the use of intelligent microneedles, including silk protein and phenylboronic acid/acrylamide matrices loaded with insulin. These needles pierce the skin, self-regulating in releasing insulin when there is a rise in skin tissue glucose levels, thus offering safe, residue-free, and sustained drug release (Starlin Chellathurai et al. 2024).

Another novel approach involved core-shell glucose oxidase and catalase-loaded PVA microneedles containing modified insulin (insulin NBC). The free insulin in these microneedles was rapidly released due to the production of high levels of H_2O_2 upon exposure to high glucose, effectively regulating blood sugar levels (Hou et al. 2023). The utilization of polymeric microneedles in illness therapy, including diabetes, has been investigated. Microneedle patches constructed from materials like alginate and hyaluronate have been found to provide superior mechanical strength and low degradation rates, hence possibly replacing conventional insulin delivery techniques. It has good mechanical strength and low degradation rates, so it can be used rather than standard insulin administration methods.

9.6.4 Acne

This is the reason that microneedle-based transdermal products have enjoyed great interest not only therapeutically but also cosmetically, as they can enhance the delivery of drugs and treat various skin disorders. A number of other products that are currently on the market and are based on microneedle technology serve these dual purposes. For example, Dermaroller®, manufactured by Derma Spark in Canada, employs a metallic microneedle array to address hair loss, stretch marks, and acne. This product also improves the permeation of medications such as minoxidil and hyaluronic acid, resulting in a variety of skin therapy benefits. Similarly, Dermapen® is a microneedle array-based device used to treat various skin disorders that include acne, stretch marks, and hair loss and enhance the permeation of topical drugs to enhance the treatment outcomes for various dermatological disorders (Halder et al. 2021). Skin lesions, acne vulgaris, and wrinkles have all been successfully treated with microneedles in cosmetic treatment applications. LiteClear® by Nanomed Skincare is an example of a product that utilizes solid silicon microneedles as a pretreatment to improve the efficacy of topically applied medications. This specific substance is intended to address skin blemishes and acne. These products are indicative of the increasing interest in the use of microneedles as noninvasive yet highly effective solutions to a diverse array of skin conditions (Zhang et al. 2023).

9.6.5 Actinic Keratosis

Microneedle-assisted photodynamic therapy utilising a single-use sterile microneedle roller with a needle length of 200 μm in 33 participants. The findings demonstrated similar effectiveness to traditional PDT, along with the advantage of less pain. It has been shown in many studies and clinical trials that microneedles can improve the effects of photodynamic therapy (PDT) (Tian et al. 2024). Microneedle-assisted PDT using a single-use sterile microneedle roller with a needle length of

200 μm in 33 volunteers. The results showed comparable efficacy to conventional PDT, with the added benefit of reduced pain. Clinical trials, such as NCT01812837 and NCT02594644 conducted by the University of California, Davis, utilized microneedles in combination with aminolevulinic acid and blue light radiation to treat actinic keratosis, effectively expediting treatment and reducing skin lesions without serious adverse effects. DUSA Pharmaceuticals, Inc., through trial NCT02632110, investigated a topical solution vehicle with microneedle pretreatment to enhance PDT efficacy. Microneedle-assisted PDT has been found to decrease the incubation time required for photosensitizers, making it a valuable method for enhancing the penetration and effectiveness of various drugs.

9.6.6 Alopecia

Alopecia, particularly alopecia areata (AA) and androgenetic alopecia (AGA), remains a challenging condition to treat, with significant psychological burdens for patients. While minoxidil and finasteride are the only FDA-approved treatments, their efficacy is often limited (Zhou et al. 2023). Microneedle (MN) patches are a new and optimistic approach to give drugs transdermally. They improve local drug delivery and patient compliance, which could lead to better therapeutic results. For instance, MN patches loaded with minoxidil (MNX) and encapsulated in a metal-organic framework (UiO-66) have been developed to improve safety and efficacy in treating alopecia areata. Additionally, a degradable MN patch containing zinc/copper dual-doped mesoporous silica nanoparticles loaded with quercetin (ZCQ) has demonstrated potential for treating AGA by promoting hair follicle regeneration through the release of quercetin, copper, and zinc ions (Zhang et al. 2020). Another innovation involves a hydrogel MN patch delivering CCL22 and IL-2 to amplify regulatory T cells locally, promoting hair regrowth and reducing inflammation in alopecia areata (Younis et al. 2024). These new findings show that MNs might be useful in treating alopecia, but more clinical studies are needed to set up standard methods and prove that they are safe and effective in the long run.

9.6.7 Hyperhidrosis

Recent advancements in the treatment of hyperhidrosis have explored several innovative approaches, including the use of botulinum toxin type A (BTX-A) and various radiofrequency-based techniques. For primary palmar hyperhidrosis, fractional CO_2 laser-assisted delivery of BTX-A has been proposed as a viable therapeutic option (Junsuwan et al. 2021). In the case of primary axillary hyperhidrosis (PAH), fractional microneedle radiofrequency (FMR) has emerged as a novel non-surgical method, effectively reducing sweat gland activity by delivering targeted radiofrequency energy through microneedles, resulting in thermal injury at the hypodermal interface. Results from clinical trials have shown that of FMR in treating refractory PAH, as seen in a 29-year-old female patient confirmed with histopathological staining, and in a prospective pilot study involving 16 Korean patients, where significant improvements were observed in patient satisfaction and sweat reduction without serious adverse events (Jung and Jin 2021).

9.6.8 Inflammatory Disease

The utilization of microneedle (MN) arrays in transdermal medication administration has been considerably broadened to address a range of skin illnesses, including inflammatory skin diseases. MNs have been utilized for conditions like atopic dermatitis (AD), psoriasis, and other chronic inflammatory skin diseases (CISDs). Atopic dermatitis, a result of skin barrier dysfunction and immune deregulation, is commonly treated with corticosteroids, though long-term use can lead to side effects such as skin atrophy, burning, and itching. Integrating corticosteroids like triamcinolone acetonide (TA) into dissolving microneedles has shown improved therapeutic outcomes with reduced side effects. For example, dissolving microneedles loaded with hyaluronic acid (HA)-encapsulated methotrexate (MTX) have been developed for psoriasis treatment, demonstrating enhanced effectiveness and reduced gastrointestinal degradation and liver metabolism compared to oral administration. Additionally, dissolving microneedles containing non-soluble cyclosporin A have been created for psoriasis therapy.

Additionally, rapidly dissolving microneedles loaded with dexamethasone-poly(lactic-*co*-glycolic acid) (PLGA) nanoparticles were designed for controlled drug release, improving patient compliance and minimizing side effects associated with conventional corticosteroid treatments (Alimardani et al. 2021). Overall, these developments in microneedle technology represent promising alternatives to traditional methods, increasing the efficacy of treatments for inflammatory skin disorders while reducing adverse effects and promoting patient comfort and adherence. Moreover, reduced irradiance photodynamic therapy (PDT) using microneedles, such as MiLD, has shown reduced histological changes and inflammation compared to higher irradiance therapies, highlighting another therapeutic advantage of microneedles in dermatological applications (Zhou et al. 2023).

9.6.9 Fungal Infection

Therapeutic-loaded transdermal microneedles provide a potential method for addressing fungal infections and surmounting obstacles linked to traditional therapy. Intravenous vancomycin (VAN), a glycopeptide antibiotic, is frequently utilized to address skin and soft tissue infections (SSTIs) induced by methicillin-resistant *Staphylococcus aureus* (MRSA); nonetheless, its effectiveness is limited by inadequate concentrations reaching the afflicted dermal regions. The topical use of VAN is ineffective because its large molecular weight impedes skin penetration. VAN-loaded microneedles with water-soluble tips are capable of penetrating thawed porcine and fresh human skin, facilitating the continuous release of a high dosage of VAN over 24 h, successfully inhibiting MRSA proliferation (Ziesmer et al. 2021).

Furthermore, small molecule pharmaceuticals like doxycycline for *Staphylococcus aureus* and *Pseudomonas aeruginosa*, along with carvacrol for multidrug-resistant bacteria, have been integrated into microneedles to enhance therapeutic efficacy. Innovative methods have also involved embedding live bacteria into microneedles. *Bacillus subtilis*-encapsulated microneedles can inhibit bacterial escape and continually release antifungal drugs that directly attach to fungal cell surface proteins, disrupting fungal cell membranes without inducing drug resistance. The microneedle-based drug delivery systems represent a novel and effective strategy for treating skin infections, offering controlled release, enhanced penetration, and reduced resistance compared to traditional therapies.

9.7 Microneedles as Nanotechnology

Nanotechnology enhances microneedles (MNs) in biomedicine by enabling the intelligent design of drug delivery systems that can control the amount and kinetics of drug release or respond to specific stimuli. The principal benefit of integrating nanomaterials into microneedles is their capacity for customization to fulfill specific roles and to react to unique microenvironments at lesion locations. These special microenvironments and landmark substances can be utilized in the design of functional nanomaterials combined with MNs for treating various diseases.

Microneedles (MNs) offer an innovative approach for osteoarthritis (OA) treatment by enabling targeted, sustained drug delivery. Glucocorticoid-loaded MNs have significantly reduced knee joint swelling and inflammation in rats, indicating their potential for long-term OA management. Meloxicam, an NSAID with poor water solubility, presents challenges for oral administration. To improve delivery, MNs encapsulating meloxicam using polyvinyl alcohol and polyvinylpyrrolidone showed comparable anti-inflammatory activity to oral tablets in rats, with enhanced safety and stability. Rapidly produced two-layer MNs further demonstrated quick drug release, high skin administration efficiency, good bioavailability, and strong anti-inflammatory effects, highlighting MNs as a promising alternative to overcome limitations of oral therapies (Zhang et al. 2022).

9.7.1 Melasma

Microneedle (MN) patches are progressively recognized as an effective treatment for melasma, a persistent hyperpigmentation disorder. Recent studies have demonstrated that hyaluronic acid (HA)-based MN patches loaded with tranexamic acid (TXA) significantly reduce epidermal thickness, melanin production, and the expression of dopachrome tautomerase (DCT) in melasma models (Zhang et al. 2024). Combining functional MN arrays with topical TXA has also proven beneficial, enhancing treatment efficacy while maintaining safety and comfort. Despite the positive outcomes, further well-designed research is needed to confirm these findings (Xu et al. 2017). Furthermore, dissolving MNs loaded with both TXA and liquorice extract have demonstrated superior efficacy, reducing melanin deposition more effectively than TXA alone and ensuring stability under stress conditions (Xing et al. 2021).

9.7.2 Osteoarthritis

Conventional drug delivery methods are frequently restricted by their inefficiency and lack of specificity, which presents a significant public health challenge in the context of bone-related disorders (Lin et al. 2023). Microneedle-based therapy offers a promising alternative by enabling targeted, programmable drug delivery in a less invasive manner. Various types of microneedles have been developed to enhance treatment efficacy for bone diseases, such as osteoporosis, with several clinical trials demonstrating their potential. For example, studies have shown that abaloparatide-coated microneedles (NCT04936984) achieve bioequivalence, while transdermal patches (NCT01674621) and solid microstructured systems (NCT04064411) improve drug delivery and patient outcomes. These advancements highlight microneedle therapy's potential as a novel strategy for bone-related disorder treatments. Dissolving microneedles offer an innovative approach for rheumatoid arthritis (RA) treatment by encapsulating drugs directly into the microneedle matrix for efficient delivery. Recent studies have advanced the use of microneedle-based therapies for various conditions. In the context of rheumatoid arthritis, neutrophil membrane-coated NSAID nanoparticles, in conjunction with hyaluronic acid-based dissolving microneedles, improved medication targeting and absorption in mice, therefore diminishing inflammation and joint damage. Developed swellable hydrogel microneedles for methotrexate administration, lowering gastrointestinal side effects, and created ultra-swellable hydrogel microneedles for colchicine delivery, demonstrating notable efficacy in alleviating joint inflammation (Sultana et al. 2023).

Microneedle-based therapies are also being explored for tendon and cartilage repair, integrating techniques like nitric oxide nanomotors and growth factors to enhance healing, as shown by Liu et al. Additionally, a "hydrogel-Chinese acupuncture" system has been innovatively designed to deliver drugs deep into tissues, improving cartilage regeneration and reducing osteoarthritis symptoms. These developments highlight the potential of microneedle-based systems for targeted, effective, and minimally invasive therapies across various medical conditions (Hou et al. 2023).

9.7.3 Scars

Microneedling (MN) is a minimally invasive cosmetic technique employed to address diverse scar types, including acne scars, hypertrophic scars, keloids, and scars arising from surgical procedures, trauma, varicella, or smallpox. It entails the use of tiny needles to induce micro-injuries in the skin, hence boosting collagen synthesis and modifying scar tissue. MN has garnered attention for its effectiveness and safety profile, particularly for darker skin types, which exhibit a less risk of post-procedure pigmentary alterations (Juhasz and Cohen 2020).

Microneedling is particularly effective for atrophic acne scars. It is a simple, minimally invasive procedure that has demonstrated positive clinical outcomes with high patient satisfaction. Multiple sessions of MN have been shown to stimulate dermal remodeling and improve the appearance of atrophic scars with minimal downtime and risk, making it a favorable option for treating acne scars, especially in skin types prone to hyperpigmentation (Sitohang et al. 2021).

Several clinical trials have investigated MN for treating scars and compared its efficacy to other modalities, such as fractional lasers, chemical reconstruction (trichloroacetic acid), radiofrequency, and silicone gel. MN has emerged as a non-ablative alternative that provides comparable or better outcomes with a reduced risk of adverse effects, making it suitable for patients concerned with aesthetic changes due to scars. MN has proven to be a promising treatment for scars, and additional research is required to further develop standardized protocols and validate its long-term efficacy (Juhasz and Cohen 2020).

9.7.4 Skin Aging

Standard microneedling techniques for treating acne and other scars typically utilize a needle length of 1.5–2 mm, but a shorter needle length of 0.5–1.0 mm is recommended for addressing aged skin and wrinkles. The efficacy of microneedling can be augmented by integrating the procedure with topical anti-aging agents such as vitamin C serum and tretinoin. Recent studies demonstrate that the integration of microneedling with conditioned media produced from human embryonic stem cell-derived endothelial precursor cells can significantly reduce wrinkles and pigmentation (Singh and Yadav 2016). Furthermore, fractional microneedling radiofrequency has been helpful in diminishing wrinkles, as evidenced by a substantial multicenter experiment. Numerous microneedling devices have been created to improve the effectiveness of these treatments, including DermaFrac, which features a small stainless steel microneedle roller combined with a serum infusion system and light-emitting devices, and Dermastamp, which employs a needle array on an electric motor unit for precise collagen induction in smaller skin regions (Table 9.4). Automated microneedling devices, like the Exceed device, have demonstrated substantial enhancement in face wrinkles, skin laxity, and texture with minimum pain, discomfort, and downtime, as evidenced by a research utilizing Phaseshift Rapid in vivo Measurement of the Skin (PRIMOS) profilometry to assess skin improvement (Ablon 2018).

Table 9.4 Applications and microneedle systems in drug delivery and diagnostics

S. no.	Application	Microneedle system	Active ingredient	Company/ manufacturer	Marketed product
1	Acne management	Microneedle patches made of poly(ionic liquid)	Salicylic acid	N/A	N/A
2	Acne scars, wrinkles, age spots	N/A	N/A	Equipmed™ Australia	Dermapen®
3	Atopic dermatitis treatment	Dissolving microneedles	Dexamethasone	N/A	N/A
4	Breast cancer screening	Microneedle patch composed of gelatin methacrylate and acrylic acid	Interstitial fluid	N/A	N/A
5	Chronic wound treatment	Microneedles composed of natural antibacterial materials	Peony leaf extract	N/A	N/A
6	Chronic wound treatment	Hyaluronic acid microneedle	Nanoparticles loaded with recombinant humanized collagen type 3 and naproxen that are composed of poly(lactic-*co*-glycolic acid)	N/A	N/A
7	Chronic wound treatment	Dressings made of zwitterionic microneedles	Zinc oxide nanoparticles and asiaticoside	N/A	N/A
8	Colorectal cancer diagnosis	Hydrogel microneedles	Interstitial fluid	N/A	N/A
9	Cosmetic applications	Handheld device	N/A	NanoPass, Inc., Israel	NanoCare
10	Cosmetic applications	Injection of liquid pharmaceutical formulations or cosmetics	N/A	AdminMed, USA	AdminPen
11	Cosmetic applications	Transdermal administration of cosmetics to deeper dermal layers	N/A	Clinical Resolution Laboratory, Inc., USA	MTS-rollers

(continued)

Table 9.4 (continued)

S. no.	Application	Microneedle system	Active ingredient	Company/ manufacturer	Marketed product
12	Cosmetic applications	Wrinkles and age spots	N/A	DermaRoller, Germany	DermaRoller®
13	Glucose determination and management of chronic diseases	A glucose measuring device with ion-conductive hollow microneedles	Interstitial fluid	N/A	N/A
14	Hypertrophic scar treatment	Dissolving hyaluronic acid microneedles	Shikonin	N/A	N/A
15	Hypertrophic scar treatment	Methacrylate gelatin/ polyethylene glycol diacrylate double-network hydrogel microneedle patch	Betamethasone	N/A	N/A
16	Hypertrophic scar treatment	Dissolving microneedle array	MiRNA-modified functional exosomes	N/A	N/A
17	Treatment for hypertrophic scarring	Dissolving microneedle patches of gelatin and starch	Losartan	N/A	N/A
18	Keloid management	Stainless solid microneedles	5-fluorouracil	N/A	N/A
19	Non-melanoma skin cancer treatment	AdminPen™ hollow microneedle array	Hypericin lipid nanocapsules	N/A	N/A
20	Psoriasis treatment	Dissolving microneedle	Triamcinolone acetonide	N/A	N/A
21	Pulmonary fibrosis treatment	Microneedle that responds to hydrogen peroxide	Integrin αvβ6-blocking antibody	N/A	N/A
22	Rheumatoid arthritis treatment	Dissolving microneedle patches	Tetramethylpyrazine	N/A	N/A
23	Skin cancer treatment	AdminPen™ hollow microneedle array	N/A	N/A	N/A

(continued)

Table 9.4 (continued)

S. no.	Application	Microneedle system	Active ingredient	Company/ manufacturer	Marketed product
24	Subcutaneous tumor treatment	Hydrothermally responsive multi-round acturable microneedle	Docetaxel	N/A	N/A
25	Topical serum containing cosmeceuticals that repair the epidermis, including hyaluronic acid, peptides, amino acids, minerals, and vitamins	N/A	N/A	AMIEA MED, Berlin, Germany	Revive™ HAP
26	Uric acid monitoring	Wearable plasmonic microneedle sensor based on microfluidics	Interstitial fluid	N/A	N/A
27	Vaccine delivery	Polymeric microneedles	Ovalbumin and CpG	N/A	N/A
28	Vaccine delivery	Hollow microneedles	Insulin	N/A	N/A
29	Vaccine delivery	Dissolving microneedle inserts	Heat-inactivated bacteria	N/A	N/A
30	Vaccine delivery	Solid pyramidal microneedle	Stabilized HIV envelope trimer immunogen and adjuvant	N/A	N/A
31	Wound healing and biomarker detection	Encoded structural color microneedle patches	Photonic crystals	N/A	N/A

9.7.5 Vaccines

Microneedles (MNs) have developed as a promising method for vaccine delivery due to their capability to induce a stronger local immune response related to traditional subcutaneous injections. MNs facilitate direct antigen presentation to dendritic cells in the skin, enhancing immunogenicity and reducing reliance on cold storage and transportation, which is a major limitation of conventional vaccines (Jung and Jin 2021). In a phase 1 clinical trial, microneedle patches for transdermal delivery of inactivated influenza vaccines elicited immunological responses comparable to intramuscular injections, while resulting in fewer adverse effects, including mild to moderate local reactions. Moreover, the use of granulocyte-macrophage colony-stimulating factor (GM-CSF) with MNs has been shown to induce a prolonged antibody response. MNs offer a minimally invasive, painless option for vaccine administration, supporting mass vaccination efforts with high patient compliance, and are being explored for their potential role in biosecurity, enabling rapid deployment and mass immunization during pandemics or bioterrorism threats (Hosangadi et al. 2020).

Microneedle vaccines represent a novel approach to immunization, particularly appealing for their potential in delivering vaccines against bioterrorism agents with reduced pain and increased compliance. Traditional vaccination methods often involve discomfort or logistical challenges, but microneedles—tiny, minimally invasive needles that penetrate only the skin's superficial layers—offer a less painful alternative. This method has shown promise across a range of vaccines:

Anthrax: Microneedles effectively deliver recombinant anthrax vaccines, potentially enhancing immune responses and reducing dose requirements compared to traditional methods.

Ebola: Microneedle-based vaccines for Ebola show potential in generating robust immune responses and maintaining stability without refrigeration, addressing both efficacy and distribution challenges (Mansoor et al. 2022).

Influenza: For influenza, microneedles offer advantages such as improved stability at room temperature and ease of mass distribution, essential for managing seasonal outbreaks and emergent strains.

Smallpox: Despite smallpox being eradicated, microneedle delivery of vaccines has demonstrated effective immune responses and long-term stability, addressing bioterrorism concerns.

Botulinum Neurotoxin: Microneedle systems for BoNT vaccines have shown efficacy comparable to traditional injections, with the added benefit of room temperature stability.

SARS-CoV-2: The microneedle approach for COVID-19 vaccines provides effective antibody production and stability without the need for cold-chain transport, offering a solution to logistical challenges faced by current vaccines. Looking forward, while microneedles have shown potential across various vaccine types, including inactivated, attenuated, recombinant, and DNA vaccines, challenges remain in manufacturing, dose accuracy, and regulatory approval. Ongoing investigation and progress are essential to resolve these challenges and enhance microneedle vaccine delivery technologies for widespread public use (McNamee et al. 2023).

9.7.6 Vitiligo

Microneedling, initially developed for skin rejuvenation, has emerged as an effective treatment for vitiligo, demonstrating notable improvements in skin pigmentation (Salloum et al. 2020). A comprehensive review of 14 studies indicates that microneedle can significantly enhance vitiligo lesions, especially when combined with topical therapies. Microneedle combined with topical treatments such as tacrolimus, 5-fluorouracil (5-FU), calcipotriol, betamethasone, and narrowband ultraviolet B (NB-UVB) has shown greater efficacy than microneedle alone (Salloum et al. 2020). The procedure facilitates the permeation of topical agents into the dermis, stimulates melanocyte activity, and then promotes neocollagenesis. A study reported that microneedle combined with 5-FU resulted in a 3.8-fold greater repigmentation response compared to microneedle alone. Similarly, microneedle with calcipotriol and betamethasone was found to be more effective than with tacrolimus (60% vs. 32%) (Starlin Chellathurai et al. 2024; Chhabra et al. 2021).

9.8 Conclusion

Microneedle array (MNA) technology is a big step forward in transdermal drug delivery. It offers less invasive ways to treat many illnesses, such as cancer, diabetes, and skin problems. Traditional therapeutic and diagnostic procedures have been revolutionized by its capacity to tailor localized therapies, decrease systemic adverse effects, and improve patient compliance. MNAs have demonstrated considerable potential, especially in the therapeutic treatment of diabetes and cancer, by enabling the direct delivery of therapeutic agents, hence increasing the effectiveness of medications with low bioavailability or frequent administration requirements. Applications in wound healing, vaccine administration, and cosmetic dermatology further highlight this technology's adaptability and potency.

Nonetheless, there are difficulties in the creation and application of MNAs. There are still many obstacles to overcome, including complicated manufacturing processes, planned medication release, safety in repeated applications, and possible restrictions when treating deeper tissue diseases. Despite these obstacles, research is still being done to improve MNA technology. Notable developments in this area include 3D printing, polymeric microneedles, and responsive microneedle systems. MNA-based drug delivery systems have enormous potential to transform therapeutic methods in a variety of medical specialties. They are a better option for drug delivery than traditional techniques because of their exceptional capacity to deliver precise dosages straight to the site of action.

9.8.1 Future Recommendation

Optimization of Drug Release Profiles: Particularly for cancer therapy and chronic illnesses like diabetes, future research should concentrate on refining the microneedle-based drug delivery systems' sequential release and dosage schedules. This would entail creating increasingly complex mechanisms that assure steady, regulated drug release.

Design of Safe and Reusable MNAs: Current restrictions on the safety of repeated applications should be addressed by advancements in hydrogel-forming and hollow microneedles. Research on materials and improved biocompatibility are required to provide safe, repetitive use without endangering patient safety.

Extension of Deeper-Lying Cancer Treatment: Future research should investigate novel approaches to deliver drugs to deeper tissues, possibly by integrating microneedles with other technologies like ultrasound or iontophoresis.

Closed-Loop Systems and Smart Microneedles: Treatment outcomes will be significantly enhanced by the development of smart microneedle devices that can self-regulate drug delivery in response to physiological inputs, like blood sugar levels in diabetic patients. More investigation into closed-loop delivery methods might ensure more accurate, patient-specific treatment plans.

Clinical Trials and Standardization: To confirm the long-term safety and effectiveness of microneedle arrays across a range of applications, future research should involve more thorough clinical trials. To ensure the wide, secure, and efficient usage of microneedle technologies in clinical practice, it is essential to standardize manufacturing and application processes.

References

Ablon G (2018) Safety and effectiveness of an automated microneedling device in improving the signs of aging skin. J Clin Aesthet Dermatol 11(8):29–34

Adepu S, Ramakrishna S (2021) Controlled drug delivery systems: current status and future directions. Molecules 26(19):5905

Akhtar N, Singh V, Yusuf M, Khan RA (2020) Non-invasive drug delivery technology: development and current status of transdermal drug delivery devices, techniques and biomedical applications. Biomed Tech (Berl) 65(3):243–272

Aldawood FK, Andar A, Desai S (2021) A comprehensive review of microneedles: types, materials, processes, characterizations and applications. Polymers 13(16):2815

Alimardani V, Abolmaali SS, Yousefi G, Rahiminezhad Z, Abedi M, Tamaddon A et al (2021) Microneedle arrays combined with nanomedicine approaches for transdermal delivery of therapeutics. J Clin Med 10(2)

Alkilani AZ, McCrudden MT, Donnelly RF (2015) Transdermal drug delivery: innovative pharmaceutical developments based on disruption of the barrier properties of the stratum corneum. Pharmaceutics 7(4):438–470

Amarnani R, Shende P (2021) Microneedles in diagnostic, treatment and theranostics: an advancement in minimally-invasive delivery system. Biomed Microdevices 24(1):4

Aung NN, Ngawhirunpat T, Rojanarata T, Patrojanasophon P, Opanasopit P, Pamornpathomkul B (2019) HPMC/PVP dissolving microneedles: a promising delivery platform to promote trans-epidermal delivery of alpha-arbutin for skin lightening. AAPS PharmSciTech 21(1):25

Avcil M, Çelik A (2021) Microneedles in drug delivery: progress and challenges. Micromachines 12(11):1321

Bao L, Park J, Bonfante G, Kim B (2022) Recent advances in porous microneedles: materials, fabrication, and transdermal applications. Drug Deliv Transl Res 12(2):395–414

Bauleth-Ramos T, El-Sayed N, Fontana F, Lobita M, Shahbazi M-A, Santos HA (2023) Recent approaches for enhancing the performance of dissolving microneedles in drug delivery applications. Mater Today 63:239–287

Bhatnagar S, Kumari P, Pattarabhiran SP, Venuganti VVK (2018) Zein microneedles for localized delivery of chemotherapeutic agents to treat breast cancer: drug loading, release behavior, and skin permeation studies. AAPS PharmSciTech 19(4):1818–1826

Brogden NK, Milewski M, Ghosh P, Hardi L, Crofford LJ, Stinchcomb AL (2012) Diclofenac delays micropore closure following microneedle treatment in human subjects. J Control Release 163(2):220–229
Cárcamo-Martínez Á, Mallon B, Domínguez-Robles J, Vora LK, Anjani QK, Donnelly RF (2021) Hollow microneedles: a perspective in biomedical applications. Int J Pharm 599:120455
Chen BZ, He MC, Zhang XP, Fei WM, Cui Y, Guo XD (2022) A novel method for fabrication of coated microneedles with homogeneous and controllable drug dosage for transdermal drug delivery. Drug Deliv Transl Res 12(11):2730–2739
Chhabra S, Chahar YS, Singh A (2021) A comparative study of microneedling combined with topical 5-fluorouracil versus microneedling alone in treatment of localized stable vitiligo. Indian J Dermatol 66(5):574
Desai VM, Priya S, Gorantla S, Singhvi G (2023) Revolutionizing therapeutic delivery with microneedle technology for tumor treatment. Pharmaceutics 15(1):14
Duarah S, Sharma M, Wen J (2019) Recent advances in microneedle-based drug delivery: special emphasis on its use in paediatric population. Eur J Pharm Biopharm 136:48–69
Gera AK, Burra RK (2022) The rise of polymeric microneedles: recent developments, advances, challenges, and applications with regard to transdermal drug delivery. J Funct Biomater 13(2):81
Glover K, Mishra D, Gade S, Vora LK, Wu Y, Paredes AJ et al (2023) Microneedles for advanced ocular drug delivery. Adv Drug Deliv Rev 201:115082
Halder J, Gupta S, Kumari R, Gupta GD, Rai VK (2021) Microneedle array: applications, recent advances, and clinical pertinence in transdermal drug delivery. J Pharm Innov 16(3):558–565
He X, Sun J, Zhuang J, Xu H, Liu Y, Wu D (2019) Microneedle system for transdermal drug and vaccine delivery: devices, safety, and prospects. Dose Response 17(4):1559325819878585
Hosangadi D, Warmbrod KL, Martin EK, Adalja A, Cicero A, Inglesby T et al (2020) Enabling emergency mass vaccination: innovations in manufacturing and administration during a pandemic. Vaccine 38(26):4167–4169
Hou X, Li J, Hong Y, Ruan H, Long M, Feng N et al (2023) Advances and prospects for hydrogel-forming microneedles in transdermal drug delivery. Biomedicines 11(8):2119
Iachina I, Eriksson AH, Bertelsen M, Petersson K, Jansson J, Kemp P et al (2023) Dissolvable microneedles for transdermal drug delivery showing skin penetration and modified drug release. Eur J Pharm Sci 182:106371
Ingrole RSJ, Gill HS (2019) Microneedle coating methods: a review with a perspective. J Pharmacol Exp Ther 370(3):555–569
Ita K (2015) Transdermal delivery of drugs with microneedles-potential and challenges. Pharmaceutics 7(3):90–105
Ito Y, Kashiwara S, Fukushima K, Takada K (2011) Two-layered dissolving microneedles for percutaneous delivery of sumatriptan in rats. Drug Dev Ind Pharm 37(12):1387–1393
Jeong HR, Lee HS, Choi IJ, Park JH (2017) Considerations in the use of microneedles: pain, convenience, anxiety and safety. J Drug Target 25(1):29–40
Juhasz MLW, Cohen JL (2020) Microneedling for the treatment of scars: an update for clinicians. Clin Cosmet Investig Dermatol 13:997–1003
Jung JH, Jin SG (2021) Microneedle for transdermal drug delivery: current trends and fabrication. J Pharm Investig 51(5):503–517
Junsuwan N, Manuskiatti W, Phothong W, Wanitphakdeedecha R (2021) Fractional CO(2) laser-assisted botulinum toxin type a delivery for the treatment of primary palmar hyperhidrosis. Lasers Med Sci 36(1):233–236
Khan S, Hasan A, Attar F, Babadaei MMN, Zeinabad HA, Salehi M et al (2021) Diagnostic and drug release systems based on microneedle arrays in breast cancer therapy. J Control Release 338:341–357
Khare N, Shende P (2021) Microneedle system: a modulated approach for penetration enhancement. Drug Dev Ind Pharm 47(8):1183–1192
Kim YC, Park JH, Prausnitz MR (2012) Microneedles for drug and vaccine delivery. Adv Drug Deliv Rev 64(14):1547–1568

Kommareddy S, Baudner BC, Bonificio A, Gallorini S, Palladino G, Determan AS et al (2013) Influenza subunit vaccine coated microneedle patches elicit comparable immune responses to intramuscular injection in Guinea pigs. Vaccine 31(34):3435–3441

Larrañeta E, Lutton REM, Woolfson AD, Donnelly RF (2016) Microneedle arrays as transdermal and intradermal drug delivery systems: materials science, manufacture and commercial development. Mater Sci Eng R Rep 104:1–32

Lee JW, Park JH, Prausnitz MR (2008) Dissolving microneedles for transdermal drug delivery. Biomaterials 29(13):2113–2124

Li J, Liu B, Zhou Y, Chen Z, Jiang L, Yuan W et al (2017) Fabrication of a Ti porous microneedle array by metal injection molding for transdermal drug delivery. PLoS One 12(2):e0172043

Liang L, Chen Y, Zhang BL, Zhang XP, Liu JL, Shen CB et al (2020) Optimization of dip-coating methods for the fabrication of coated microneedles for drug delivery. J Drug Deliv Sci Technol 55:101464

Lin Z, Zheng K, Zhong J, Zheng X (2023) Advances in microneedle-based therapy for bone disorders. Biomed Pharmacother 165:115013

Luo X, Yang L, Cui Y (2023) Microneedles: materials, fabrication, and biomedical applications. Biomed Microdevices 25(3):20

Lyu S, Dong Z, Xu X, Bei HP, Yuen HY, James Cheung CW et al (2023) Going below and beyond the surface: microneedle structure, materials, drugs, fabrication, and applications for wound healing and tissue regeneration. Bioact Mater 27:303–326

Mansoor I, Eassa HA, Mohammed KHA, Abd El-Fattah MA, Abdo MH, Rashad E et al (2022) Microneedle-based vaccine delivery: review of an emerging technology. AAPS PharmSciTech 23(4):103

McAlister E, Kirkby M, Domínguez-Robles J, Paredes AJ, Anjani QK, Moffatt K et al (2021) The role of microneedle arrays in drug delivery and patient monitoring to prevent diabetes induced fibrosis. Adv Drug Deliv Rev 175:113825

McNamee M, Wong S, Guy OJ, Sharma S (2023) Microneedle technology for potential SARS-CoV-2 vaccine delivery. Expert Opin Drug Deliv 20:799

Mdanda S, Ubanako P, Kondiah PPD, Kumar P, Choonara YE (2021) Recent advances in microneedle platforms for transdermal drug delivery technologies. Polymers 13(15):2405

Nazary Abrbekoh F, Salimi L, Saghati S, Amini H, Fathi Karkan S, Moharamzadeh K et al (2022) Application of microneedle patches for drug delivery; doorstep to novel therapies. J Tissue Eng 13:20417314221085390

Nguyen HX, Nguyen CN (2023) Microneedle-mediated transdermal delivery of biopharmaceuticals. Pharmaceutics 15(1):277

O'Mahony C, Sebastian R, Tjulkins F, Whelan D, Bocchino A, Hu Y et al (2023) Hollow silicon microneedles, fabricated using combined wet and dry etching techniques, for transdermal delivery and diagnostics. Int J Pharm 637:122888

Omidian H, Dey Chowdhury S (2024) Swellable microneedles in drug delivery and diagnostics. Pharmaceuticals (Basel) 17(6):791

Park J-H, Allen MG, Prausnitz MR (2005) Biodegradable polymer microneedles: fabrication, mechanics and transdermal drug delivery. J Control Release 104(1):51–66

Pere CPP, Economidou SN, Lall G, Ziraud C, Boateng JS, Alexander BD et al (2018) 3D printed microneedles for insulin skin delivery. Int J Pharm 544(2):425–432

Pires LR, Vinayakumar K, Turos M, Miguel V, Gaspar J (2019) A perspective on microneedle-based drug delivery and diagnostics in paediatrics. J Pers Med 9(4):49

Roxhed N, Samel B, Nordquist L, Griss P, Stemme G (2008) Painless drug delivery through microneedle-based transdermal patches featuring active infusion. IEEE Trans Biomed Eng 55(3):1063–1071

Ruby PK, Pathak S, Aggarwal D (2014) Critical attributes of transdermal drug delivery system (TDDS)—a generic product development review. Drug Dev Ind Pharm 40(11):1421–1428

Sabri AHB, Anjani QK, Utomo E, Ripolin A, Donnelly RF (2022) Development and characterization of a dry reservoir-hydrogel-forming microneedles composite for minimally invasive delivery of cefazolin. Int J Pharm 617:121593

Sadeqi A, Kiaee G, Zeng W, Rezaei Nejad H, Sonkusale S (2022) Hard polymeric porous microneedles on stretchable substrate for transdermal drug delivery. Sci Rep 12(1):1853
Salih ARC, Farooqi HMU, Amin H, Karn PR, Meghani N, Nagendran S (2024) Hyaluronic acid: comprehensive review of a multifunctional biopolymer. Future J Pharm Sci 10(1):63
Salloum A, Bazzi N, Maalouf D, Habre M (2020) Microneedling in vitiligo: a systematic review. Dermatol Ther 33(6):e14297
Seetharam AA, Choudhry H, Bakhrebah MA, Abdulaal WH, Gupta MS, Rizvi SMD et al (2020) Microneedles drug delivery systems for treatment of cancer: a recent update. Pharmaceutics 12(11):1101
Shriky B, Babenko M, Whiteside BR (2023) Dissolving and swelling hydrogel-based microneedles: an overview of their materials, fabrication, characterization methods, and challenges. Gels 9(10):806
Singh A, Yadav S (2016) Microneedling: advances and widening horizons. Indian Dermatol Online J 7(4):244–254
Sitohang IBS, Sirait SAP, Suryanegara J (2021) Microneedling in the treatment of atrophic scars: a systematic review of randomised controlled trials. Int Wound J 18(5):577–585
Srinivasan PN, Singaravelu R (2017) Solid silicon microneedles for drug delivery applications. Int J Adv Manuf Technol 93:407
Starlin Chellathurai M, Mahmood S, Mohamed Sofian Z, Wan Hee C, Sundarapandian R, Ahamed HN et al (2024) Biodegradable polymeric insulin microneedles—a design and materials perspective review. Drug Deliv 31(1):2296350
Sultana N, Ali A, Waheed A, Jabi B, Yaqub Khan M, Mujeeb M et al (2023) Dissolving microneedle transdermal patch loaded with risedronate sodium and ursolic acid bipartite nanotransfersomes to combat osteoporosis: optimization, characterization, in vitro and ex vivo assessment. Int J Pharm 644:123335
Tian R, Wang X, Li Y, Zhang L, Wen X (2024) Application of microneedling in photodynamic therapy: a systematic review. Photodiagn Photodyn Ther 46:104016
Tucak A, Sirbubalo M, Hindija L, Rahić O, Hadžiabdić J, Muhamedagić K et al (2020) Microneedles: characteristics, materials, production methods and commercial development. Micromachines 11(11):961
Waghule T, Singhvi G, Dubey SK, Pandey MM, Gupta G, Singh M et al (2019) Microneedles: a smart approach and increasing potential for transdermal drug delivery system. Biomed Pharmacother 109:1249–1258
Wang B, Pei J, Xu S, Liu J, Yu J (2023) Recent advances in mRNA cancer vaccines: meeting challenges and embracing opportunities. Front Immunol 14:1246682
Wu T, You X, Chen Z (2022) Hollow microneedles on a paper fabricated by standard photolithography for the screening test of prediabetes. Sensors (Basel) 22(11):4253
Xie X, Pascual C, Lieu C, Oh S, Wang J, Zou B et al (2017) Analgesic microneedle patch for neuropathic pain therapy. ACS Nano 11(1):395–406
Xing M, Wang X, Zhao L, Zhou Z, Liu H, Wang B et al (2021) Novel dissolving microneedles preparation for synergistic melasma therapy: combined effects of tranexamic acid and licorice extract. Int J Pharm 600:120406
Xu Y, Ma R, Juliandri J, Wang X, Xu B, Wang D et al (2017) Efficacy of functional microarray of microneedles combined with topical tranexamic acid for melasma: a randomized, self-controlled, split-face study. Medicine 96(19):e6897
Xu J, Danehy R, Cai H, Ao Z, Pu M, Nusawardhana A et al (2019) Microneedle patch-mediated treatment of bacterial biofilms. ACS Appl Mater Interfaces 11(16):14640–14646
Younis N, Puigmal N, Kurdi AE, Badaoui A, Zhang D, Morales-Garay C et al (2024) Microneedle-mediated delivery of immunomodulators restores immune privilege in hair follicles and reverses immune-mediated alopecia. Adv Mater 36(31):e2312088
Zahn J, Talbot N, Liepmann D, Pisano A (2000) Microfabricated polysilicon microneedles for minimally invasive biomedical devices. Biomed Microdevices 2:295–303
Zhang Y, Yu J, Kahkoska AR, Wang J, Buse JB, Gu Z (2019) Advances in transdermal insulin delivery. Adv Drug Deliv Rev 139:51–70

Zhang T, Sun B, Guo J, Wang M, Cui H, Mao H et al (2020) Active pharmaceutical ingredient poly(ionic liquid)-based microneedles for the treatment of skin acne infection. Acta Biomater 115:136–147

Zhang W, Zhang W, Li C, Zhang J, Qin L, Lai Y (2022) Recent advances of microneedles and their application in disease treatment. Int J Mol Sci 23(5)

Zhang T, Sun B, Ding W, Zhang C, Yin X, Wang B et al (2023) Combining rapid degrading microneedles with slow-released drug delivery system for the treatment of alopecia areata. Chem Eng J 471:144351

Zhang C, Song X, Ma C, Wang M, Deng H, Yuan M et al (2024) Hyaluronic acid dissolving microneedle patch loaded with tranexamic acid for melasma treatment. Int J Biol Macromol 270(Pt 2):132255

Zheng Z, Diaz-Arévalo D, Guan H, Zeng M (2018) Noninvasive vaccination against infectious diseases. Hum Vaccin Immunother 14(7):1717–1733

Zheng Y, Ling Z, Li Z, Zhao P, Wen X, Qu F et al (2023) A rapidly dissolvable microneedle patch with tip-accumulated antigens for efficient transdermal vaccination. Macromol Biosci 23(12):e2300253

Zhou Y, Jia L, Zhou D, Chen G, Fu Q, Li N (2023) Advances in microneedles research based on promoting hair regrowth. J Control Release 353:965–974

Ziesmer J, Tajpara P, Hempel NJ, Ehrström M, Melican K, Eidsmo L et al (2021) Vancomycin-loaded microneedle arrays against methicillin-resistant staphylococcus aureus skin infections. Adv Mater Technol 6(7):2001307

Minimally Invasive Drug Delivery Systems for Delivery of Biological Macromolecules

10

Subhajit Dutta, Shanmugam Ramaswamy, L. Priyanka Dwarampudi, Malakapogu Ravindra Babu, S. P. Dhanabal, Mirunalini Gobinath, and Parikshit Roychowdhury

Abstract

Painless methods that provide alternatives to oral, nasal, ocular, rectal, or pulmonary routes for therapeutic drug delivery are referred to as minimally invasive drug administration. Polyvinylpyrrolidone polymer with varying molecular weights was used to create microneedles. The polymer contained three model molecules with increasing MW: fluorescein sodium and fluorescein isothiocyanate–dextrans. Biomacromolecule transdermal distribution without needles is still a difficulty. This study develops a transdermal delivery platform utilizing nontoxic fluorocarbon-modified chitosan (FCS) to facilitate the noninvasive, highly effective distribution of biomacromolecules, such as antigens and antibodies. The resulting nanocomplexes demonstrate efficient transdermal penetration through both transappendageal and intercellular pathways. Compared to intravenous injection, noninvasive transdermal administration of immune checkpoint inhibitor antibodies decreases systemic toxicity and strengthens immune responses against melanoma in female mice. In order to regulate glucose levels

S. Dutta · S. Ramaswamy (✉)
TIFAC CORE in Herbal Drugs, Department of Pharmacognosy, JSS College of Pharmacy, JSS Academy of Higher Education and Research, Ooty, Tamil Nadu, India

L. P. Dwarampudi · S. P. Dhanabal · M. Gobinath
Department of Pharmacognosy, JSS College of Pharmacy, JSS Academy of Higher Education and Research, Ooty, Tamil Nadu, India

M. R. Babu
Department of Pharmacy, School of Medical and Allied Sciences, Galgotias University, Greater Noida, Uttar Pradesh, India

P. Roychowdhury
Department of Pharmaceutics, JSS College of Pharmacy, JSS Academy of Higher Education and Research, Ooty, Tamil Nadu, India

M. R. Babu et al. (eds.), *Transdermal Applications of Minimally Invasive Drug Delivery Systems*, https://doi.org/10.1007/978-981-96-6275-3_10

in the blood and time, there has been a great deal of research interest in creating new insulin administration technologies that can also improve patient compliance by using noninvasive delivery methods. This chapter's focus is on different, less intrusive medication delivery methods, such as biological macromolecules. This chapter also discusses several smart stimuli-responsive techniques, kinds of minimally invasive medication delivery, case studies, and applications. Lastly, a thorough discussion of the benefits, drawbacks, and implications of each minimally invasive method for delivering biological macromolecules is presented.

10.1 Introduction

Minimally invasive drug administration refers to painless techniques that offer substitute routes for therapeutic drug delivery via oral, nasal, ocular, rectal, or pulmonary. The most common method of treating illnesses has been oral delivery. Still, some barriers—like the pulmonary and blood barrier—affect how well medications are delivered to their intended locations. The brain-to-brain barrier (BBB) is a sophisticated anatomical and physiological membrane preventing the entry of chemicals into the brain. As such, it is very difficult to deliver medications to the brain efficiently. Similarly, when it comes to pulmonary medication delivery, there are three main obstacles. Developments on medical science create benefits that render respiratory delivery a great option for medication administration treatments. One formulation that shows great promise for getting beyond obstacles to medication distribution is nanoformulation. As prospective feasible formulations that increase therapeutic efficiency, nanocarriers are crucial research targets for preclinical and clinical use (Zhang et al. 2022).

In order to facilitate the minimally invasive drug delivery systems for delivery of biomacromolecule drugs, particularly proteins, a range of physical enhancement devices, such as thermal ablation, electroporation, cavitational ultrasound, and microneedles, as well as novel chemical enhancers like membrane-penetrating peptides, have been developed. These approaches still have a number of drawbacks, despite the possibility of using them to deliver therapeutic proteins among other macromolecules. Membrane-penetrating peptides, for instance, have been shown to facilitate transdermal administration of tiny proteins, like insulin, but at the expense of an inadequate delivery efficiency. They also continue to be ineffectual for proteins with enormous molecular weights. In recent times, transdermal delivery has made extensive use of microneedles, or patches with several tiny needles. These patches have demonstrated considerable promise for delivering influenza and insulin. Microneedle patch production and quality assurance would be challenging, nevertheless, particularly if biomacromolecule payloads were used. Furthermore, the microneedles may still cause some skin injury, which could raise the risk of infection (Nazary Abrbekoh et al. 2022).

Owing to the previously mentioned factors as well as the quick development of pharmaceutical and medical technologies, the twenty first-century pharmaceutical development trend has already started to change from focusing only on creating new medications to creating novel drug delivery systems in addition to new drugs. Natural macromolecules with significant medicinal and pharmacological potential include lipids, proteins, peptides, and carbohydrates. Numerous human illnesses have been treated using these macromolecules as biomedicines. In this sense, biomedicine is a subfield of medicine that applies physiological and biological concepts to clinical settings. It concentrates on those aspects of the biological and chemical sciences that have application in the medical field. Following from the previous discourse, the current chapter's goal is to recognize and emphasize the importance of delivering the biological macromolecules by using minimally invasive drug delivery systems. But the main barriers included in this system are mucus layer and epithelial barrier (Molnar and Gair 2015).

10.2 Biological Macromolecules

10.2.1 Types of Macromolecules (Mandelkern 2012)

Macromolecules	Formula	Monomer	Uses	Examples
Nucleic acids	CHONP	Nucleotides	Genetic data	DNA, RNA
Carbohydrates	C:H:O	Monosaccharides	Structure and energy storage	Glucose, starch, fructose, cellulose, glycogen
Lipids	C:H:O	Fatty acid and glycerol	Water repulsion, chemical messengers, energy storage, and protection	Oil, beeswax, cholesterol, butter
Proteins	CHON	Amino acids	Receptors, storage, enzyme, transport, structural, contractile, defensive	Hormones, enzymes

10.2.2 Applications of Macromolecules

10.2.2.1 Application in Material Science

Due to the recent realization that microchip-grade silicon supplies are unlikely to be sufficient to meet the increasing demand for data storage, researchers are exploring potential alternatives to the current data storage medium. This is necessary because new data is created every day. Given that nature has evolved over the previous billion years to optimize the method of keeping a full genome within these biopolymers and that DNA sequencing has become a hot topic for decades, it should come as no surprise that DNA represents a few of the alternatives. As an alternative, the synthesis of longer, homogeneous macromolecules can also be used to expand the capacity of data storage. Sequencing, or the accurate and error-free readout of this data, is as important as the level of molecules' data storage, which is a feat in and of itself. In nature, interactions between molecules to produce higher-order structures are ubiquitous. A number of theoretical models explaining and forecasting the behavior of tiny amphiphilic compounds based on supramolecular interactions and molecular geometry have been established via comprehensive studies of their molecular assembly. Additionally, block copolymers may exhibit a propensity to form distinct morphologies upon assembly. This phenomenon is caused by a conflict of the various components, which separate to lower the interfacial energy (Yuan et al. 2024). The building blocks of life as we know it are biomacromolecules with a specific monomer sequence. Although tremendous progress has been made in recent decades, folding protein structure prediction from amino acid sequence is still difficult. Because DNA nanotechnology was developed due to the predictable nature of this base pairing, nucleic acid chains were assembled to create sophisticated nanoscale structures through thoughtful and deliberate sequence design. By utilizing a small number of building components, biomacromolecules can execute a wide range of impressive and complicated tasks, including molecular recognition, transportation, storage of data, and catalysis, through the folding process into secondary, tertiary, and quaternary structures. Enzymes have the extraordinary ability to catalyze all chemical processes that are essential to life, which is one of their most fascinating characteristics. Because of their precise control over their core structure, these biomacromolecules fold into secondary structures that give rise to their distinctive properties. Catalytic pockets, which are made up of several reactive groups condensed in a small space, are actually created when these three-dimensional structures expand. Developing novel asymmetric catalysts with a variety of uses could be highly beneficial, given the majority of building blocks utilized in the manufacturing of pharmaceuticals require exact control over the stereochemistry (Aksakal et al. 2021).

10.2.2.2 Application in Life Science

Carbohydrates are essential to numerous biological processes in living things, such as adhesion, inflammation, immunological responses, and cell identification and growth regulation. The requirement for extreme cell type selectivity in the carbohydrate-cell interaction, particularly in biomedical treatments, is one of the primary challenges in order to avoid unfavorable reactions. Despite the fact that antibiotic-resistant bacteria have greatly increased human longevity in the last century, they nevertheless pose one of the biggest hazards to human health (Chevolot et al. 2009). These peptides are quite short; in fact, 88% of the 3180 reported structures have a maximum of 50 amino acids (Marshall and Arenas 2003).

10.2.2.3 Other Applications

Meier and colleagues used conjugated that is rod-like pentamers derived from oligo(phenylene ethynylene)s to apply the sequence-definition idea of iterative synthesis. The resultant macromolecules varied more significantly in their hydrodynamic volume and thermal conductivity than in their photophysical characteristics, which were only marginally different. The investigation might be expanded to include longer sequences than a pentamer, which might further affect the photophysical characteristics. Notwithstanding the positive outcomes, new synthetic approaches must be created in order to employ completely conjugated macromolecules in practical electronic applications due to their extremely low overall yields (3.2%) (Hahn et al. 2021).

10.3 Minimally Invasive Delivery Techniques

10.3.1 Transdermal Patches and Devices for Drug Delivery

Many TDDS patches have been developed based on design considerations arising from various formulation factors and related physical-chemical and envisioned biological qualities, including adhesives and excipient characteristics. These patches include single- and multiple-layered devices, vapor patches, drug and polymer matrix characteristic-based items, and reservoir-based delivery systems (Hahn et al. 2021). A transdermal patch is made up of a liner that comes into direct touch with the medication, prevents drug loss during storage, and needs to be taken off before usage. Following the liner's removal, a film of adhesive remains, acting as both a platform for the patch to stick to the skin and an adhesive for the patch's component parts. The membrane provides the required support for the patch and loaded medication while regulating drug release from the container and complicated patches to the skin and throughout the circulatory system over a predetermined period of time. The backing membrane isolates the external surroundings from the TDDS. In addition to acting as the substance that adheres the patch to the surface of the skin, an adhesive also operates as the solution that contains the medication. The biggest problem with this procedure is the minimal amount of medication in the glue, which alters the adhesive's adherence. Therefore, to provide adequate, continuous doses of the drug, either the width of the patch needs to be raised or the patient needs to reapply the patch multiple times if substantial amounts of the drug are to be supplied. To address this problem, a variety of pharmaceutical-grade, biocompatible, and biodegradable materials—also known as enhancers—have been used with the formulation matrix to improve the drug's absorption and capacity to permeate the skin from the TDDS device (Wong et al. 2023). Enhancers facilitate the drug's entry into the bloodstream with a transdermal patch that is applied topically and contains the exact dosage. The rate of drug release in diffusion-based transdermal drug delivery systems is controlled by the polymeric membrane employed in the sheathing system. This rate is dependent on various factors such as adhesive properties, permeability coefficient, membrane thickness, and polymer properties. The drug compartment is totally submerged in the metallic plastic seal, with the polymeric membrane controlling the release of the drug in line with the following expression:

$$\text{The drug release rate}\left(\frac{dq}{dt}\right)=\frac{C_R}{\left(\frac{1}{Pm}\right)+\left(\frac{1}{Pa}\right)}$$

where *Pm* and *Pa* stand for the membrane's and the adhesive's respective permeability coefficients and C_R stands for the drug concentration. As a result, the membrane and adhesive permeability control the release of the drug, even at high drug concentrations in the reservoir (Ramadon et al. 2022).

10.3.2 Techniques for Improving Payload Delivery

10.3.2.1 Improvements to Structure-Based Delivery

Drug delivery enhancement strategies that are based on structure or chemistry make use of structural variety, changes, and the addition of novel structural scaffolds or entities throughout the TDDS's fabrication and formulation stages. Both natural and synthetic polymers have been widely used to reach desired delivery levels, and chemical boosters have been added to provide a net improved delivery experience (Mansour et al. 2023).

10.3.2.2 Transformations to Prodrug Strategies

The parent drug molecule's structural alterations are used in the prodrug approach's chemical modification process. Most of the time, it converts the drug into a lipophilic entity, a bioreversible compound that is biotransformed into the target drug or its active structural form within the biological system. This method makes it easier and faster for the drug's new derivative to pass through the epidermal barrier. The method gives the parent drug molecule the proper physicochemical, pharmacological, and pharmacokinetic properties, which improves the drug's permeability, therapeutic value, and efficacy in the transdermal drug delivery system. The semisynthetic prodrug conversion alters the bioactivity of the drug and, depending on the circumstance, can convert an active drug into an inactive version that will either be converted back to the active state by the body when it reaches the site of action or be modified by skin enzymes. The transition is simple for drug candidates containing hydroxyl, carboxyl, amine, and thiol groups. The most common conversions produce derivatives of ester, amide, and methoxyl, which convert hydrophilic drugs into skin-permeable, biocompatible lipophilic substances. Changing the number of hydrophilic groups—mainly hydroxyl and carboxyl—to their ester, amide, or ether derivatives is another way to control the quantity of lipophilicity added to the original therapeutic molecule (N'Da 2014). PEGylation, alkyl chain lengthening, and total chemical conversion to a new form are more methods for creating prodrugs. The drug candidate undergoes a modification to alter its physicochemical characteristics in order to maximize transdermal dispersion. The prodrug is converted back into the original drug by the skin's enzymes. However, permeability-controlling factors influenced through the pharmacokinetic, physiological/biological, and physicochemical features of the resultant prodrug must be considered when converting generally hydrophilic medications into comparatively more lipophilic ones (Markovic et al. 2020).

10.3.2.3 Improvements in Electromechanical, Physical, and Acoustic Delivery

The inability to increase drug transport, rate variation, poor amount of delivery, skin irritation, limited efficacy, safety issues, and undesired levels of skin problems for easier delivery were some of the reasons why passive enhancers needed to be replaced. These include an array of chemical enhancers that can quickly and reliably penetrate deeper into the skin to deliver medications in substantially higher concentrations. An active method of permeability enhancement is the application of external energy to drive drug movement across the skin or to physically disrupt SC (Stratum Corneum), which is the outermost layer of the skin. This allows relatively large amounts of medicines and biomacromolecules to be delivered. The technique helps to shorten the interval between the administration of TDDS and the start of the treatment at the site and provides greater control over the delivery. Transdermal medication delivery device performance has reached new heights thanks to improvement strategies based on electromechanical, physical, and acoustical principles (Zaid Alkilani et al. 2015).

10.3.2.4 Iontophoresis

With the use of tiny electrodes that remain in contact with the drug and apply a limited value, short-time, and localized milliampere (0.1–1.0 mA/cm^2) current to the skin, the drug can be applied to both the entire and restricted area beneath the skin layers. Delivery of pilocarpine might be used as a demonstration to cause perspiration for diagnosing cystic fibrosis. The iontophoretic administration of lidocaine is a successful illustration of this type of delivery for a rapid onset of anesthesia. Inflammation, cystic fibrosis, and acne are a few ailments where iontophoretic administration has been effectively used. Medicinal substances such as lidocaine, retinoids and steroids, pilocarpine, ketoprofen, and anticholinergics have been administered using this method to provide better, longer-lasting dosages at higher levels of drug release (Yu et al. 2021). Nevertheless, it was discovered that iontophoresis was not carrying HMW drug molecules larger than 7 kDa. Nevertheless, this was made feasible by the employment of a family of chemical permeation-enhancing compounds and liposomal formulations containing insulin and BSA delivered via microneedle delivery. Additionally, the method gave the roots of hair and nails higher dosages of the medication. An hour-long, single-use transdermal patch with a rapid beginning of therapeutic activity that delivered sumatriptan was approved by the FDA to treat migraines. A 2 mA current allowed for 3 h of uninterrupted delivery. Iontophoresis does not involve mechanical penetration; yet, it does have very slight, transient impacts on the skin's structure. The drug delivery through this technique is controlled by a number of factors, including the pH of the drug solution, the drug's molecular size, hydrophilicity, preferred electrode type (Ag/AgCl), current strength (0.5 mA/cm^2), application time, types of currents, etc. Iontophoresis is also utilized for insulin delivery, blood glucose monitoring, and cystic fibrosis diagnosis. Recently delivered include ribonuclease A, cytochrome *c*, human basic fibroblast growth factor, and other HMW biomacromolecular products greater than 12 kDa (Vikelis et al. 2015).

10.3.2.5 Tape Stripping

This method, which removes the skin layer, SC, involves applying sticky tape repeatedly. Considerations include the patient's age, the depth of the epidermal layer, the skin site, the amount and makeup of the lipid substances at the site of the skin, the skin's surface pH and the level of moisture, the force used to remove the tape, the length of time the tape was present, and the tape's effective adhesion, which all affect how well the tape removes the skin layer and then delivers the medication (Olesen et al. 2019).

10.3.2.6 Electroporation

This technique applies high-voltage, 5–500 V, millisecond electrical pulses to the skin to increase permeability, which in turn creates tiny pores in the SC, aiding in the drug's diffusion. Electrodes placed closely apart introduce electrical pulses for safe and painless delivery. This technique is extremely safe and painless for giving HMW drugs, including oligonucleotide, antiangiogenic metaargidin peptides, and the negatively charged anticoagulant heparin, as well as LMW substances, such as doxorubicin, mannitol, timolol, or calcein. The method also helps with gene transfer through a two-step process that entails making the skin permeable first and then electrophoresis. The administration of macromolecular medications has been significantly improved by the use of microneedle electroporation. Nonetheless, the drawbacks include limited distribution; severe disruption of cells, which can infrequently cause denaturation of protein treatments; damage to heat-labile drugs; and cell death, as well as addition of biomacromolecular therapeutic elements (Escobar-Chávez et al. 2009).

10.3.2.7 Microneedles

Microneedles combine the transdermal patch, or TDDS device, with the pattern-using, convenient, and effective features of traditional injection needles. One of the least invasive forms of TDDS, microneedles operate by piercing the SC without causing pain. Because microneedles create superficial, hydrophilic channels through the layers of the skin, they enable the infusion of more types and higher quantities of drugs. When the pores are blocked by a liquid or film, they remain open for 72 h, but this can be extended to a week. Medication is administered by hollow needle, or using sophisticated manufacturing techniques, drugs can be rendered biodegradable to be stored in the body's tissues to break down on their own. Additionally, vaccinations and medication and vaccine self-administration have been performed with microneedles. Additionally, they have been successfully employed to deliver macromolecular substances like proteins and antigens, LMW heparins, insulin, and some vaccinations, as well as tiny compounds, nanoparticles, and fluid extractions. The method offered lidocaine-coated microneedles a quick onset of 1 min and prolonged delivery for up to 90 min. When inserted into the skin, polyvinylpyrrolidone microneedles made by photopolymerizing *N*-vinylpyrrolidone solution combined with lyophilized vaccinations quickly dissolved. Vaccination using microneedle patches was more effective than intramuscular injections. The procedure allowed for a 1-min rapid start and up to a 90-min delivery period for microneedles coated with lidocaine. Microneedles made from polyvinylpyrrolidone, produced through the photopolymerization of *N*-vinylpyrrolidone solution combined with lyophilized vaccines, disintegrated swiftly upon injection into the skin. Intranasal injections were not as successful as microneedle patches for vaccination. Still, the technique is a helpful substitute for traditional needle-based immunization and can be used to stop pandemic outbreaks and minimize the hazards associated with reusing hypodermic needles. Like a hypodermic syringe and needle, hollow microneedles inactively provide the drug through a process known as "poke and flow," wherein the medicinal solution travels via the hole of the needle. On the other hand, an active delivery system needs a driving force generated by a small pump or gas pressure to apply the drug formulation as a gel or patch at the desired location. With "coat and poke" method, tiny needles laden with medication puncture the skin. The porosity hydrogel-forming microneedles are dissolved using the "poke and release" approach, which allows the medication to diffuse into the bloodstream. Soluble microneedle patches are successfully transferring LMW and HMW drugs in vitro and in vivo by employing the "poke and release" method (Escobar-Chávez et al. 2009; Tucak et al. 2020; Donnelly et al. 2010; Jeong et al. 2021).

10.3.3 Payload Delivery Devices Based on Velocity

Micro-mechanical device delivery improvements are a latecomer to the TDDS delivery improvement landscape. The development of methodologies appeared to be impacted by the requirement for mass delivery in greater quantities and at a faster rate. Velocity-based delivery systems were made possible by the necessity for sufficient quantity, fast-paced distribution, and ability to treat a large population in a short amount of time while maintaining the drug's therapeutic effect and control (Escobar-Chávez et al. 2009; Jeong et al. 2021; Rajalakshmi et al. 2023).

10.3.3.1 Jet Injectors

Velocity-based delivery systems were made possible by the necessity for sufficient quantity, fast-paced distribution, as well as the capacity to quickly treat a sizable population while preserving the medication's therapeutic impact and management. Pneumatic injection, air cannon, and jet gun are some other names for it. It can provide insulin to individuals on an individual basis and is quick and simple to use for mass vaccination campaigns. The Biojector 2000 is a commercially available injector that uses a spring-powered device to deliver prepared medication subcutaneously and intramuscularly in a 1 mm zone. This needle-free, ergonomic syringe system is safe to use, is ideal for single-dose administration, and may be used for various injection purposes. The method has two drawbacks: the nozzle may become contaminated and the jet stream from the prior delivery may reappear in the subsequent dosage for a multipurpose device. Consequently, although the product has been discontinued for some purposes, cautious use is suggested. A different injector device called Intraject ® uses electroporation to administer hydrophilic medicines through the application of low-voltage electrical pulses. It is a pre-filled, single-use, disposable, needle-free pharmaceutical device. The procedure makes tiny holes in the skin through which hydrophilic medications, peptides, and proteins are delivered. The

Implaject® method is painless, quick, dependable, and easy to use. Using a needle-free injection device, this technology has been used to deliver vaccinations, insulin, painkillers, and other proteins. The Crossjet® device, which is also needle-free, is composed of a formulation portion that functions as a syringe, a gas generator for utilizing chemical energy, with a control valve with vents made of polycarbonate. It has been applied to administer human growth hormones, insulin, and Imitrex®, a medication that treats migraines and cluster headaches. The apparatus burns chemicals, which gives it an unpleasant smell (Trimzi and Ham 2021; Trucillo 2021).

10.3.3.2 Powder Jet Devices

The device is a rapid injection that delivers powdered medication at a pace of 600–900 m/s using a fast speed helium gas push. The medication is delivered through cassettes that are sandwiched between two membranes, rupturing the membranes and pushing pressure and gas through the skin in the process. Altering the penetration depth involves adjusting the acceleration of the drug particle, which is solid, within the gas. This gadget causes the least amount of skin damage because it does not use needles and does not bleed, in contrast to other physical and acoustic-based devices. The technique has a quick and long-lasting release capacity and is site-specific (Choi et al. 2023).

10.3.3.3 Needleless Shots

These are reusable instruments that transfer very viscous formulations of drugs over the skin using forces like electrophoresis, Lorentz, shock waves, and high-pressure gas flow. The method prevents needle-related injuries and has addressed the drawbacks associated with traditional shot needle/syringe applications (Nagpal and Ravi 2015).

10.3.4 Controlled Heat-Based Devices and Those Assisted by Heat

Various methods, such as thermally assisted devices, radiofrequency and laser applications, and medicated tattoos, have been documented to improve the delivery of multiple medications (Li et al. 2022).

10.3.4.1 Magnetophoresis

As the applied strength of the magnetic field increased, so did the influence on the drug substance's diffusion flux. The use of a defined intensity magnetic field improves the drug's penetration into the systemic circulation. Lidocaine was administered at three different magnetic field strengths (30, 150, and 300 mT) using a magnetophoretic TDDS patch. Medication molecules are transported more easily thanks to the phenomenon known as magnetokinesis. Furthermore, the drug's octanol/water partition coefficient increased from 13.80 to 25.94 at 300 mT using magnetophoresis. Studies conducted in vivo revealed superior cutaneous bioavailability compared to standard nonmagnetic patches (Murthy et al. 2010).

Thermally assisted drug delivery systems and devices function by the temperature-regulated process that enhances absorptivity and blood circulation at the skin's blood vessel levels, hence improving medication distribution. Chemoreactive oxidative reactions generated by a micro-unit inside the device deliver the necessary quantity and relay the temperature at the designated time. Using a regulated heat application of 43 °C, the amount of medication delivered from a nicotine patch placed on the upper arm of ten healthy nonsmokers increased by up to 13 times. Similar results were observed when high ambient temperature nitroglycerine patches were subjected to. The controlled heat-aided drug delivery system facilitates the absorption of medication by the skin by applying heat. The controlled heat-aided drug delivery systems consist of a tiny heating component that generates heat with restricted duration and intensity by using oxidation reactions and operating on a chemoreactive basis (Hao et al. 2016).

10.3.4.2 Medicated Tattoos

This method makes use of straightforward tattoos that are filled with the necessary medication formulas. Aside from the actual tattooing procedure, it is painless and simple to remove. The technique offers a simple approach to administer drugs. As part of the observable technique of qualitative delivery monitoring, the tattoo's validity time is determined by its color. The hue's colorants play a role in determining the delivery levels. Among the examples are acetaminophen, vitamin C med-tats®, and antibiotics. Lithography and silkscreen printing methods are used to create med-tats®, which are tattoos with multiple layers. The medications dissolve in a carefully regulated fashion after being encapsulated as hydrogel, liposomes, and microemulsion. There are additional options for digital tattoos (Brown et al. 2024).

10.3.4.3 Laser Ablation

In order to generate ablation without causing damage to subsequent skin layers, the skin's ROI is exposed to a laser beam. Drug distribution among lipophilic and hydrophilic substances is enhanced by the removal of the SC. Optimizing delivery is aided by the specific elimination of SC without causing harm to the tissues beneath and the creation of microchannels and water evaporation caused by laser ablation. The elements that determine the efficiency of the approach include the employment of a regulated wavelength, the pulse's duration and penetration strength, the tissue thickness at the site, the pulse's energy, the tissue absorption coefficient, the duration of contact, the pulse repetition rate, and the total number of pulses. Moreover, monochromatic laser beams are an option. It is also useful for administering HMW medications. In human volunteers, a laser pretreatment showed the technique's efficacy in delivering lidocaine quickly and initiating its therapeutic activity quickly (Brown et al. 2024).

10.3.4.4 Thermal Ablation Using Radiofrequency

Using a tiny, needle-shaped microelectrode inserted into the skin, a high-frequency AC (100 kHz) is delivered to create permeation channels in the SC by means of vigorous vibrations, hence ablation of the skin's ROI. Topically applied hydrophilic medicines and macromolecular compounds can be achieved at high frequencies (100–500 kHz) (Lee et al. 2011).

10.3.4.5 Raising and Scraping of the Skin

To improve the penetration of the topically administered medication, this technique entails the direct removal or modification of the skin's outermost layer. Generally speaking, the method—also referred to as micro-scissuining—erodes the skin's impervious outer layer by utilizing tiny, sharp metal grains to form microchannels. In order to make tiny openings for the medicine to enter the skin, metallic, sharp microparticles can also cause skin scratches. There are, however, few concerns regarding safety, and research has been done on the subject of damage to the skin safety during immunotherapy (Hmingthansanga et al. 2022).

10.4 Nanotechnology in MIDDS

In the development of the MIDDS technique, the solid, liquid, and liquid crystalline phase nanoscale devices have been used and 10–50 nm-sized polymeric micelles containing the immunosuppressive calcineurin inhibitor tacrolimus were added to increase the drug's penetration. Tacrolimus lipid nanoparticles in a size range of 20–100 nm demonstrated improved skin targeting. Nano-lipid carriers were used to encapsulate fluticasone propionate. Electrically charged polymeric chitosan nanoparticles, measuring 230 nm, were used to transport hydrocortisone and hydroxytyrosol through the skin. The skin-specific delivery resulted in greater skin selectivity and decreased systemic uptake. Reduced medication demand was demonstrated by nano-delivery. Products such as retinoids, tazarotene, and tretinoin, when encapsulated in polymeric nanocarriers, have been reported to facilitate transdermal drug delivery using nanoscale devices. Nanoparticles coated with 4 nm methotrexate have been utilized to enhance medication absorption by keratinocytes and skin penetration. Additionally, the in vitro penetration of drug-loaded lipid-based nanocarriers was higher than that of non-nano-delivery, and sodium carbonate-mediated nanogels with a diameter of 100 nm also showed improved drug permeation. Additionally, 73 nm solid lipid nanocarriers and cyclosporin-A polymeric nanoparticles have been used to deliver dermatitis. Silver ions loaded with antimicrobials, lipid-complexed at 200 nm, and electrostatically bonded have been administered for atopic dermatitis. Zinc oxide nanoparticles, nC60 fullerenes, 200 nm lipid-core capsules containing clobetasol propionate, 200 nm chitosan nanoparticles containing ketoprofen, and 200–270 nm liquid crystalline nanodispersions containing interleukin (IL)-6 siRNA are a few examples of nano-based skin delivery methods. Successful nano-delivery applications include transdermal immunizations, antiparkinsonian as well as anti-hypertensive medications, various chemotherapeutic agents, and even targeted gene knockdown using siRNA administration to the skin, which could lead to the development of personalized therapeutics (Liu et al. 2023; Mishra et al. 2018).

10.5 Biocompatibility and Safety Concerns

Since the beginning of MIDDS development and commercialization, a number of safety concerns have surfaced. Among the notable issues are those related to immunological safety and sterility of the devices; electromechanical device burns; MIDDS, device, and product preparation issues during in-process manufacturing; adherence to cGMP; health risks associated with long-term continuous use of patches and devices; and the use of patches during MRI procedures that result in burns. Even though some equipment and patches contain metal, finding alternatives for devices that cause problems and burns and removing the metallic part from MRI patches are recommended. Major injuries from metallic elements in devices and electrical injuries are a few of the concerns. Recently, concerns about the safety of MIDDS delivery have been raised in addition to skin toxicity of nanoparticles and other nanocarriers (Siswomihardjo 2016; Huzum et al. 2021).

10.6 Case Studies

Research is underway to enhance the targeting capability of lipid nanoparticles (LNPs) through the use of PEG lipids with distinct termini to conjugate antibodies to the surface of carriers. By loading baicalin into SLN PEGylated in PEG-maleimide, researchers saw that this modified PEG lipid was connected to an OX26 antibody that targets thiolated brain cells via the maleimide-thiol reaction. Following conjugation, the cells' targeting capacity increased. The process of creating PEGylated cubosome-based DDSs that exhibit selectivity toward the EGFR, or epidermal growth factor receptor, was also facilitated by this conjugation procedure. The EDC-NHS process was used in a study to combine tumor-targeting peptides with PEG-NLCs exposed to carboxyl acids. They created potent NLCs that target breast cancer. The study attached BS(PEG)5 and PEG-amine to LNPs, therefore conjugating TLNk, a protein possessing keratinocyte targeting potential. Because of their improved targeting abilities, the modified LNPs exhibited greater rates of skin regeneration in mice with burned skin than the LNPs (Marques et al. 2023; Hald Albertsen et al. 2022).

Poloxamers and poloxamines, two block copolymers (BCPs) made from polyethylene and polypropylene oxides, have been applied to LNP surface modification in addition to PEGylation. Because of their low cytotoxicity and special structure-based amphiphilic capabilities, they are used as excipients in a variety of medications. BCPs are one of these excipients that, like PEGylation, helps to lengthen their period of residence in the body, improve carrier support, and increase LNP bioavailability (Seo et al. 2023).

Research has indicated that BCPs, particularly LNPs with a poloxamer surface modification, enhance the capacity of delivery vehicles to target cells by interacting with plasma proteins. Researchers examined the rates of adhesion between different apolipoproteins (apo) and stabilized SLNs using low molecular weight poloxamers. This validates apo E's strong adsorption, which facilitates uptake over the BBB. Poloxamer 235 was shown to have a remarkably low rate of apo CII adsorption, which hindered the activity of the apo E receptor. Furthermore, poloxamer 235's unimer blocks the bovine brain endothelial cells' glycoprotein P efflux pump, which causes an increase in drug buildup in the cells. Therefore, the best delivery system for brain-targeting DDSs may be surface-modified SLNs containing poloxamer 235 (Hald Albertsen et al. 2022). According to research findings, surface modification using unique polymers can increase the bioavailability of LNPs. In addition to converting MnO_4 to MnO_2, cationic polyelectrolyte poly(allylamine hydrochloride) acts as a shield to stabilize LNPs through electrostatic repulsion during the formation process. However, the polyelectrolyte-coated MnO_2-LNPs are fragile and hazardous in cell medium or saline due to their small particle sizes and positively charged properties. In order to prevent colloidal MnO_2 from interacting with H_2O_2 too fast and to improve the hypoxia of the tumor microenvironment (TME), two hybrid LNPs were synthesized and incorporated into matrices that were either hydrophilic terpolymer/protein-MnO_2 (TMD) or hydrophobic polymer/lipid-MnO. The generated hybrid LNPs were nontoxic to tissue after 7 days of IV administration, and in comparison to PMD-LNPs, they exhibited noticeably superior colloidal properties and biocompatibility in both in vitro and in vivo settings. LMD-LNPs demonstrated excellent tumor formation and preservation at an average pH level of 7.4, as well as an effective reduction of hypoxia as well as HIF-1α in tumors that were solid due to their lower reactivity (Ashfaq et al. 2023).

By altering their surface, particular lipids can give LNPs different activities. The main purpose of cholesterol has been to act as an assistant lipid to improve the integrity of LNP membranes. Research has indicated that cholesterol not only provides stability to lipophilic particles (LNPs) but also serves other specific purposes. Lipid nanoparticle absorption by hepatocytes and BBB permeability are both enhanced by the significant cholesterol content of LNP. Moreover, cholesterol enhances the ability of lipid nanoparticles to load drugs (Jung et al. 2022).

Representative surfactant protein oleosin has improved membrane stability and a distinct amphipathic topological structure. Therefore, research is being done on surface modification with oleosin to give liposomes and LNPs stability and usefulness. They suggested using soybean oleosin-coated liposomes to deliver the lipophilic medication luteolin. Oleosin increased the liposome membrane's structural stability and tolerance to a variety of external stresses (such as temperature, pH, and ions) in the previously stated study. GFP-hybridized LNPs encapsulated in genetically modified oleosin (GM-oleosomes) are used to target breast cancer with the lipophilic anticancer medication carmustine. They combined the HER2 antibody, which is unique to the SK-BR-3 cell type, with the GFP-LG protein to create an extremely functional LNP that could be tracked and targeted without the need for chemical treatment. In vivo experiments on mice confirmed that the GM-oleosome selectively delivered drugs to target cancer cells. All things considered, the previously reported findings indicate that oleosin and other naturally occurring surfactant proteins are useful surface modifiers for producing high-functionality LNPs (Suzuki and Kamiya 2023).

10.7 Future Perspectives

Different MIDDS may be used in the future for anticancer, theragnostic, and diagnostic purposes in the treatment of different kinds of cancer. In the near future, clinical trials involving human volunteers and/or applied research should be started to confirm the safety and effectiveness of thermoresponsive prodrugs loaded with anticancer agent(s) in people. Furthermore, the present idea of using combined cancer along with genome therapy via MIDDS to target both passive and aggressive tumors may be seen as a potential future treatment option for a variety of cancers. Human clinical trials should be conducted on MIDDS for biological macromolecules in the future to demonstrate the pH-responsive medications' safety and efficacy in treating various malignancies and other biological applications. In the future, it will also be necessary to address the requirements for the achievement of targeting, which include appropriate size, charged surface, stealth effects, sterilization, and resemblance of the cell membrane of the MIDDS of macromolecules/nanomedicines. For many years to come, the current analysis might provide support for the efforts of scientists studying medication delivery and oncogenesis. Finally, the attention will be on the subsequent steps that the field of alternative nanochemistry hopes to take with the goal to eventually use green chemistry to safely treat a range of illnesses through nanomedicine on a big scale (Debela et al. 2021).

The role of artificial intelligence in biological macromolecule creation and MIDDS is becoming more and more significant. With the increasing advancement of AI algorithms, which can leverage larger and more accurate datasets, researchers will be able to make faster and more informed conclusions. According to one study, AI might produce more than 50 novel treatments in 10 years, saving $50 billion just in preclinical expenses. To take advantage of these prospects, numerous pharmaceutical corporations are increasingly collaborating with AI startups. Over $5.2 million was invested in AI for the study of drugs (Bhardwaj et al. 2022).

10.8 Conclusion

Although none of the noninvasive LMWH delivery methods have been shown to be secure or efficient for clinical use, they all show promise for improving patient compliance in the therapy of thrombosis. Notably, one review weakness is the inability to critically analyze and compare the bioavailability data (Qi et al. 2004). The significant variations in operational practices are the cause of this constraint. Studies conducted in vitro have shown that MNs could quickly dissolve into the ocular regions to build a depot there. They also demonstrated a greater degree of high molecular weight model molecule penetration and sustained release from the tissues. This holds significant value in terms of delivering high MW anti-VEGF medications in a minimally invasive way. Furthermore, it was discovered that the substance utilized to create MNs was biocompatible with retinal cells. The viability of employing this technology to deliver biological macromolecules that are active requires more research (Sivadasan et al. 2023; Akhter et al. 2022; Hou et al. 2019; Gade et al. 2024).

References

Akhter MH, Ahmad I, Alshahrani MY, Al-Harbi AI, Khalilullah H, Afzal O et al (2022) Drug delivery challenges and current progress in nanocarrier-based ocular therapeutic system. Gels 8(2):82

Aksakal R, Mertens C, Soete M, Badi N, Du Prez F (2021) Applications of discrete synthetic macromolecules in life and materials science: recent and future trends. Adv Sci 8(6):2004038

Ashfaq R, Rasul A, Asghar S, Kovács A, Berkó S, Budai-Szűcs M (2023) Lipid nanoparticles: an effective tool to improve the bioavailability of nutraceuticals. Int J Mol Sci 24(21):15764

Bhardwaj A, Kishore S, Pandey DK (2022) Artificial intelligence in biological sciences. Life 12(9):1430

Brown E, Barry T, Mai T, Harvey J (2024) Clinical experience of a tattoo application device. Tech Innov Patient Support Radiat Oncol 30:100254

Chevolot Y, Vidal S, Laurenceau E, Morvan F, Vasseur JJ, Souteyrand E (2009) Carbohydrates as recognition receptors in biosensing applications. In: Recognition receptors in biosensors. Springer, Berlin, pp 275–341

Choi D, Lee Y, Lin ZH, Cho S, Kim M, Ao CK et al (2023) Recent advances in triboelectric nanogenerators: from technological progress to commercial applications. ACS Nano 17(12):11087–11219

Debela DT, Muzazu SG, Heraro KD, Ndalama MT, Mesele BW, Haile DC et al (2021) New approaches and procedures for cancer treatment: current perspectives. SAGE Open Med 9:20503121211034366

Donnelly RF, Raj Singh TR, Woolfson AD (2010) Microneedle-based drug delivery systems: microfabrication, drug delivery, and safety. Drug Deliv 17(4):187–207

Escobar-Chávez JJ, Bonilla-Martínez D, Villegas-González MA, Revilla-Vázquez AL (2009) Electroporation as an efficient physical enhancer for skin drug delivery. J Clin Pharmacol 49(11):1262–1283

Gade S, Glover K, Mishra D, Sharma S, Guy OJ, Donnelly R et al (2024) Hollow microneedles for ocular drug delivery. J Control Release 371:43–66

Hahn D, Schneider R, Foitzik E, Meier M (2021) A practical and efficient synthesis of uniform conjugated rod-like oligomers. Macromol Rapid Commun 42:e2000735

Hald Albertsen C, Kulkarni JA, Witzigmann D, Lind M, Petersson K, Simonsen JB (2022) The role of lipid components in lipid nanoparticles for vaccines and gene therapy. Adv Drug Deliv Rev 188:114416

Hao J, Ghosh P, Li SK, Newman B, Kasting GB, Raney SG (2016) Heat effects on drug delivery across human skin. Expert Opin Drug Deliv 13(5):755–768

Hmingthansanga V, Singh N, Banerjee S, Manickam S, Velayutham R, Natesan S (2022) Improved topical drug delivery: role of permeation enhancers and advanced approaches. Pharmaceutics 14(12):2818

Hou A, Quan G, Yang B, Lu C, Minglong C, Yang D et al (2019) Rational design of rapidly separating dissolving microneedles for precise drug delivery by balancing the mechanical performance and disintegration rate. Adv Healthc Mater 8:1900898

Huzum B, Puha B, Necoara RM, Gheorghevici S, Puha G, Filip A et al (2021) Biocompatibility assessment of biomaterials used in orthopedic devices: an overview (review). Exp Ther Med 22(5):1315

Jeong WY, Kwon M, Choi HE, Kim KS (2021) Recent advances in transdermal drug delivery systems: a review. Biomater Res 25:24

Jung HN, Lee SY, Lee S, Youn H, Im HJ (2022) Lipid nanoparticles for delivery of RNA therapeutics: current status and the role of in vivo imaging. Theranostics 12(17):7509–7531

Lee JW, Gadiraju P, Park JH, Allen MG, Prausnitz MR (2011) Microsecond thermal ablation of skin for transdermal drug delivery. J Control Release 154(1):58–68

Li S, Kim Y, Lee JW, Prausnitz MR (2022) Microneedle patch tattoos. iScience 25(10):105014

Liu L, Zhao W, Ma Q, Gao Y, Wang W, Zhang X et al (2023) Functional nano-systems for transdermal drug delivery and skin therapy. Nanoscale Adv 5(6):1527–1558

Mandelkern L (2012) An introduction to macromolecules. Springer Science & Business Media, Berlin. 171 p

Mansour A, Romani M, Acharya AB, Rahman B, Verron E, Badran Z (2023) Drug delivery systems in regenerative medicine: an updated review. Pharmaceutics 15(2):695

Markovic M, Ben-Shabat S, Dahan A (2020) Prodrugs for improved drug delivery: lessons learned from recently developed and marketed products. Pharmaceutics 12(11):1031

Marques AC, Costa PC, Velho S, Amaral MH (2023) Lipid nanoparticles functionalized with antibodies for anticancer drug therapy. Pharmaceutics 15(1):216

Marshall SH, Arenas G (2003) Antimicrobial peptides: a natural alternative to chemical antibiotics and a potential for applied biotechnology. Electron J Biotechnol 6(3):271–284

Mishra V, Bansal KK, Verma A, Yadav N, Thakur S, Sudhakar K et al (2018) Solid lipid nanoparticles: emerging colloidal nano drug delivery systems. Pharmaceutics 10(4):191

Molnar C, Gair J (2015) 2.3 biological molecules. 14 [cited 2024 Sep 29]. https://opentextbc.ca/biology/chapter/2-3-biological-molecules/

Murthy SN, Sammeta SM, Bower C (2010) Magnetophoresis for enhancing transdermal drug delivery: mechanistic studies and patch design. J Control Release 148(2):197–203

N'Da DD (2014) Prodrug strategies for enhancing the percutaneous absorption of drugs. Molecules 19(12):20780–20807

Nagpal D, Ravi A (2015) Needle free drug delivery systems. Int J Pharm Investig

Nazary Abrbekoh F, Salimi L, Saghati S, Amini H, Fathi Karkan S, Moharamzadeh K et al (2022) Application of microneedle patches for drug delivery; doorstep to novel therapies. J Tissue Eng 29(13):20417314221085390

Olesen CM, Fuchs CSK, Philipsen PA, Hædersdal M, Agner T, Clausen ML (2019) Advancement through epidermis using tape stripping technique and reflectance confocal microscopy. Sci Rep 9:12217

Qi Y, Zhao G, Liu D, Shriver Z, Sundaram M, Sengupta S et al (2004) Delivery of therapeutic levels of heparin and low-molecular-weight heparin through a pulmonary route. Proc Natl Acad Sci U S A 101(26):9867–9872

Rajalakshmi P, Halith SM, Salam SM, Monisha P, Muhilarasi S, Murugan D et al (2023) Review on transdermal drug delivery system. Int J Pharm Sci Rev Res 83(2):49–57. http://globalresearchonline.net/ijpsrr/v83-2/08.pdf

Ramadon D, McCrudden MTC, Courtenay AJ, Donnelly RF (2022) Enhancement strategies for transdermal drug delivery systems: current trends and applications. Drug Deliv Transl Res 12(4):758–791

Seo Y, Lim H, Park H, Yu J, An J, Yoo HY et al (2023) Recent progress of lipid nanoparticles-based lipophilic drug delivery: focus on surface modifications. Pharmaceutics 15(3):772

Siswomihardjo W (2016) Biocompatibility issues of biomaterials. In: Biomaterials and medical devices. Springer, Cham, pp 41–65

Sivadasan D, Sultan MH, Alqahtani SS, Javed S (2023) Cubosomes in drug delivery—a comprehensive review on its structural components, preparation techniques and therapeutic applications. Biomedicines 11(4):1114

Suzuki M, Kamiya K (2023) Cell-sized asymmetric phospholipid-amphiphilic protein vesicles with growth, fission, and molecule transportation. iScience 26(3):106086
Trimzi MA, Ham YB (2021) A needle-free jet injection system for controlled release and repeated biopharmaceutical delivery. Pharmaceutics 13(11):1770
Trucillo P (2021) Drug carriers: classification, administration, release profiles, and industrial approach. Processes 9(3):470
Tucak A, Sirbubalo M, Hindija L, Rahić O, Hadžiabdić J, Muhamedagić K et al (2020) Microneedles: characteristics, materials, production methods and commercial development. Micromachines 11(11):961
Vikelis M, Spingos KC, Rapoport AM (2015) The iontophoretic transdermal system formulation of sumatriptan as a new option in the acute treatment of migraine: a perspective. Ther Adv Neurol Disord 8(4):160–165
Wong WF, Ang KP, Sethi G, Looi CY (2023) Recent advancement of medical patch for transdermal drug delivery. Medicina (Mex) 59(4):778
Yu YQ, Yang X, Wu XF, Fan YB (2021) Enhancing permeation of drug molecules across the skin via delivery in nanocarriers: novel strategies for effective transdermal applications. Front Bioeng Biotechnol 9:646554
Yuan J, Wang S, Yang J, Schneider KH, Xie M, Chen Y et al (2024) Recent advances in harnessing biological macromolecules for wound management: a review. Int J Biol Macromol 266:130989
Zaid Alkilani A, McCrudden MTC, Donnelly RF (2015) Transdermal drug delivery: innovative pharmaceutical developments based on disruption of the barrier properties of the stratum corneum. Pharmaceutics 7(4):438–470
Zhang YB, Xu D, Bai L, Zhou YM, Zhang H, Cui YL (2022) A review of non-invasive drug delivery through respiratory routes. Pharmaceutics 14(9):1974

Applications of Minimally Invasive Drug Delivery Systems

11

Latatai Nandkumar Patil, Vancha Harish,
Malakapogu Ravindra Babu, and Shriyansh Srivastav

Abstract

In recent years, minimally invasive drug delivery systems (MIDDS) have gained enormous attention of the researchers as they provide creative ways to achieve accurate and effective therapeutic effects while reducing procedural risks and patient pain. The concepts of MIDDS were used in the treatment of various diseases such as cardiovascular diseases, central nervous system-related disorders, respiratory diseases, cancer, and diabetes. The delivery of therapeutic agents in all the above cases utilizes invasive procedures. So, there is a need in developing the potential systems that bypass all types of limitations of traditional oral or intravenous routes and improve both patient compliance and bioavailability of therapeutic agent. The system that has all the required qualities includes transdermal, nose to brain, inhalational, and nanomedicines. The present chapter begins by specifying the primary mechanisms of key MIDDS such as intranasal, hydrogel, nanoparticles, nose to brain, and transdermal systems. Further the chapter also explores the inclusion of nanotechnology that allows for controlled and targeted drug delivery. The clinical effectiveness and safety profiles of the system were highlighted by some case studies related to cancer, diabetes, pain control, CNS disorders, and cardiovascular disorders. MIDDS play a critical role in tackling issues related to chronic diseases by providing solutions that lessen systemic adverse effects and enhance treatment outcomes. Lastly, to satisfy the changing demands of patients and healthcare systems throughout the world, this

L. N. Patil · V. Harish (✉)
Department of Pharmaceutics, School of Pharmaceutical Sciences, Lovely Professional University, Phagwara, Punjab, India
e-mail: harish.24838@lpu.co.in

M. R. Babu · S. Srivastav
Department of Pharmacy, School of Medical and Allied Sciences, Galgotias University, Greater Noida, Uttar Pradesh, India

M. R. Babu et al. (eds.), *Transdermal Applications of Minimally Invasive Drug Delivery Systems*, https://doi.org/10.1007/978-981-96-6275-3_11

chapter highlights the revolutionary potential of MIDDS in transforming healthcare delivery, with an emphasis on overcoming present constraints and encouraging innovation.

Keywords

Microneedles · Transdermal patches · Nanomedicine · Hydrogels

11.1 Introduction

The development of MIDDS in recent years has significantly transformed the medical field and provided a new architect for the delivery of therapeutic drugs to the target effectively without patient discomfort. These systems provide various benefits over other conventional systems to patients by decreasing the invasiveness and enhancing the therapeutic efficacy (Kulkarni 2009). MIDDS was utilized in the management of various chronic diseases such as cancer and diabetes and pain because of its ability to deliver the drugs in targeted, controlled, and patient-friendly manner (Khafagy et al. 2007). The most important benefit of MIDDS was its capability to transport drugs with less side effects, enhanced absorption, higher bioavailability and targetability, and improved patient compliance by using novel drug delivery systems, which offers site specificity and enhanced therapy over conventional delivery systems (Vllasaliu 2021). The drug from these systems gets released at certain intervals up on predetermined triggering of temperature, pH, and biomarkers. This assures that right doses of drugs are administered with less possible over- or under-dosing and less doses.

There are various MIDDS, out of which nose to brain, implantable, and transdermal delivery of drugs have gained more popularity because they offer continuous release (implantable) and absorption of drugs through the skin (transdermal patch) (이영식 2020; Akhtar et al. 2020). MIDDS are specifically useful in the treatment of all chronic illness including diabetes (Gupta et al. 2009) and cardiovascular disease (Huynh 2017) and chronic pain (Priya and Singhvi 2022) as they require long-term delivery of drugs. Diabetes treatment has been greatly improved by wearable infusion devices and insulin pumps, which provide patients with greater flexibility and ease in monitoring and managing their medicine supply (Gupta et al. 2009).

Another intriguing area of MIDDS is the use of nanotechnology, namely, in the development of nanoparticles and nanocarrier system. These highly competent carriers can deliver drugs directly to the target region, boosting the effectiveness of treatment, particularly in conditions like cancer (Patra et al. 2018). The tailored drug delivery systems were developed to deliver anticancer drugs directly to the target (cancer cells) by reducing the collateral damage to the healthy tissues that traditional chemotherapy usually causes. They not only improve the efficacy of the treatment but also reduce undesirable side effects such as hair loss, nausea, and immunological suppression.

The present chapter mainly deals with the pioneering MIDDS techniques such as nose to brain and microneedle patches, which provide a more efficient, effective, and painless substitute for traditional injections. They have attracted the attention due to their potential to increase patient compliance by administering vaccinations and other treatments via the skin without the need for needles. This practice is particularly true for older and younger patients who might be hesitant to undergo conventional needle-based treatments. In addition to these therapeutic applications, MIDDS is gaining traction in the domains of wound healing and regenerative medicine. Biodegradable drug delivery methods such as hydrogels are being used to directly release growth factors or antibiotics into wound sites to promote tissue repair and reduce the risk of infection. These systems are not only effective but also lessen patient suffering since they are often designed to dissolve or break down naturally in the body, eliminating the need for surgery or the risk of foreign body rejection.

11.2 Advantages and Limitations

MIDDS is a popular choice for individuals and medical professionals due to its various advantages. One significant benefit is that these systems typically require fewer incisions or none at all than conventional treatments, which leads to quicker recovery periods and less postoperative discomfort (Muankaew and Loftsson 2018; Baryakova et al. 2023). Additionally, MIDDS provide more precise and controlled drug delivery, which can enhance treatment efficacy and reduce side effects (Jimmy and Jose 2011). This can be quite beneficial when long-term drug administration is required for therapies like cancer therapy or chronic pain management (Hariharan and Tran 2023). Furthermore, MIDDS employs smaller devices, which are often safer for patients because of their decreased risk of infection and related complications (Jimmy and Jose 2011). Less invasive methods can minimize hospital stays and related expenses, because they eliminate the need for more intrusive operations and their consequences (Vllasaliu 2021). Most of the minimally invasive systems such as transdermal patches and insulin pens are user-friendly, due to which they allow patients to apply themselves. The comfort provided by them can enhance the overall quality of life.

These methods do have certain drawbacks, though. The intricacy of design and manufacturing is a major disadvantage that may result in increased production costs and restricted supply. Advanced technology is necessary for these systems' accuracy, but not all medical settings may have access to it. Furthermore, not every medicine type may be appropriate for MIDDS, particularly those that require fast, high-concentration release or that must be administered in big amounts. Another difficulty is making sure the medication gets to the right place at a high enough concentration, particularly for medications with low bioavailability. Last but not least, continual research and close patient monitoring are necessary to properly understand the long-term consequences of minimally invasive drug delivery systems on tissue and organ function (Zhang et al. 2022).

11.3 Types of MIDDS

MIDDS encompass various types, viz., oral, transdermal, nasal, rectal, and pulmonary delivery systems, each tailored for specific medical needs and therapeutic applications (Fig. 11.1). Nanotechnology plays a key role in the development of these MIDDS that bypass all the limitations associated with the conventional delivery systems.

11.3.1 Oral Drug Delivery Systems

Oral drug administration is the most convenient and commonly used method for delivering medications. However, it faces significant challenges such as poor drug absorption and degradation caused by the acidic environment of the stomach and the presence of gastrointestinal microorganisms. The drugs' bioavailability and efficacy may be lessened by these factors, especially for sensitive substances like proteins and peptides. Advanced formulations are frequently needed to improve stability and absorption in order to solve these problems (Park et al. 2011).

11.3.2 Transdermal Drug Delivery Systems (TDDS)

The TDDS is a significant advancement in pharmaceutical technology that offers a less invasive way to dispense drugs. TDDS enables the controlled delivery of

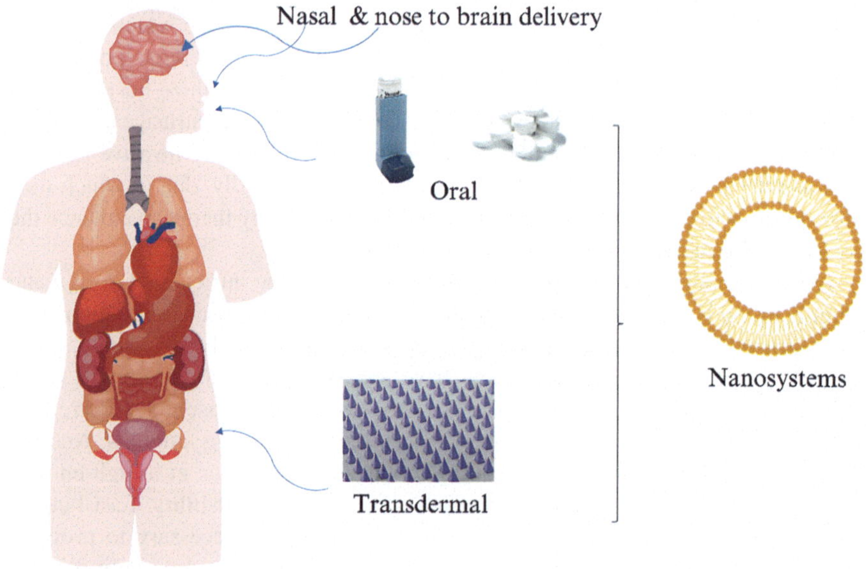

Fig. 11.1 Types of MIDDS

pharmaceuticals through the skin directly into the systemic circulation by avoiding the GIT and hepatic first-pass metabolism. This strategy ensures improved bioavailability and sustained plasma drug levels while reducing the possibility of GI distress and systemic side effects as compared to oral and injectable treatments. The invasive nature of TDDS increases patient compliance particularly for individuals who require chronic medication, because it eliminates the discomfort and anxiety that are occasionally associated with needles and other various dosage forms (Kanikkannan et al. 2000).

Moreover, modern TDDS employs cutting-edge technology like microneedles, iontophoresis, and nanocarriers to get beyond the skin's natural barrier function and deliver a greater range of drugs including macromolecules like peptides and proteins. These techniques are particularly helpful for drugs with limited therapeutic indices or those that need exact dose. The user-friendly nature, extended duration of action, and potential for on-demand drug release from TDDS further position it as a crucial element in the development of advanced patient-centered treatment options. Therefore, transdermal systems are shining example of a paradigm shift toward effective, simple, and less intrusive drug delivery methods.

11.3.2.1 Microneedles

Microneedles are a novel and less invasive way to administer drugs through the skin, specifically to the top dermis or epidermis (Fig. 11.2). In contrast to conventional injections, which pierce deeply into tissues or the circulation, microneedles provide a quick and painless method of delivering therapeutic substances without the need to deal with difficulties like needle fear and medication deterioration in the digestive tract (Prausnitz et al. 2004). Microneedles are tiny needles in micron size

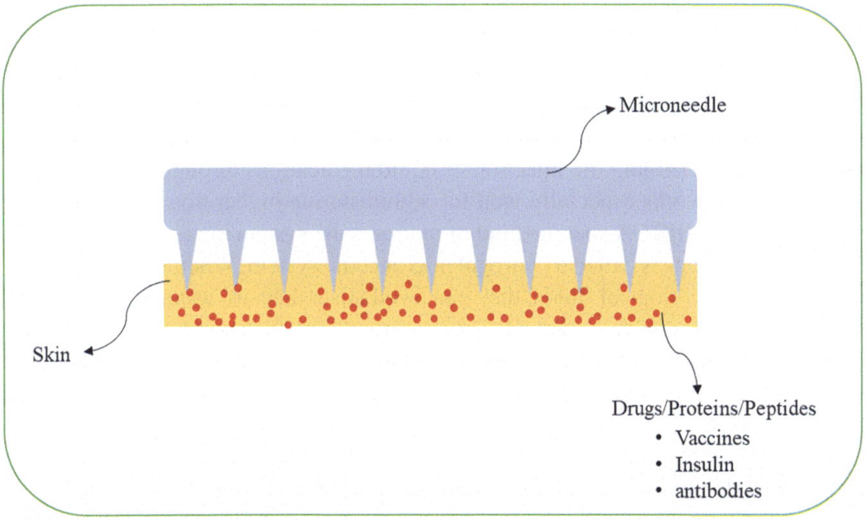

Fig. 11.2 Delivery of proteins and peptides using microneedles

that improve drug permeability by forming tiny reversible routes through the stratum corneum of the skin. Usually, these microstructures are made with lengths under 1 mm (Larrañeta et al. 2016). Recent developments in drug delivery systems using microneedles have shown their efficacy in the range of medical settings. The use of dissolving microneedles to treat skin cancer is a well-known example. The research conducted by Lan et al. demonstrated the development of cisplatin-loaded microneedle to target melanoma cells. They prepared by using biodegradable polymers, which dissolve when they come in contact with the skin, thereby releasing the drug in controlled fashion directly in the cancer site. This kind of approach not only enhances the delivery of therapeutic agent but also reduces the systemic side effects that are associated with the conventional therapies (Lan et al. 2018). Similarly, researchers have explored microneedle for vaccine delivery, and it has shown promising results and produced strong immune responses on par with conventional injections, and it is more patient-acceptable and caused less discomfort (Mansoor et al. 2022). The complete dissolution of the microneedles after use reduced the risks associated with needlestick injuries and medical waste. Microneedles have also been extensively researched for chronic disease management. Zhao et al. (2022) created hydrogel microneedles for insulin delivery in diabetic patients (Chen et al. 2019). These microneedles swell upon insertion into the skin and release insulin in response to glucose levels, providing controlled and sustained drug release.

Solid microneedles are used to create microchannels in the skin, and the drug is applied afterward, either through a patch or other topical formulations (poke-and-patch method). Solid microneedles are often used for pretreatment to increase skin permeability (Kim et al. 2012). **Coated microneedles** are coated with the drug on their surface. Once inserted into the skin, the coating dissolves, releasing the drug into the targeted area. This type allows precise dosing but is limited by the amount of drug that can be coated onto the needles (Ita 2015). **Dissolving microneedles** are made from biocompatible and biodegradable materials that dissolve after being inserted into the skin. The drug is either incorporated within the needle structure or within the material, and as the microneedles dissolve, the drug is released into the skin (Lee et al. 2008). Despite being somewhat smaller, hollow microneedles resemble regular needles. They offer a direct method for administering medications into the skin by enabling the injection of liquid medications through a hollow tube. This technique works especially well for administering higher dosages of medication (Kim et al. 2012). A specific kind of microneedle system made using hydrogel-forming ingredients is called a hydrogel microneedle system. When hydrogels come into touch with biological fluids, they can swell since they are water-absorbent and extremely biocompatible polymers. This feature makes them perfect for developing microneedles that may expand, collect interstitial fluid from the skin, and enable controlled medication administration or sensing applications (Courtenay et al. 2020).

Microneedle's less invasive nature improves patient experience by providing a number of benefits for drug administration. One of the key benefits is their pain-free application; due to their small size, microneedles penetrate only the outer layers of the skin and do not reach the deeper, nerve-rich areas, resulting in a painless delivery method (Kim et al. 2012). Additionally, many microneedle patches are designed for self-administration, allowing patients to apply them independently without the need for healthcare professional intervention (Prausnitz et al. 2004). This ease of use not only empowers patients but also contributes to improved compliance, making microneedles a preferred option, particularly for those with chronic conditions that necessitate regular medication (Bariya et al. 2011).

Microneedles penetrate the stratum corneum, which is the main barrier for transdermal drug delivery, and reach the viable epidermis. This creates microchannels that allow the passage of drugs either through diffusion (solid and coated microneedles) or through active delivery (hollow microneedles). In the case of dissolving microneedles, the matrix of the needle itself degrades and releases the drug into the skin (Aldawood et al. 2021). **Peptide and protein drugs**, which typically degrade in the gastrointestinal tract, can be delivered via microneedles (Moreira et al. 2019). Drugs like insulin, painkillers, or chemotherapy agents are being explored for microneedle delivery (Li et al. 2022). Microneedles have gained popularity in the skincare industry for anti-aging treatments and the delivery of substances like hyaluronic acid (Yélamos et al. 2023).

11.3.2.2 Transdermal Patches

Transdermal patches are adhesive patches that deliver drugs through the skin into the bloodstream. These patches offer a noninvasive, controlled, and sustained method of drug delivery (Alkilani et al. 2015). Transdermal patches provide noninvasive, bypass the gastrointestinal tract (no first-pass metabolism), reduce dosing frequency, and provide steady drug release. Various types of drugs are available in the form of transdermal patch such as nicotine patches, pain management (fentanyl patches), hormone therapy (estrogen patches), and cardiovascular treatments (nitroglycerin patches) (Alkilani et al. 2015).

Transdermal patches have evolved significantly over the last decade, providing a noninvasive method for delivering drugs through the skin, offering benefits like sustained release, improved patient compliance, and bypassing first-pass metabolism. Nicotine patches are the most often used transdermal patches. Efficiency and user convenience have been the main goals of the recent developments. Dale et al.'s (1995) study looked at how high-dose nicotine transdermal patches affected heavy smokers. Comparing higher-dose patches (21 and 42 mg) to conventional lower-dose patches, the study discovered that the former increased smoking cessation rates among heavy smokers. Although fentanyl transdermal patches have been used to treat chronic pain, more recent studies have concentrated on enhancing the system's effectiveness and safety. A new fentanyl transdermal patch was created by Bahrami et al. (2024) using a temperature-sensitive polymer that adjusted drug delivery according to the patient's skin temperature. By enabling regulated fentanyl release during elevated body temperatures (such as during exercise), this invention decreased the danger of overdosing while preserving efficient pain management under a variety of circumstances. Recent studies have concentrated on enhancing the safety and effectiveness of fentanyl patches, which have been used to treat chronic pain. Estradiol patches are extensively researched for hormone replacement therapy, especially for women with menopause. Matrix technology was used by

Kopper et al. (2009) to develop estradiol patch and reported the enhanced drug delivery consistently than conventional reservoir systems. Women on hormone replacement therapy saw superior therapeutic results, because of the patch's enhanced adherence, less skin irritations, and more consistent hormone levels. Development of transdermal patch of rivastigmine resulted in enhanced patient compliance by reducing the GI side effects in Alzheimer's patient. In another study, Cummings et al. (2007) developed rivastigmine patches with enhanced patch adhesion and skin permeability that led to better cognitive outcomes in Alzheimer's patients than its oral administration. They also reduced dosing frequency and enhanced steady-state concentration of the drug that minimized the unwanted effects and maximized overall efficiency.

11.3.3 Inhalation Devices

Inhalational drug delivery devices are a cutting-edge technique in the realm of minimally invasive therapeutic techniques that provide a number of advantages over other traditional techniques such as oral and injectable systems (Fig. 11.3). The respiratory tract's large surface area and rich vascular network were utilized by these systems to either target specific lung area or enable the rapid and efficient absorption of lungs directly into the systemic circulation. Drug delivery methods that are inhaled will bypass the GIT and hepatic first-pass metabolism, increasing bioavailability and providing a faster start of action (Anselmo et al. 2019). This special feature of them is particularly helpful for ailments that require immediate treatment, such as asthma, chronic obstructive pulmonary disease (COPD), and acute respiratory infections (Kew et al. 2014; Fu et al. 2022).

Technological advancements have led to the development of inhalation devices like nebulizers, pressurized metered-dose inhalers (pMDIs), and dry powder inhalers (DPIs), which are tailored to specific medications and patient needs. These devices are easy to use and noninvasive and ensure accurate dosing, improving patient compliance. Additionally, biologic drugs, including peptides and proteins, can now be inhaled, expanding their use beyond respiratory conditions to treat systemic diseases such as cardiovascular diseases, diabetes, and CNS disorders (Labhasetwar 2005).

Furthermore, inhalable delivery systems are gaining recognition for their role in personalized medicine, offering targeted treatments with minimal systemic side effects. Nanotechnology has enhanced their effectiveness by enabling nanoparticle-based carriers for controlled and sustained drug release. Despite their advantages, challenges such as drug stability, formulation optimization, and device compatibility remain under investigation. Therefore, these systems hold significant promise as a patient-focused minimally invasive approach to modern medicine (Ahmad 2022).

Turpeinen et al. (2021) developed novel dry powder inhaler (DPI) using electrostatic charge technology to improve lung deposition of fluticasone propionate and salmeterol. Additionally, smart inhalers have advanced COPD management by

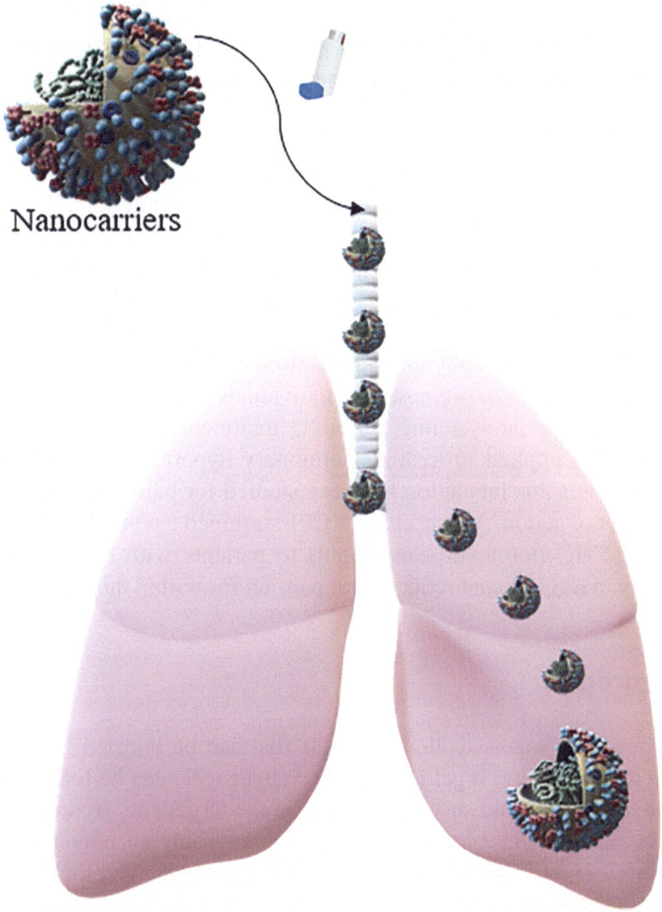

Fig. 11.3 Nanocarrier-based pulmonary drug delivery system

integrating real-time monitoring and feedback system. The smart inhaler developed by Chan et al. (2021) has attracted the patients more as it is equipped with a Bluetooth sensor that tracks drug usage and administration. By using this device, patients showed better treatment adherence, leading to reduced exacerbations and hospitalizations. Nebulizers, which have historically been used to treat cystic fibrosis, have also undergone advancements to increase effectiveness and shorten treatment durations. In contrast to jet nebulizers, Coates et al.'s study of a vibrating mesh nebulizer that administered tobramycin to patients with cystic fibrosis showed that the method created smaller aerosol particles enabling deeper lung penetration and shorter treatment durations. Patients' quality of life and lung function increased as a result of this development (Coates et al. 2011).

Inhalation devices have been extensively researched during COVID-19 pandemic for the delivery of vaccine. In one study, Heida and his group administered vaccine containing adenovirus through nasal inhaler for the treatment of COVID-19 and reported that this technique has elicited greater mucosal immunity in both the upper and lower respiratory tracts, providing enhanced protection compared to conventional injectable vaccines (Heida et al. 2022).

Inhalation devices are also extensively utilized to deliver antibiotics for the treatment of lung infections. Almangour et al. (2020) developed an inhalable colistin powder for patients with multidrug-resistant bacterial lung infections. This formulation achieved high local concentrations of colistin in the lungs while reducing systemic exposure and associated side effects. The study demonstrated that inhaled colistin was effective in treating infections caused by multidrug-resistant pathogens, including *Pseudomonas aeruginosa*, in individuals with cystic fibrosis, thereby lowering the reliance on systemic antibiotic treatments. Additionally, inhalation devices have been applied in treating pulmonary hypertension Olschewski et al. (2000). Lastly, cannabis inhalation has been studied for pain management, providing a rapid onset of action. Almog et al. (2020) examined a portable cannabis inhaler that delivered THC (tetrahydrocannabinol) to patients with chronic neuropathic pain, resulting in a significant reduction in pain scores within minutes of inhalation.

11.3.4 Hydrogels

Hydrogels are soft, water-swollen materials that can be injected into the body in liquid form and then form a gel in situ. These hydrogels can be loaded with drugs and provide a localized and sustained release at the target site. Hydrogels are made from biocompatible and biodegradable polymers such as polyethylene glycol (PEG), alginate, and hyaluronic acid. Minimally invasive nature, localized delivery, sustained and controlled drug release, and reduced systemic toxicity give it more advantage over others. It is having wide application in the tissue engineering, wound healing, cancer therapy (localized delivery to tumors), and drug delivery for bone regeneration (Peppas 2000). Tanga et al. (2023) developed a thermoresponsive injectable hydrogel based on poly(*N*-isopropylacrylamide) (PNIPAAm) for controlled drug release. The hydrogel showed a significant release of loaded drugs at physiological temperatures, making it suitable for applications in localized drug delivery. Zhao et al. (2010) investigated an injectable calcium phosphate-alginate hydrogel-umbilical cord mesenchymal stem cell paste for bone tissue engineering. Fan et al. (2019) designed a biodegradable injectable hydrogel that encapsulates chemotherapeutic agents for localized cancer treatment. The hydrogel effectively targeted tumor sites and released the drug over an extended period. Salehi et al. (2023) presented a smart injectable hydrogel that responds to environmental stimuli (pH and temperature), enabling on-demand drug release. The hydrogel showed promise for applications in personalized medicine.

11.4 Applications of MIDDS

11.4.1 Chronic Disease Management

Minimally invasive drug delivery technologies have become vital tools in chronic disease management. They provide more effective treatment options, improve patient compliance, reduce side effects, and enhance the overall quality of life for individuals suffering from chronic diseases. Diabetes mellitus, particularly type 1 and type 2, requires lifelong management. Traditional insulin injections are painful and inconvenient, leading to poor compliance in some patients. Minimally invasive drug delivery technologies offer novel solutions for better glycemic control. Microneedles, especially dissolving microneedles, can deliver insulin through the skin painlessly. They provide a sustained release of insulin, which can help maintain blood sugar levels more consistently than traditional injections (Zhang et al. 2021). Desmospray, a nasal desmopressin solution, significantly improved drug absorption compared to its oral form. While oral desmopressin shows minimal GI absorption (0.08–0.16% bioavailability), nasal route of administration enhanced absorption up to 20% offering a more efficient and effective administration route (Ozsoy et al. 2009). CP024, a nasal powder of human growth hormone, demonstrated in Phase 1 trials the ability to elevate insulin-like growth factor levels comparable to those achieved with subcutaneous injections of existing marketed products. This suggests its potential as a noninvasive alternative for hormone therapy delivery (Lewis et al. 2015). Continuous glucose monitoring (CGM) systems use minimally invasive sensors that are inserted under the skin. They continuously monitor glucose levels and can transmit real-time data to mobile devices. Advanced CGMs are now coupled with insulin pumps to automate insulin delivery (Moser et al. 2010). Inhalable insulin devices (e.g., Afrezza) allow patients to inhale insulin powder into their lungs, providing a noninvasive alternative to injections. This method has been shown to offer faster absorption and action compared to subcutaneous injections (Goldberg and Wong 2015). A powdered intranasal glucagon-like peptide-1 formulation was tested for blood glucose regulation showing 2.7% bioavailability compared to subcutaneous injection. Dose-dependent insulin levels increased without significant toxicity. Though safe overall, mild side effects like nausea occurred in some patients, confirming the safety and feasibility of intranasal administration (Ueno et al. 2014). In 1925, Gansslen demonstrated the potential of pulmonary drug delivery by showing that inhaled insulin could lower blood glucose levels in five diabetic patients. This study marked an early milestone in exploring noninvasive methods for administering therapeutic agents (Gänsslen 1925). There were noninvasive glucose monitoring devises, some optical techniques for transdermal glucose monitoring, and nanomaterial-based diabetes sensing devises or tests that help the patient to monitor glucose without any discomfort (Vashist 2012; Sieg et al. 2005; Makaram et al. 2014). These improve patient compliance, more stable glycemic control, and less frequent blood sugar fluctuations. Oral-lyn™, developed by Generex Biotechnology, is an oral insulin spray using RapidMist™ technology for convenient self-administration. Insulin is delivered as a tasteless mist to the buccal mucosa, with absorption directly proportional to the dose. It offers a faster onset and shorter action duration compared to subcutaneous insulin (Bansal et al. 2014). Amogh et al. developed a biodegradable chitosan-based polymeric patch for buccal insulin delivery, incorporating ionic liquids or deep eutectic solvents to enhance drug transport. The system demonstrated sustained insulin release, with serum levels stabilizing after 3 h. This platform shows potential for delivering injectable drugs noninvasively via the buccal route (Vaidya and Mitragotri 2020). Yiyue et al. developed buccal delivery system for insulin-loaded deformable nanovesicle (DNV)-loaded gel system with improved storage stability using a temperature-sensitive gelatin-based gel (IPC-DNV-TSG). Stable at 4 °C for 3 months, it maintained a hypoglycemic effect. While IPC-DNV-TSG showed slightly lower cellular uptake than IPC-DNVs, overall transport and cell viability remained high with no significant mucosal irritation observed (Guo et al. 2022).

Chronic pain affects millions of people worldwide and can severely reduce the quality of life. Minimally invasive drug delivery technologies provide more efficient and targeted pain relief, reducing the need for oral medications and their associated systemic side effects. Transdermal patches are widely used for chronic pain management, particularly for delivering opioids like fentanyl. The patch provides continuous drug delivery, reducing the need for frequent dosing and maintaining steady pain control levels (Hadley et al. 2013). Microneedle patches have been designed to administer local anesthetics directly to the target area, avoiding the gastrointestinal system and minimizing the chance of systemic side effects (Priya and Singhvi 2022). For individuals with chronic pain, devices like intrathecal pumps can be surgically implanted to deliver medications such as morphine directly into the spinal fluid. These methods are particularly advantageous for managing cancer-related pain or severe neuropathic pain (Jonas et al. 2015; De Andres et al. 2022).

Cardiovascular diseases such as hypertension, heart failure, and atherosclerosis necessitate prolonged management. MIDDS can enhance the treatment of the above diseases effectively by releasing drugs in a controlled and sustained fashion. In recent years, clonidine, an antihypertensive drug, was delivered using transdermal patch and achieved steady-state concentration without peak and valley effect, which is associated with oral delivery (Prisant and Elliott 2003). Nanoparticles are used for the delivery of anti-inflammatory and lipid-lowering agents in conditions such as atherosclerotic plaques directly to the site of plaque formation (Liu et al. 2022). Hydrogels are used to deliver growth factors directly to the damaged heart tissue. These hydrogels also help in improving heart function in post-myocardial infraction and tissue regeneration (Peña et al. 2018).

Cancer treatment often requires targeted and controlled drug delivery to reduce side effects and improve the therapeutic index of chemotherapeutic agents. MIDDS can be tailored to achieve this, providing localized drug delivery and reducing systemic toxicity. Nanoparticles can be engineered to carry chemotherapy drugs directly to the tumor site, sparing healthy tissues. Liposomal doxorubicin is a well-known example of nanoparticle-based chemotherapy, which reduces the cardiotoxicity commonly associated with doxorubicin (Jonas et al. 2015). Microneedle patches are being developed for localized delivery of chemotherapeutic agents in skin cancers such as melanoma. This method allows the direct targeting of tumor cells with minimal systemic exposure (Heo et al. 2023). Implantable drug delivery systems, such as the Gliadel wafer used in brain cancer, release chemotherapy agents directly into the tumor site over time. This allows for a higher concentration of drugs at the tumor while minimizing systemic side effects (Jonas et al. 2015).

11.4.2 Management of Central Nervous System (CNS) Disorders

Drug molecules are still not efficiently transported to the CNS, despite the fact that neurological and psychiatric conditions including brain tumors, epilepsy, Parkinson's disease, Alzheimer's disease, multiple sclerosis, and depression are becoming more common worldwide (Zhang et al. 2022; Islam et al. 2020). Managing neurological disorders often involves long-term drug therapy. Drugs can be directly injected into the parenchyma or cerebrospinal fluid by intrathecal or intraparenchymal infusions, respectively, but both methods are intrusive and inappropriate for long-term care (Bahadur et al. 2020; Misra and Pathak 2023). Additionally, untargeted delivery of therapeutic drugs may also seriously harm the glial cells and neurons. MIDDS have been developed to enhance the effectiveness and convenience of treatments for these complex conditions (Botti et al. 2021). Nanoparticles (Md et al. 2014), nanogels (Picone et al. 2018), liposomes (Kurano et al. 2022), and solid lipid nanoparticles (Crowe and Hsu 2022) are used to deliver drugs across the blood-brain barrier, which is otherwise difficult to penetrate with conventional drugs. Nanocarriers can transport therapeutic agents like dopamine, large proteins and peptides, antioxidants, or neurotrophic factors directly to the brain by passing the potential protective barrier, i.e., the blood-brain barrier (Liu et al. 2022; Dhuria et al. 2010). Inhalable formulations of drugs like levodopa provide a noninvasive route to deliver drugs quickly to the brain (Islam et al. 2020). However, the mechanism of drug delivery to the brain through this route is not fully demonstrated, but most of the researchers have reported that drugs enter CNS through the trigeminal and olfactory nerves present in the nose. These systems help manage "off" periods in Parkinson's patients, where oral medications are less effective (Grosset et al. 2020). There were certain studies that demonstrated the ligand-based nanocarrier system to deliver drug to the brain through nasal route (Costa et al. 2019; Fan et al. 2018). The research conducted by Sridhar et al. revealed that delivery of selegiline by loading in chitosan nanoparticles through the nose has enhanced the concentrations in the brain by 20 folds in comparison to oral route. The C_{max} was also higher with intranasal thermosensitive gel than solution (Chang et al. 2017). Chitosan glutamate nanoparticles loaded with rasagiline were prepared by Mittal et al., and they found enhanced concentrations in CNS through nasal route of administration than its naïve form (Mittal et al. 2016). Deep brain stimulation devices are also reported in some researches, but they are not a drug delivery method but represent a minimally invasive option that can complement drug therapies. They are used in conditions like Parkinson's and dystonia to modulate brain function (Groiss et al. 2009).

Microemulsion formulation was used by Jogini et al. to deliver tacrine hydrochloride, an FDA-approved drug for Alzheimer's treatment through nasal route. She used Carbopol 934P as mucoadhesive agent to increase the contact time of microemulsion and increase absorption of tacrine hydrochloride in the brain. The biodistribution of the drug formulation was enhanced along with brain selectivity by nasal route than intravenous administration (Jogani et al. 2008). Qian et al. delivered HLS-3 through intranasal administration to treat Alzheimer's disease. The study revealed that HLS-3 has enhanced CNS effects with minimal peripheral side effects (Qian et al. 2018). Microneedles were also developed for nasal delivery of therapeutic agents such as huperzine A and hyaluronic acid for Alzheimer's disease treatment (Ruan et al. 2024). Therefore, we can conclude that nose-brain delivery of therapeutic agents may be the best choice for treating chronic CNS disorders with minimal invasiveness.

Haque and his team developed alginate nanogel formulation loaded with venlafaxine, an antidepressant drug with inherent hydrophilicity and short elimination half-life. Haque research demonstrated the increased permeation of venlafaxine through porcine nasal mucosa. The in vivo studies also revealed significant increase in drug concentration in the brain (743 ng/mL) than intravenous route (383 ng/mL). In another study, they developed alginate nanoparticles and evaluated by forced swim test and reported enhancement in the behavioral analysis parameters (Haque et al. 2014). Dange et al. developed nanostructure lipid carriers of venlafaxine for intranasal delivery (Dange et al. 2014).

Woensel et al. developed chitosan nanoparticles loaded with siRNA for nasal delivery to inhibit Gal-1 in glioblastoma. These nanoparticles preserved siRNA effectiveness, rapidly targeting murine and human glioblastoma cells, reducing tumor migration. Intranasal administration in mice enabled fast distribution to the nasal mucosa, olfactory bulb, and hindbrain demonstrating CNS delivery efficiency (Van Woensel et al. 2016). Ferreira et al. delivered CTX and α-cyano-4-hydroxycinnamic acid by developing PLGA and oligomeric chitosan-based mucoadhesive nanoparticles to the brain through nasal delivery and treated glioblastoma (Ferreira et al. 2020). Jain et al. developed amiloride-loaded mucoadhesive nanoemulsion for intranasal delivery to the brain. These formulations were nontoxic to sheep nasal mucosa and showed safety for antiepileptic drug administration. However, further pharmacokinetic and pharmacodynamic studies are needed to confirm their therapeutic efficacy and clinical potential (Bonferoni et al. 2019). Gonclaves et al. developed a thermosensitive gel containing levetiracetam for intranasal delivery, achieving a bioavailability of 107.44%, indicating effective systemic absorption. The gel, composed of pluronic F-127 and carbopol, formed a stable aerosol suitable for nasal mucosa application. Nonclinical studies showed no toxicity, presenting for epilepsy treatment (Gonçalves et al. 2019).

11.4.3 Applications in Pediatrics

Pediatric population presents unique challenges in drug delivery due to their distinct physiological and metabolic characteristics. Minimally invasive drug delivery systems are increasingly being adapted to meet the needs of these groups, ensuring safer and more effective treatment. Pediatric patients often require tailored drug delivery systems due to their size, developmental stage, and varying levels of compliance. Minimally invasive and user-friendly delivery methods are essential for effective treatment in children. Liquid formulations are commonly used in pediatric medicine because they can be easily dosed and adjusted. Oral syringes and dropper bottles ensure accurate dosing and are preferable for younger children who may have difficulty swallowing tablets or capsules (Batchelor and Marriott 2015). Microneedles designed for pediatric use can deliver vaccines or medications painlessly. For instance, microneedle patches for vaccine delivery are being explored as a less intimidating option compared to traditional needles (Duarah et al. 2019). **Transdermal patches** are used for chronic conditions in children, such as ADHD or chronic pain. They offer the advantage of providing steady medication delivery over time and reducing the need for frequent dosing (Golhen et al. 2023). For respiratory conditions like asthma, inhalers with spacer devices are designed to enhance drug delivery to the lungs in young children. Nebulizers can also be used, though they are less convenient (Gillette et al. 2016).

11.4.4 Applications in Geriatrics

Geriatric patients often face challenges such as diminished organ function, polypharmacy, and cognitive decline. MIDDS plays a crucial role in these patients as it requires special attention on efficacy and safety. The frequency of dosing can be reduced in many ways, viz., controlled, extended release formulations to help the elderly patients suffering from diverse chronic conditions. They also improve treatment adherence by providing constant drug levels (Romley et al. 2020). For instance, TDDS especially **patches** are utilized for the patients who have problem in swallowing pills. Patches for hormones, analgesics, or antihypertensives provide a steady dose of medication and are easy to apply and remove (Vadivelu and Hines 2008). **Smart pill dispenser devices** can help manage complex medication regimens by providing reminders and tracking adherence. They are particularly useful for elderly patients who may struggle with remembering to take their medications (Faisal et al. 2023). For elderly patients with respiratory conditions, inhalers equipped with spacers can improve medication delivery and reduce the effort required to use the device (Hagmeyer et al. 2023).

11.4.5 Applications in Special Populations

Special populations, including patients with cognitive impairments and disabilities or those undergoing complex treatments, require additional considerations in drug delivery. For patients with dementia or other cognitive impairments, simple and easy-to-use drug delivery systems are essential. Solutions include blister packs that simplify pill-taking or liquid formulations that can be administered through a feeding tube if necessary (Lee et al. 2018). For individuals with physical disabilities, drug delivery systems should be designed for ease of use. This includes devices with larger, more accessible controls or assistive technologies that facilitate the administration of medications (Lee 2017). Patients undergoing complex treatments, such as chemotherapy or insulin therapy, benefit from integrated systems that combine monitoring and drug delivery. Examples include insulin pumps with continuous glucose monitoring or infusion pumps for chemotherapy (Diabetes Control and Complications Trial (DCCT) 1987).

11.5 Conclusion

The progression from conventional to advanced drug delivery systems, especially minimally invasive approaches, represents a transformative shift in healthcare. Traditional drug delivery often lacks precision and ease of use, which can limit patient adherence and therapeutic efficacy. Minimally invasive technologies, such as microneedles, nanoparticles, transdermal patches, inhalation devices, and hydrogels, address these limitations by offering patient-friendly, targeted, and efficient methods of drug administration. These technologies are particularly beneficial in managing chronic diseases, as well as in pediatric, geriatric, and other specialized care populations, enhancing both safety and patient outcomes. As a result, MIDDS not only improve treatment accuracy and patient comfort but also exemplify the potential for more personalized and accessible healthcare solutions across diverse medical needs.

References

Ahmad A (2022) Pharmacological strategies and recent advancement in nano-drug delivery for targeting asthma. Life 12(4):596

Akhtar N, Singh V, Yusuf M, Khan RA (2020) Non-invasive drug delivery technology: development and current status of transdermal drug delivery devices, techniques and biomedical applications. Biomed Eng Tech 65(3):243–272

Aldawood FK, Andar A, Desai S (2021) A comprehensive review of microneedles: types, materials, processes, characterizations and applications. Polymers (Basel) 13(16):2815

Alkilani AZ, McCrudden MTC, Donnelly RF (2015) Transdermal drug delivery: innovative pharmaceutical developments based on disruption of the barrier properties of the stratum corneum. Pharmaceutics 7(4):438–470

Almangour TA, Alenazi B, Ghonem L, Alhifany AA, Aldakheel BA, Alruwaili A (2020) Inhaled colistin for the treatment of nosocomial pneumonia due to multidrug-resistant Gram-negative

bacteria: a real-life experience in tertiary care hospitals in Saudi Arabia. Saudi Pharm J 28(8):1009–1013
Almog S, Aharon-Peretz J, Vulfsons S, Ogintz M, Abalia H, Lupo T et al (2020) The pharmacokinetics, efficacy, and safety of a novel selective-dose cannabis inhaler in patients with chronic pain: a randomized, double-blinded, placebo-controlled trial. Eur J Pain 24(8):1505–1516
Anselmo AC, Gokarn Y, Mitragotri S (2019) Non-invasive delivery strategies for biologics. Nat Rev Drug Discov 18(1):19–40
Bahadur S, Pardhi DM, Rautio J, Rosenholm JM, Pathak K (2020) Intranasal nanoemulsions for direct nose-to-brain delivery of actives for CNS disorders. Pharmaceutics 12(12):1230
Bahrami F, Psikuta A, Rossi RM, Dommann A, Defraeye T (2024) Exploring the thermally-controlled fentanyl transdermal therapy to provide constant drug delivery by physics-based digital twins. Eur J Pharm Sci 200:106848
Bansal M, Bansal S, Kumria R (2014) The RapidMist™ system for buccal delivery of insulin. In: das Neves J, Sarmento B (eds) Mucosal delivery of biopharmaceuticals: biology, challenges and strategies. Springer US, Boston, pp 423–436. https://doi.org/10.1007/978-1-4614-9524-6_19
Bariya SH, Gohel MC, Mehta TA, Sharma OP (2011) Microneedles: an emerging transdermal drug delivery system. J Pharm Pharmacol 64(1):11–29
Baryakova TH, Pogostin BH, Langer R, McHugh KJ (2023) Overcoming barriers to patient adherence: the case for developing innovative drug delivery systems. Nat Rev Drug Discov 22(5):387–409
Batchelor HK, Marriott JF (2015) Formulations for children: problems and solutions. Br J Clin Pharmacol 79(3):405–418
Bonferoni MC, Rossi S, Sandri G, Ferrari F, Gavini E, Rassu G et al (2019) Nanoemulsions for "nose-to-brain" drug delivery. Pharmaceutics 11(2):84
Botti G, Dalpiaz A, Pavan B (2021) Targeting systems to the brain obtained by merging prodrugs, nanoparticles, and nasal administration. Pharmaceutics 13(8):1144
Chan AHY, Pleasants RA, Dhand R, Tilley SL, Schworer SA, Costello RW et al (2021) Digital inhalers for asthma or chronic obstructive pulmonary disease: a scientific perspective. Pulm Ther 7(2):345–376
Chang Y, Wang L-B, Li D, Lei K, Liu S-Y (2017) Efficacy of rasagiline for the treatment of Parkinson's disease: an updated meta-analysis. Ann Med 49(5):421–434
Chen G, Yu J, Gu Z (2019) Glucose-responsive microneedle patches for diabetes treatment. J Diabetes Sci Technol 13(1):41–48
Coates AL, Denk O, Leung K, Ribeiro N, Chan J, Green M et al (2011) Higher tobramycin concentration and vibrating mesh technology can shorten antibiotic treatment time in cystic fibrosis. Pediatr Pulmonol 46(4):401–408
Costa C, Moreira JN, Amaral MH, Lobo JMS, Silva AC (2019) Nose-to-brain delivery of lipid-based nanosystems for epileptic seizures and anxiety crisis. J Control Release 295:187–200
Courtenay AJ, McAlister E, McCrudden MTC, Vora L, Steiner L, Levin G et al (2020) Hydrogel-forming microneedle arrays as a therapeutic option for transdermal esketamine delivery. J Control Release 322:177–186
Crowe TP, Hsu WH (2022) Evaluation of recent intranasal drug delivery systems to the central nervous system. Pharmaceutics 14(3):629
Cummings J, Winblad B (2007) A rivastigmine patch for the treatment of Alzheimer's disease and Parkinson's disease dementia. Expert Rev Neurother 7(11):1457–1463
Dale LC, Hurt RD, Offord KP, Lawson GM, Croghan IT, Schroeder DR (1995) High-dose nicotine patch therapy. Percentage of replacement and smoking cessation. JAMA 274(17):1353–1358
Dange SM, Kamble MS, Bhalerao KK, Chaudhari PD, Bhosale AV, Nanjwade BK et al (2014) Formulation and evaluation of venlafaxine nanostructured lipid carriers. J Bionanoscience 8(2):81–89
De Andres J, Hayek S, Perruchoud C, Lawrence MM, Reina MA, De Andres-Serrano C et al (2022) Intrathecal drug delivery: advances and applications in the management of chronic pain patient. Front Pain Res (Lausanne) 3:900566

Dhuria SV, Hanson LR, Frey WH II (2010) Intranasal delivery to the central nervous system: mechanisms and experimental considerations. J Pharm Sci 99(4):1654–1673
Diabetes Control and Complications Trial (DCCT): results of feasibility study. The DCCT Research Group. Diabetes Care 1987 10(1):1–19.
Duarah S, Sharma M, Wen J (2019) Recent advances in microneedle-based drug delivery: special emphasis on its use in paediatric population. Eur J Pharm Biopharm 136:48–69
Faisal S, Samoth D, Aslam Y, Patel H, Park S, Baby B et al (2023) Key features of smart medication adherence products: updated scoping review. JMIR aging 6:e50990
Fan Y, Chen M, Zhang J, Maincent P, Xia X, Wu W (2018) Updated progress of nanocarrier-based intranasal drug delivery systems for treatment of brain diseases. Crit Rev Ther Drug Carr Syst 35(5):433–467
Fan D-Y, Tian Y, Liu Z-J (2019) Injectable hydrogels for localized cancer therapy. Front Chem 7:675
Ferreira NN, Granja S, Boni FI, Prezotti FG, Ferreira LMB, Cury BSF et al (2020) Modulating chitosan-PLGA nanoparticle properties to design a co-delivery platform for glioblastoma therapy intended for nose-to-brain route. Drug Deliv Transl Res 10(6):1729–1747. https://doi.org/10.1007/s13346-020-00824-2
Fu Y, Chapman EJ, Boland AC, Bennett MI (2022) Evidence-based management approaches for patients with severe chronic obstructive pulmonary disease (COPD): a practice review. Palliat Med 36(5):770–782
Gänsslen M (1925) Über Inhalation von Insulin. Klin Wochenschr 4(2):71
Gillette C, Rockich-Winston N, Kuhn JA, Flesher S, Shepherd M (2016) Inhaler technique in children with asthma: a systematic review. Acad Pediatr 16(7):605–615
Goldberg T, Wong E (2015) Afrezza (insulin human) inhalation powder: a new inhaled insulin for the management of type-1 or type-2 diabetes mellitus. P T 40(11):735–741
Golhen K, Buettcher M, Kost J, Huwyler J, Pfister M (2023) Meeting challenges of pediatric drug delivery: the potential of orally fast disintegrating tablets for infants and children. Pharmaceutics 15(4):1033
Gonçalves J, Bicker J, Gouveia F, Liberal J, Oliveira RC, Alves G et al (2019) Nose-to-brain delivery of levetiracetam after intranasal administration to mice. Int J Pharm 564:329–339
Groiss SJ, Wojtecki L, Südmeyer M, Schnitzler A (2009) Deep brain stimulation in Parkinson's disease. Ther Adv Neurol Disord 2(6):20–28
Grosset DG, Dhall R, Gurevich T, Kassubek J, Poewe WH, Rascol O et al (2020) Inhaled levodopa in Parkinson's disease patients with OFF periods: a randomized 12-month pulmonary safety study. Parkinsonism Relat Disord 71:4–10
Guo Y, Yang Y, Xu Y, Meng Y, Ye J, Xia X et al (2022) Deformable nanovesicle-loaded gel for buccal insulin delivery. Pharmaceutics 14:2262
Gupta J, Felner EI, Prausnitz MR (2009) Minimally invasive insulin delivery in subjects with type 1 diabetes using hollow microneedles. Diabetes Technol Ther 11(6):329–337
Hadley G, Derry S, Moore RA, Wiffen PJ (2013) Transdermal fentanyl for cancer pain. Cochrane Database Syst Rev 2013(10):CD010270
Hagmeyer L, van Koningsbruggen-Rietschel S, Matthes S, Rietschel E, Randerath W (2023) From the infant to the geriatric patient-strategies for inhalation therapy in asthma and chronic obstructive pulmonary disease. Clin Respir J 17(6):487–498
Haque S, Md S, Sahni JK, Ali J, Baboota S (2014) Development and evaluation of brain targeted intranasal alginate nanoparticles for treatment of depression. J Psychiatr Res 48(1):1–12
Hariharan A, Tran SD (2023) Localized drug delivery systems: an update on treatment options for head and neck squamous cell carcinomas. Pharmaceutics 15(7):1844
Heida R, Hinrichs WL, Frijlink HW (2022) Inhaled vaccine delivery in the combat against respiratory viruses: a 2021 overview of recent developments and implications for COVID-19. Expert Rev Vaccines 21(7):957–974
Heo J, Jeon EY, Joo KI, Cha HJ (2023) Locoregional melanoma therapy by tissue adhesive microneedle patch-assisted trans-tumoral delivery of anticancer drug. Biotechnol Bioprocess Eng 28(3):473–482

Huynh K (2017) Minimally invasive delivery of engineered cardiac patches for heart repair. Nat Rev Cardiol 14(10):568. https://doi.org/10.1038/nrcardio.2017.137

Islam SU, Shehzad A, Ahmed MB, Lee YS (2020) Intranasal delivery of nanoformulations: a potential way of treatment for neurological disorders. Molecules 25(8):1929

Ita K (2015) Transdermal delivery of drugs with microneedles-potential and challenges. Pharmaceutics 7(3):90–105

Jimmy B, Jose J (2011) Patient medication adherence: measures in daily practice. Oman Med J 26(3):155–159

Jogani VV, Shah PJ, Mishra P, Mishra AK, Misra AR (2008) Intranasal mucoadhesive microemulsion of tacrine to improve brain targeting. Alzheimer Dis Assoc Disord 22(2):116–124

Jonas O, Landry HM, Fuller JE, Santini JT, Baselga J, Tepper RI et al (2015) An implantable microdevice to perform high-throughput in vivo drug sensitivity testing in tumors. Sci Transl Med 7(284):284ra57

Kanikkannan N, Kandimalla K, Lamba SS, Singh M (2000) Structure-activity relationship of chemical penetration enhancers in transdermal drug delivery. Curr Med Chem 7(6):593–608

Kew KM, Dias S, Cates CJ (2014) Long-acting inhaled therapy (beta-agonists, anticholinergics and steroids) for COPD: a network meta-analysis. Cochrane Database Syst Rev 2014(3):CD010844

Khafagy E-S, Morishita M, Onuki Y, Takayama K (2007) Current challenges in non-invasive insulin delivery systems: a comparative review. Adv Drug Deliv Rev 59(15):1521–1546. https://www.sciencedirect.com/science/article/pii/S0169409X07001913

Kim Y-C, Park J-H, Prausnitz MR (2012) Microneedles for drug and vaccine delivery. Adv Drug Deliv Rev 64(14):1547–1568

Kopper NW, Gudeman J, Thompson DJ (2009) Transdermal hormone therapy in postmenopausal women: a review of metabolic effects and drug delivery technologies. Drug Des Devel Ther 2:193–202

Kulkarni VS (2009) Handbook of non-invasive drug delivery systems: science and technology. Elsevier, Amsterdam

Kurano T, Kanazawa T, Ooba A, Masuyama Y, Maruhana N, Yamada M et al (2022) Nose-to-brain/spinal cord delivery kinetics of liposomes with different surface properties. J Control Release 344:225–234. https://www.sciencedirect.com/science/article/pii/S0168365922001432

Labhasetwar V (2005) Nanotechnology for drug and gene therapy: the importance of understanding molecular mechanisms of delivery. Curr Opin Biotechnol 16(6):674–680

Lan X, She J, Lin D-A, Xu Y, Li X, Yang W-F et al (2018) Microneedle-mediated delivery of lipid-coated cisplatin nanoparticles for efficient and safe cancer therapy. ACS Appl Mater Interfaces 10(39):33060–33069

Larrañeta E, McCrudden MTC, Courtenay AJ, Donnelly RF (2016) Microneedles: a new frontier in nanomedicine delivery. Pharm Res 33(5):1055–1073

Lee H (2017) The internet of things and assistive technologies for people with disabilities. In: Internet of things and advanced application in healthcare. IGI Global, Hershey, pp 32–65

Lee JW, Park J-H, Prausnitz MR (2008) Dissolving microneedles for transdermal drug delivery. Biomaterials 29(13):2113–2124

Lee L, Patel T, Molnar F, Seitz D (2018) Optimizing medications in older adults with cognitive impairment: considerations for primary care clinicians. Can Fam Physician 64(9):646–652

Lewis AL, Jordan F, Patel T, Jeffery K, King G, Savage M et al (2015) Intranasal human growth hormone (hGH) induces IGF-1 levels comparable with subcutaneous injection with lower systemic exposure to hGH in healthy volunteers. J Clin Endocrinol Metab 100(11):4364–4371. https://doi.org/10.1210/jc.2014-4146

Li X, Zhao Z, Zhang M, Ling G, Zhang P (2022) Research progress of microneedles in the treatment of melanoma. J Control Release 348:631–647

Liu P, Chen G, Zhang J (2022) A review of liposomes as a drug delivery system: current status of approved products, regulatory environments, and future perspectives. Molecules 27(4):1372

Makaram P, Owens D, Aceros J (2014) Trends in nanomaterial-based non-invasive diabetes sensing technologies. Diagnostics 4:27–46

Mansoor I, Eassa HA, Mohammed KHA, Abd El-Fattah MA, Abdo MH, Rashad E et al (2022) Microneedle-based vaccine delivery: review of an emerging technology. AAPS PharmSciTech 23(4):103

Md S, Haque S, Fazil M, Kumar M, Baboota S, Sahni JK et al (2014) Optimised nanoformulation of bromocriptine for direct nose-to-brain delivery: biodistribution, pharmacokinetic and dopamine estimation by ultra-HPLC/mass spectrometry method. Expert Opin Drug Deliv 11(6):827–842

Misra SK, Pathak K (2023) Nose-to-brain targeting via nanoemulsion: significance and evidence. Colloids Interfaces 7(1):23

Mittal D, Md S, Hasan Q, Fazil M, Ali A, Baboota S et al (2016) Brain targeted nanoparticulate drug delivery system of rasagiline via intranasal route. Drug Deliv 23(1):130–139

Moreira AF, Rodrigues CF, Jacinto TA, Miguel SP, Costa EC, Correia IJ (2019) Microneedle-based delivery devices for cancer therapy: a review. Pharmacol Res 148:104438

Moser EG, Crew LB, Garg SK (2010) Role of continuous glucose monitoring in diabetes management. Av en Diabetol 26(2):73–78

Muankaew C, Loftsson T (2018) Cyclodextrin-based formulations: a non-invasive platform for targeted drug delivery. Basic Clin Pharmacol Toxicol 122(1):46–55

Olschewski H, Ghofrani HA, Schmehl T, Winkler J, Wilkens H, Höper MM et al (2000) Inhaled iloprost to treat severe pulmonary hypertension. An uncontrolled trial. German PPH Study Group. Ann Intern Med 132(6):435–443

Ozsoy Y, Gungor S, Cevher E (2009) Nasal delivery of high molecular weight drugs. Molecules 14:3754–3779

Park K, Kwon IC, Park K (2011) Oral protein delivery: current status and future prospect. React Funct Polym 71(3):280–287

Patra JK, Das G, Fraceto LF, Campos EVR, del Rodriguez-Torres MP, Acosta-Torres LS et al (2018) Nano based drug delivery systems: recent developments and future prospects. J Nanobiotechnol 16(1):1–33

Peña B, Laughter M, Jett S, Rowland TJ, Taylor MRG, Mestroni L et al (2018) Injectable hydrogels for cardiac tissue engineering. Macromol Biosci 18(6):e1800079

Peppas N (2000) Hydrogels in pharmaceutical formulations. Eur J Pharm Biopharm 50(1):27–46

Picone P, Sabatino MA, Ditta LA, Amato A, San Biagio PL, Mulè F et al (2018) Nose-to-brain delivery of insulin enhanced by a nanogel carrier. J Control Release 270:23–36

Prausnitz MR, Mitragotri S, Langer R (2004) Current status and future potential of transdermal drug delivery. Nat Rev Drug Discov 3(2):115–124

Prisant LM, Elliott WJ (2003) Drug delivery systems for treatment of systemic hypertension. Clin Pharmacokinet 42(11):931–940

Priya S, Singhvi G (2022) Microneedles-based drug delivery strategies: a breakthrough approach for the management of pain. Biomed Pharmacother 155:113717

Qian S, He L, Wang Q, Wong YC, Mak M, Ho C-Y et al (2018) Intranasal delivery of a novel acetylcholinesterase inhibitor HLS-3 for treatment of Alzheimer's disease. Life Sci 207:428–435

Romley JA, Xie Z, Chiou T, Goldman D, Peters AL (2020) Extended-release formulation and medication adherence. J Gen Intern Med 35(1):354–356

Ruan S, Li J, Ruan H, Xia Q, Hou X, Wang Z et al (2024) Microneedle-mediated nose-to-brain drug delivery for improved Alzheimer's disease treatment. J Control Release 366:712–731. https://www.sciencedirect.com/science/article/pii/S0168365924000178

Salehi S, Naghib SM, Garshasbi HR, Ghorbanzadeh S, Zhang W (2023) Smart stimuli-responsive injectable gels and hydrogels for drug delivery and tissue engineering applications: a review. Front Bioeng Biotechnol 11:1104126

Sieg A, Guy RH, Delgado-Charro MB (2005) Noninvasive and minimally invasive methods for transdermal glucose monitoring. Diabetes Technol Ther 7(1):174–197. https://doi.org/10.1089/dia.2005.7.174

Tanga S, Aucamp M, Ramburrun P (2023) Injectable thermoresponsive hydrogels for cancer therapy: challenges and prospects. Gels 9(5):418

Turpeinen A, Eriksson P, Happonen A, Husman-Piirainen J, Haikarainen J (2021) Consistent dosing through the salmeterol-fluticasone propionate easyhaler for the management of asthma and chronic obstructive pulmonary disease: robustness analysis across the easyhaler lifetime. J Aerosol Med Pulm Drug Deliv 34(3):189–196

Ueno H, Mizuta M, Shiiya T, Tsuchimochi W, Noma K, Nakashima N et al (2014) Exploratory trial of intranasal administration of glucagon-like peptide-1 in Japanese patients with type 2 diabetes. Diabetes Care 37(7):2024–2027. https://doi.org/10.2337/dc13-0690

Vadivelu N, Hines RL (2008) Management of chronic pain in the elderly: focus on transdermal buprenorphine. Clin Interv Aging 3(3):421–430

Vaidya A, Mitragotri S (2020) Ionic liquid-mediated delivery of insulin to buccal mucosa. J Control Release 327:26–34. https://www.sciencedirect.com/science/article/pii/S0168365920304211

Van Woensel M, Wauthoz N, Rosière R, Mathieu V, Kiss R, Lefranc F et al (2016) Development of siRNA-loaded chitosan nanoparticles targeting Galectin-1 for the treatment of glioblastoma multiforme via intranasal administration. J Control Release 227:71–81

Vashist SK (2012) Non-invasive glucose monitoring technology in diabetes management: a review. Anal Chim Acta 750:16–27. https://www.sciencedirect.com/science/article/pii/S0003267012004606

Vllasaliu D (2021) Non-invasive drug delivery systems. Pharmaceutics 13(5):611

Yélamos O, Geller S, Tokez S (2023) Skin cancer special issue in skin health and disease. Ski Heal Dis 3(2):e224

Zhang N, Zhou X, Liu L, Zhao L, Xie H, Yang Z (2021) Dissolving polymer microneedles for transdermal delivery of insulin. Front Pharmacol 12:719905

Zhang Y-B, Xu D, Bai L, Zhou Y-M, Zhang H, Cui Y-L (2022) A review of non-invasive drug delivery through respiratory routes. Pharmaceutics 14:1974

Zhao L, Weir MD, Xu HHK (2010) An injectable calcium phosphate-alginate hydrogel-umbilical cord mesenchymal stem cell paste for bone tissue engineering. Biomaterials 31(25):6502–6510

Zhao J, Xu G, Yao X, Zhou H, Lyu B, Pei S et al (2022) Microneedle-based insulin transdermal delivery system: current status and translation challenges. Drug Deliv Transl Res 12(10):2403–2427

이영식 (2020) Implantable and minimally-invasive drug delivery system using soft bioelectronics for diagnosis and treatment. 서울대학교 대학원

Industrial Perspective of Minimally Invasive Drug Delivery System

12

Bhumika Parashar and Rishabha Malviya

Abstract

Minimally invasive drug delivery systems (MIDDS) have emerged as transformative tools in modern medicine, addressing the limitations of traditional administration methods such as oral and injectable routes. These innovative systems, including microneedles, nanocarriers, and implantable devices, provide targeted, efficient, and patient-friendly solutions. Technologies like 3D printing and bioresponsive materials have enhanced drug delivery precision, while digital integration supports personalized healthcare. By overcoming barriers such as poor bioavailability, systemic toxicity, and patient noncompliance, MIDDS are reshaping drug administration across various therapeutic areas, from chronic diseases to oncology. This article aims to explore the advancements in minimally invasive drug delivery systems (MIDDS), including microneedles, nanocarriers, and implantable devices, emphasizing their industrial applications and innovations for personalized healthcare solutions. The industrial perspective on MIDDS highlights its transformative role in pharmaceuticals by offering alternatives to traditional drug administration methods. Microneedle technology, combined with advances in materials science and fabrication techniques, has enabled painless and efficient drug delivery for vaccines, insulin, and cancer therapies. Nanocarriers provide targeted delivery and controlled release, enhancing efficacy while minimizing systemic toxicity. Integration with digital health tools, such as wearable devices and AI-driven systems, further supports personalized medicine and remote patient management. Continued research and cross-disciplinary collaboration are crucial to overcoming current limitations and optimizing MIDDS for clinical and industrial use. Minimally invasive drug delivery systems (MIDDS) demonstrate significant potential to improve drug administra-

B. Parashar · R. Malviya (✉)
Department of Pharmacy, School of Medical and Allied Sciences, Galgotias University, Greater Noida, Uttar Pradesh, India

M. R. Babu et al. (eds.), *Transdermal Applications of Minimally Invasive Drug Delivery Systems*, https://doi.org/10.1007/978-981-96-6275-3_12

tion, enhance patient compliance, and achieve targeted therapeutic outcomes. Key technologies, such as microneedles, nanocarriers, and implantable devices, have revolutionized drug delivery with innovations like 3D printing, bioresponsive systems, and digital integration. These systems offer precise control, reduced side effects, and broader applicability across various medical fields. However, challenges like manufacturing complexity, regulatory barriers, and cost must be addressed for widespread adoption.

Keywords

Microneedles · Nanocarriers · Implantable devices · Personalized medicine · Digital health integration · 3D printing · Bioresponsive systems · Controlled drug release

12.1 Introduction

The event commemorates the 10th anniversary of the regulatory endorsement of the "Natpara case," the inaugural instance in which quantitative systems pharmacology (QSP) impacted a regulatory determination. The FDA Office of Clinical Pharmacology employed a calcium homeostasis quantitative systems pharmacology model to substantiate their request for a post-marketing clinical trial aimed at investigating how alternate dosage methods would mitigate an adverse event. Combination therapies: QSP models elucidate therapeutic interventions at the cellular level, connecting the target molecule to pharmacodynamic (PD) biomarkers via a network of biological events. This enables the simulation of pharmacological effects on multiple targets within a singular model. The benefit of QSP models is that, although the pharmacological action is additive at the cellular level, synergistic effects may arise at the systemic (patient) level. Considering the myriad permutations of dosages, dosing regimens, and timings for double and triple drug administrations, QSP simulations can assist in prioritizing the drug combinations for clinical use. This approach can also be utilized to assess a company's asset distinction from the standard of care or competing medicines (Venkatakrishnan and van der Graaf 2022). Conventional administration methods, including subcutaneous injection and oral ingestion, are extensively utilized for the delivery of the majority of medications. Transdermal drug delivery reduces needle anxiety and pain compared to traditional approaches, hence improving patient compliance. Moreover, transdermal distribution generally necessitates reduced drug dosages compared to oral administration, and it can bypass the detrimental effects associated with the erratic absorption and metabolism of pharmaceuticals in the gastrointestinal system (Ramadon et al. 2020). In 1976, Gerstel patented the notion of microscale arrays for the immediate penetration of the stratum corneum. At that time, the absence of manufacturing technology rendered the production and commercialization of microneedles unfeasible. The advent of microelectronic instruments in the semiconductor sector throughout the 1990s rekindled scientists' interest in the production of microneedles. Since that time, a variety of materials have been employed in the manufacturing of microneedles, including ceramics, stainless steel, polymers, silicon, silk, and sugars, among others. The first fabrication technique for microneedles is MEMS technology, which has subsequently led to the development of laser cutting, photolithography, wet and dry etching, micro-molding, and 3D printing (Kulkarni et al. 2022). The process of additive manufacturing, also known as 3D printing, is a modern technique that fabricates objects layer by layer based on computer-aided design models. This technology may generate intricate structures, enhanced precision, and substantial drug loading capacity, rendering it very effective for delivering customized medications and medical devices that successfully address the limitations of traditional dosage forms (Cutrim et al. 2021). The nanocarrier exhibited exceptional loading efficiency and pH sensitivity, facilitating a regulated release of the therapeutic payload in both the acidic tumor milieu (pH 5.4) and the physiological normal milieu (pH 7.4), thereby augmenting its therapeutic efficacy. The cytotoxicity of the created system was assessed on the A549 cell line to determine the efficacy of the cancer therapy. The L929 cell line was analyzed to investigate any potential unfavorable effects induced by the nanocarrier. The varied physiological conditions within the human body catalyze the development and investigation of nanocarriers with advanced physicochemical properties. These nanocarriers are engineered to surmount impediments and barriers, enabling them to precisely target and treat specific regions impacted by diseases (Xu et al. 2020). The intricacy of diabetes arises from the interaction of genetic, environmental, and nutritional lifestyle factors. Effective interventions necessitate a thorough understanding of the patient, encompassing genetics, demographics, health history, family history, lifestyle patterns, dietary habits, and real-time health data, along with an awareness of the patient's knowledge regarding their condition. Nonetheless, current methodologies frequently encounter challenges related to tailoring, complication prevention, patient awareness, and adherence concerns (Rizvi and Saleh 2018).

12.2 Microneedle Array

Recently, there has been considerable focus on minimally invasive transdermal medication delivery systems utilizing microneedles. Microneedles comprise a patch embedded with many needles of micrometer length, surpassing the thickness of the skin's stratum corneum. As illustrated in Fig. 12.1, upon insertion into the skin, microneedles effectively traverse the stratum corneum, enabling the direct administration of pharmaceuticals to the epidermis and dermis, which are then absorbed into the circulatory system via microvessels and lymphatic capillaries (Donnelly and Larrañeta 2019).

Microneedles (MNs) are micron-scale devices that facilitate the transdermal delivery of drugs by penetrating the stratum corneum. MNs include many components, including ceramics, polymers, and metals, designed for the epidermal and intradermal administration of vaccinations and bioactive substances. The principal

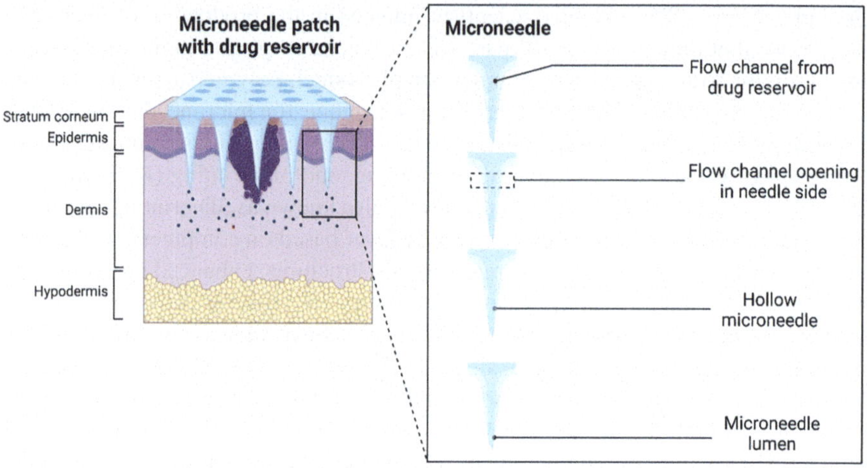

Fig. 12.1 Microneedles for transdermal DDS

method of MN drug delivery entails creating a micron-sized conduit by compromising the epidermal barrier. One of the primary obstacles to drug delivery by microneedles is the stratum corneum, the outermost layer of the skin. To enhance transdermal medication administration, various approaches have been utilized, including sonophoresis, iontophoresis, and electroporation technologies. Magnetic nanoparticles have been employed to enhance the drug concentration at the site of delivery (Ligon et al. 2017). MNs are broadly endorsed by patients and healthcare professionals. There exist several categories of microneedles: solid microneedles, dissolvable microneedles, coated microneedles, hydrogel-based microneedles, hollow microneedles, and polymeric microneedles. The categorization of MNs is illustrated in Fig. 12.2. The base may be selected based on the delivery requirements during fabrication. Recently, MNs have attracted significant attention as an innovative TDD approach due to their unique benefits, which encompass low invasiveness, self-administration, and painless delivery attributed to their micron-sized nature. Microneedles (MNs) can effectively administer diverse agents, including tiny pharmacological molecules, macromolecules (such as peptides, proteins, vaccines, and nucleic acids), and nanoparticles, facilitating their passage through the subcutaneous layer to achieve direct entry into the systemic circulation (Ligon et al. 2017).

12.2.1 3D-Printed MN for Delivery of Insulin

3D-printed microneedle arrays including spear and pyramid shapes were fabricated using a commercial SLA printer and a biocompatible Class 1 polymer. Subsequently, insulin-sugar films were added to the 3D-printed arrays using inkjet printing. Inkjet printing is employed to modify the surface of the MN by uniformly applying physiologically relevant components. It produces patterns with diverse geometries. It

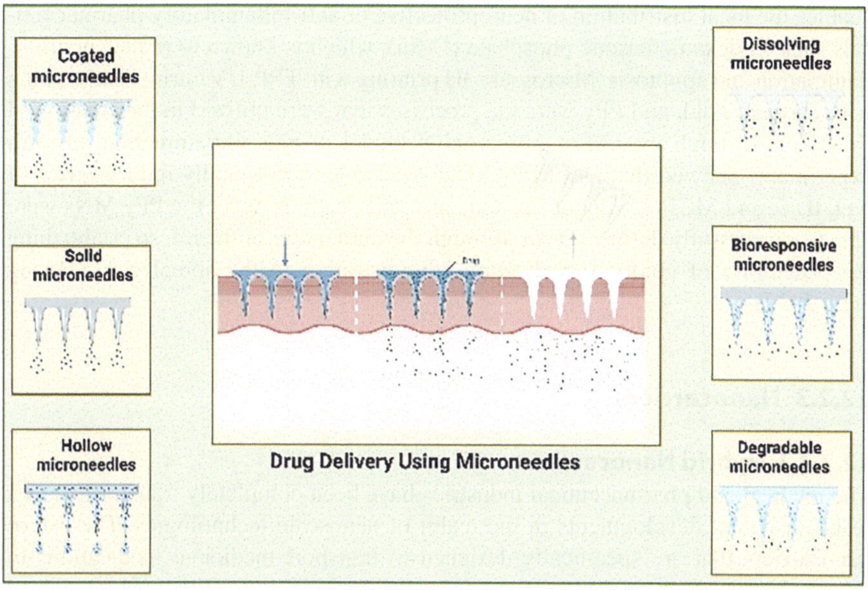

Fig. 12.2 Different types of microneedles

could mitigate the drawbacks associated with dip coating techniques such as immersion, retraction, and drying phases (Boehm et al. 2011). Mannitol, trehalose, and xylitol served as insulin carriers to preserve insulin action before the deposition of active films on the surface of the microneedles. Both in vivo and in vitro experiments demonstrated that coated microneedle systems rapidly released insulin (Economidou et al. 2019).

12.2.2 3D-Printed MNs for Electronically Controlled Transdural Drug Release

Subsequent injury mechanisms such as neuroinflammation, excitotoxicity, and ischemia contribute to continued damage after a spinal cord injury (SCI). Inflammation and cytotoxicity can cause additional harm to neural cells after a SCI has occurred. Excess neurotransmitters, free radicals, and reactive oxygen species are the causes of excitotoxicity, which ultimately results in the death of cells in the vicinity of the damage site. Free radicals, such as nitric oxide that is generated by inflammatory microglia, have the potential to cause harm to surrounding tissues that have not been affected, which can lead to secondary neuron loss lasting several weeks after the injury. The academics came up with a solution to this problem by developing a biocompatible polymer called polypyrrole (PPy), which is conductive and may be used for electronically controlled medication delivery through implanted electrodes. For transdural medication delivery, a PPy MN array was manufactured. This array enables the local distribution of neuroprotective or anti-inflammatory pharmaceuticals such as dexamethasone phosphate (Dexa), which is known to reduce neuroinflammation and apoptosis. Microscale 3D printing with TPP, UV curing, and coating with chrome, gold, and PPy were the processes that were utilized in the creation of the MNs. Through the use of a microglial model of neuroinflammation, in vitro experiments showed that Dexa PPy MNs were able to drastically reduce levels of NO, IL-6, and MCP-1 without inducing any cell death. In vivo, the PPy MNs were able to successfully deliver (Dexa) through the dura mater of the rat, so establishing the possibility of epidural medication administration to the spinal cord (Huang et al. 2022).

12.2.3 Nanocarriers

12.2.3.1 Hybrid Nanocarriers

The medical and pharmaceutical industries have been completely transformed as a result of recent developments in the realm of nanoscale technologies. The use of nanocarriers that are specifically designed to transport medicinal medications to specific locations within the human body is a promising solution that can be achieved through the application of nanotechnology (Ghasemiyeh and Mohammadi-Samani 2019). Nanocarriers have attracted a lot of attention in the field of cancer therapies since they possess distinctive qualities and advantages that standard drug delivery methods do not possess. Enhanced drug solubility, controlled release capabilities, prolonged circulation in the body, and the capacity to target specific areas are all characteristics of these carriers, which are made possible by the utilization of the nanoscale size range. There have been a great number of papers that have brought attention to the accomplishments that have been made in the field of pH-responsive drug carriers, highlighting their potential to improve the therapeutic effectiveness of pharmaceuticals (Zauška et al. 2022). In this context, the selection of appropriate nanocarriers is of utmost importance throughout the process of encapsulation. This is because it has an impact on targeted distribution, controlled release, and, ultimately, the bioavailability of active components. As potential therapeutic solutions, polymer-based hybrid materials stand out as a promising choice among the many available options. These hybrid materials have already demonstrated an outstanding commitment to addressing and offering answers to critical difficulties in important areas such as human health, the environment, and energy (Beňová et al. 2021). Despite this, their design, performance, and applicability in the real world could still use some improvement (Sabir et al. 2021). A great number of articles have brought attention to the accomplishments that have been made in the field of pH-responsive drug carriers, suggesting that these carriers have the potential to improve the therapeutic efficacy of medications. In one particular study, for instance, the primary objective was to develop a microemulsion that was based on polyvinylpyrrolidone (PVP) and had the capability of transporting the anticancer medicine doxorubicin. The purpose of this endeavor was to improve the treatment of cancer. The

Table 12.1 Characteristics and applications of pH-responsive nanocarriers for targeted drug delivery

Parameter	Details
Components of nanocarrier	Polyacrylic acid (PAA), polyvinylpyrrolidone (PVP), and molybdenum disulfide (MoS_2) (Aboudzadeh et al. 2020)
Drug used	5-Fluorouracil (5-FU)
Key properties	pH-responsive, enhanced drug loading (46%) and encapsulation efficiency (87.25%) (Hamzehlou and Aboudzadeh 2021)
Particle size	~315 nm (average), PDI: 0.27
Surface charge	Zeta potential: −33.5 mV
Release profile	pH 5.4—33% at 12 h, 49% at 24 h; pH 7.4—15% at 12 h, 31% at 24 h (Jin et al. 2014)
Cytotoxicity (MTT assay)	Significant cancer cell death in A549 cells; minimal effects on L929 normal cells
Techniques used	FT-IR, XRD, FE-SEM, DLS, UV-Vis spectrophotometry
Morphology	Quasi-spherical particles (Zafar et al. 2021)
Synthesis method	Hydrothermal synthesis of MoS_2; double emulsion (water-oil-water) for drug encapsulation (Valencia-Lazcano et al. 2023)
Applications	Lung cancer treatment; targeted drug delivery with reduced side effects on normal tissues (Matusiak et al. 2021)
Advantages	High stability, pH-sensitive release, synergistic effects of MoS_2 and 5-FU, customizable with targeting ligands (Shakouri et al. 2023)

polyvinylpyrrolidone/polyacrylic acid/molybdenum disulfide nanomaterial is one of the most promising nanomaterials that has been exploited in the construction of an effective nanocarrier (Table 12.1) (Prosheva et al. 2019).

12.2.3.2 Functionalize Exosomes

Exosomes are vesicular structures that originate from endolysosomes and range in size from 30 to 150 nm. They are discharged by cells into extracellular fluids such as blood, cerebrospinal fluid (CSF), urine, saliva, and other fluids. In their aqueous core, they are made up of double-layered lipid membranes that are interspersed with transmembrane proteins. These proteins encapsulate nucleic acids, proteins, or bioactive compounds (Kalluri and LeBleu 2020). The invagination of the plasma membrane into the cytosol, which results in the production of an intraluminal vesicle (ILV), is the first step in the process of exosome development (Kim et al. 2020). To generate multivesicular bodies (MVBs), various proteins that are found on intraluminal vectors (ILVs) are recycled back into the plasma membrane, and new cargos are injected into them. Exosomes are released into the cytoplasm when MVBs fuse with either lysosome, which causes them to be destroyed, or with the plasma membrane, which causes them to fuse (Li et al. 2023). The combination of ultracentrifugation and density gradient centrifugation is the technique that is utilized the most frequently for the objective of isolating exosomes. This approach demands a significant volume of samples to be isolated, in addition to being laborious and time-consuming all at the same time. Because of this, cutting-edge techniques such as ultrafiltration or affinity-based capture of exosomes are utilized. These techniques are not only efficient but also produce a significant quantity of exosomes (Lombardo et al. 2020).

Chemical Functionalization of Exosomes

There is an abundance of amine/carboxylic acid-ended phospholipids and transmembrane proteins on the surface of exosomes. These proteins may be functionalized by targeting moiety by an immediate chemical process (Tian et al. 2018). When it comes to chemical functionalization strategies, click chemistry is preferred over traditional chemical reactions because the latter use toxic solvents and reaction conditions that compromise the exosomes' integrity (Salunkhe et al. 2020). Physical forces are utilized in click chemistry reactions, which are rapid and carried out in aqueous solvents and produce a high yield of functionalization. Physical pressures such as sonication, extrusion, and freeze-thaw cycles are employed to disrupt the integrity of the membrane (Chitkara and Mittal 2020). Another method involves the use of functionalized liposomes in a hydrophobic insertion technique. A hybridized liposome-exosome system with targeting ligands is created when physically damaged exosomes are exposed to functionalized liposomes. Internal component release during the procedure is a major drawback of this strategy (Ahlschwede et al. 2019).

Biological Functionalization of Exosomes

A common method to biologically functionalize a target is to employ the machinery of exosome-producing cells to construct a transmembrane protein coupled with a targeting ligand. Exosomes including the transmembrane proteins Lamp, GPI, CD63, CD9, CD81, etc. can serve as a foundation for functionalization. The genes for the transmembrane protein and the targeting ligand are fused in chimeric genes that are carried by expression vectors (plasmids/viruses) into cells that can produce exosomes. These chimeric genes are processed by the cells that produce exosomes, which then result in a transmembrane protein with a targeting ligand expressed on its surface (Chen et al. 2016).

12.2.3.3 Functionalize Lipid Nanoparticles

A variety of methods exist for functionalizing LNPs, one of which is surface modification with target ligands such as cell-penetrating peptides, exogenous proteins, or antibodies. Popular targeting ligands for targeted nucleic acid delivery include CPPs such as rabies virus glycoprotein (RVG) peptide and penetratin (PEN) peptide, as well as exogenous proteins such as transferrin (Tf) and lactoferrin (Lf) (Huang et al. 2007).

Functionalization with Exogenous Proteins

Several in vitro and in vivo experiments showed that Tf-liposomal formulation increased the transport of drugs with poor brain penetrability to the brain (Chowdhury and Chowdhury 2023). The multifunctional lactoferrin (Lf) protein is a member of the transferrin family and a cationic iron-binding protein with anticarcinogenic and anti-inflammatory characteristics, among others. Lf interacts with many different things, one of which is the LDL receptor-related protein, which helps get to the brain utilizing receptor-mediated endocytosis (Gothwal et al. 2023). Potential for brain targeting is suggested by the fact that Lf expression is considerably elevated in both neurons and glia in AD. As a ligand for targeted delivery to the brain, a study developed Lf-liposomes. Compared to commonly manufactured liposomes, this formulation showed a better absorption profile in primary brain capillary endothelial cells, a remarkable ability to traverse the blood-brain barrier in vitro, a greater accumulation of the formulation in the brain, and a decreased cellular toxicity. It has also been demonstrated that surface Lf effectively carries nerve growth factor (NGF) as Lf/NGF-liposomes across the blood-brain barrier (BBB), preventing the SK-N-MC cell degeneration through Aβ-induced neurotoxicity (Wang et al. 2014).

Functionalization with Antibodies

A novel set of antibodies that specifically target receptor-mediated transcytosis (RMT) has recently been demonstrated to bind to the blood-brain barrier (BBB) in slices of brain tissue from both mice and humans. An effective RMT antibody-targeted liposomal system using antibodies was developed to improve brain-specific delivery. In vitro cell absorption of modified liposomes was demonstrated. Brain accumulation was ten times more in the pralidoxime-loaded antibody-liposome group compared to the liposomal control group, according to in vivo assessments. These findings suggest that the antibody liposomal formulation could be a useful synthetic vehicle for delivering medications to the central nervous system (Ye et al. 2022). Using Tf as a targeted ligand also has its limits, such as the possibility of blocking iron uptake due to the saturation of Tf-receptors. One potential solution is to investigate antibodies that attach to Tf-receptors through sites other than Tf to circumvent this problem. Without affecting Tf-binding in rats, the anti-Tf-receptor monoclonal antibody OX26 interacts with the extracellular domain of the Tf-receptor. Drug carriers conjugated with OX26 can cross the blood-brain barrier (BBB) in the same way as Tf. Conjugated with streptavidin and OX26, the neurotrophin bFGF, a powerful neuroprotective agent, was found to enhance brain absorption. Neuroimaging in Alzheimer's disease can be accomplished by intravenous injection of radio-iodinated Aβ (125I-Aβ40) and subsequent linking to OX26 using a streptavidin-biotin linker (Wong et al. 2019).

12.2.4 Transdermal Drug Delivery Systems

The transfer of the drug through the skin is referred to as transdermal drug delivery (TDD), and it is an appealing, noninvasive systemic route that is an alternative to parenteral and oral drug delivery methods. TDD provides painless drug delivery, eliminates the need for first-pass metabolism, improves patient compliance, and allows for self-administration (Aich et al. 2022) and minimally intrusive medication therapy. Furthermore, this approach improves the bioavailability of the medications, which in turn leads to an increase in both the local and systemic therapy of the drugs. TDD and other dosage forms, such as transdermal and iontophoretic patches, which have been clinically accepted and sold, were investigated by the researchers (Ita 2015). It is vital to discuss the layers of the skin to gain an understanding of how the medicine can penetrate the skin. Three layers make up the human skin: the dermis, the epidermis, and the hypodermis. The epidermis is composed of several layers, including the stratum granulosum, the stratum corneum, the stratum lucidum, the stratum spinosum, and the stratum basale, for example. Because of its high hydrophobicity and the presence of dead keratinized cells and corneocytes, the stratum corneum can block the administration of the majority of medications through the skin. When applied to the skin, it functions as a barrier, protecting a variety of substances. As a result, smaller particles with a low molecular weight can be easily pierced, but larger molecules require specialized procedures (Jeong et al. 2021).

12.2.4.1 Microelectronic Patches

When it involves transdermal drug administration, microneedle patches take use of their minimally invasive nature to successfully penetrate the surface layer of the skin, which in turn facilitates the quick absorption and release of the drug. For example, the incorporation of flexible surface acoustic wave technology with microneedle patches has increased the capability of transdermal delivery of macromolecular medications (Zhang et al. 2024a). In addition, a lipid nanocapsule-microneedle array patch loaded with fenretinide has been developed as a novel delivery strategy for the chemoprevention of breast cancer. This method improves the local distribution of the drug, which in turn increases the drug's bioavailability. A proof-of-concept study has also been successful in fabricating a three-layer dissolving microneedle (TDMN) to improve the bioavailability and brain distribution of rivastigmine, a medication that is used for the treatment of Alzheimer's disease (Apolinário et al. 2024). A few microneedle patches have been designed in a biomimetic manner, while others have incorporated sophisticated response therapies. In the field of ocular surgery, for instance, a microneedle patch that is inspired by the mouthparts of mosquitoes has been developed to provide local anesthetic. Polyvinylpyrrolidone (PVP) and hyaluronic acid (HA) are the two components that make up this patch. It can rapidly break down and release lidocaine, which results in efficient ocular analgesia (Ilyas et al. 2024). Additionally, a unique double-barbed microneedle was developed to treat tendinopathy in another recent study. This particular microneedle is equipped with nanocapsules that contain the JAK/STAT inhibitor WP1066, which promotes the self-renewal, migration, and stemness of tenocyte stem/progenitor cells (TSPCs), hence alleviating the symptoms of tendinopathy. In addition, a core-shell microneedle (CSMN) patch has been developed to deliver tannic acid-magnesium (TA-Mg) complex and extracellular vesicles from *Lactobacillus druckerii* (LDEVs) sequentially. To resist pathogen overburden and reduce reactive oxygen species (ROS), this CSMN@TA-Mg/LDEV patch initially releases TA-Mg when it is placed at the site of infection. This helps to facilitate the transition to the proliferative phase. The subsequent effect of the prolonged release of LDEVs is that it stimulates the activity of keratinocytes and fibroblasts, encourages the formation of blood vessels, and controls the accumulation of collagen (Zhang et al. 2024b).

Microneedle Patch for Skin Photoaging Treatment

For thousands of years, people have been applying substances to their skin to achieve medicinal results. The potential of transdermal drug delivery as an alternative to oral and subcutaneous administration has not yet been completely realized, although it plays a major role in the practice of medicine. There has been a consistent rise in the utilization of transdermal drug delivery systems for the surface administration of low-dose, lipophilic, and low-concentration medicines in clinical settings (Yang et al. 2024). On the other hand, because of the high molecular size of biological macromolecules and vaccines like insulin, parathyroid hormone, and influenza vaccine, there is not yet a transdermal delivery system that is effective in delivering these substances. The restricted number of medications that can be supplied through the transdermal method is one of the most significant obstacles that transdermal delivery must overcome today. Successful transdermal medications often have a molecular weight of only a few hundred daltons, with an octanol-water partition coefficient that favors lipids, and require daily doses of milligrams or less. This is because the current delivery systems are not as effective as they once were. The transport of hydrophilic medicines, peptides, and big molecules (such as DNA or small interfering RNA) through the transdermal route provides a unique set of problems (Gu et al. 2020). As a consequence of this, scholars have started to take notice of microneedle technology. They are attempting to make use of this technology to achieve high-efficiency transdermal delivery of medications that are difficult to administer. At this time, the prevention or treatment of skin photoaging includes not only the use of sunscreen regularly and the application of skincare products like vitamin A derivatives, particularly products containing retinol, but also the utilization of clinical cosmetic procedures (Liang et al. 2020). Laser or radiofrequency treatments, photodynamic therapy, and the transdermal delivery of bioactive substances like platelet-rich plasma (PRP) using microneedle rollers are some examples of these types of treatments (El-Domyati et al. 2018), which is an effort to encourage the restoration of the skin that has been photoaged. Researchers are always looking at ways to repair skin photoaging that are both more effective and more convenient, such as the use of microneedle patches among other options (Lyons et al. 2018).

Microneedle Patches Deliver Extracellular Vesicles (EVs)

Hydrogels in microneedle patches have been investigated to improve drug delivery. ADSC-EVs can be delivered to the skin using roller microneedles. ADSC-EVs had immunomodulatory and anti-photoaging properties, but the skin barrier prevented their absorption (Sun et al. 2022). Therefore, MN rollers have been used to administer EVs subcutaneously, which induce microscopic lesions that start the wound healing process and cause neocollagenesis. To promote photoaging in SKH-1 mice, persistent UV light and several back treatments were used. One side received MN alone or MN + EVs, whereas the other was untreated. Compared to the untreated side, MN alone or MN + EVs reduced epidermal thickness and enhanced skin barrier function. The MN + EVs group had fewer wrinkles, higher collagen density, and better collagen fiber organization. On the third day post-treatment, MN + EVs had reduced CD11b + cell infiltration than MN. These findings indicate that MN therapy alone improves photoaging skin's epidermal structure and function, while ADSC-EVs accelerate inflammatory recovery and collagen content (Akers et al. 2013). Microneedle rollers have usage constraints and operational methods that make long-term use difficult. Scholars are investigating the use of microneedle patches to load EVs, which can be chemically generated to carry nucleic acids like mRNA, siRNA, etc. for drug delivery (Pomatto et al. 2021). Some studies used gene regulation like mRNA therapy to overcome collagen's enormous molecular weight. EVs containing COL1A1 mRNA and microneedle patches were used to cure skin photoaging in mice. They created EVs with collagen-expressing mRNA plasmids using cellular nanoporation (CNP) technology. These COL1A1-EVs successfully transfect cells and increase collagen protein expression. Injection alone improves skin photoaging. However, loading them onto microneedle patches prolongs and uniformizes drug release, enhancing therapeutic effects (Jin et al. 2022).

12.2.5 Implantable Devices

New bioelectronic implanted devices are changing physiological signal recording and closed-loop neuromodulation in a wide range of disorders across organs. The rising prevalence of chronic illnesses, from neurological to metabolic disorders, prompted this transformation. Global mortality from chronic diseases accounts for 71% of all deaths, underscoring the need for novel treatments (Banatvala and Bovet 2023). The growing prevalence of neurological illnesses worldwide highlights bioelectronic applications' potential. The fast improvement of biomedical technologies and an aging population will increase demand for chronic disease management. This review examines implantable devices on nervous, cardiovascular, respiratory, digestive, urinary, and musculoskeletal systems for physiological signal recording and neuromodulation therapy, a major electroceutical advancement. Bioelectronic implanted devices are also promising for neuromodulation therapy (Shin et al. 2019). Many disorders can be treated using neuromodulation, which alters nerve activity with electrical, visual, or chemical stimuli. Chronic pain, mobility difficulties, and psychiatric diseases are treated using DBS and SCS. A novel optical neuromodulation approach, optogenetics, is being investigated for treating neurological illnesses and modifying nervous system neuronal functioning. Chemical neuromodulation with targeted medication delivery is promising for Parkinson's disease, hormone abnormalities, and chronic pain. Neuromodulation therapy can also treat arrhythmias, chronic gastrointestinal diseases, urine incontinence, and mood disorders like depression and bipolar disorder (Table 12.2) (Zhao et al. 2023).

Table 12.2 Organs targeted by implantable devices

Organ system	Organ	Function	Signal type	Modality	Clinical application
Nervous system	Brain	Recording	Mechanical, electrical	Intracranial pressure, electrical monitoring	Bioelectronic devices record intracranial pressure and monitor neural circuits for studying brain injuries and conditions like Parkinson's and epilepsy. They also enable tumor monitoring and neurotransmitter imbalance assessments for disorders such as schizophrenia, aiding in early diagnosis and treatment (Pu et al. 2024)
		Neuromodulation	Electrical, optical, chemical	Electrical stimulation, optogenetics, drug delivery	Neuromodulation technologies provide targeted stimulation for managing epilepsy, Parkinson's disease, chronic pain, and mood disorders. Optogenetics allows precision control of behavioral and neural functions, while drug delivery systems offer site-specific treatments for schizophrenia and anxiety disorders. These interventions are critical for improving patients' quality of life and enabling neural rehabilitation post-injury (Yoon et al. 2022)
	Spine	Recording, neuromodulation	Electrical	Electrical monitoring, electrical stimulation	Devices enable the restoration of motor functions post-spinal cord injuries through electrical monitoring and stimulation. They manage conditions like spasticity, neuropathic pain, and nerve recovery during rehabilitation. These systems are particularly useful in enhancing mobility and sensory functions for patients with severe spinal trauma (Zhou et al. 2019)
Cardiovascular system	Heart	Recording	Mechanical, electrical, optical	ECG, hemodynamic monitoring, oxygen saturation	Devices record mechanical, electrical, and optical signals for diagnosing and managing arrhythmias, ischemia, and cardiac failure. They also monitor oxygen saturation to detect hypoxia in critical conditions. These devices play a pivotal role in continuous cardiovascular health assessment and early detection of life-threatening cardiac events (Ouyang et al. 2023)
		Neuromodulation	Electrical, optical	Cardiac pacing, optogenetics	Cardiac pacing devices and optogenetic systems help manage bradycardia and heart failure by providing precise electrical or optical stimulation. Adaptive cardiac pacing ensures appropriate heart rates, improving the survival and functionality of patients with severe cardiac conditions (Harland et al. 2022)

(continued)

Table 12.2 (continued)

Organ system	Organ	Function	Signal type	Modality	Clinical application
Respiratory system	Lungs	Recording	Thermal, electrical	Temperature, EMG	Bioelectronic implants measure lung temperature and EMG signals to monitor homeostasis and detect laryngeal paralysis. These systems aid in the early diagnosis of respiratory insufficiencies and restore vocal cord functions, improving breathing and speech outcomes for patients recovering from laryngeal trauma or disorders (Kathe et al. 2022)
Digestive system	Stomach, intestines	Recording, neuromodulation	Chemical, optical	Serotonin monitoring, endoscopy	Implantable devices provide real-time monitoring of serotonin levels for managing bowel diseases and optical sensors for diagnosing gastric cancers. Neuromodulation systems restore normal motility in patients with gastroparesis and chronic intestinal pseudo-obstruction, offering targeted therapeutic solutions that alleviate symptoms like bloating and pain (Chen et al. 2022)
Urinary system	Kidney, bladder	Recording, neuromodulation	Mechanical, thermal, electrical	Deformation sensors, electrical stimulation	These devices monitor bladder deformation and kidney temperature to diagnose graft rejection and manage neurogenic bladder conditions. Neuromodulation interventions control bladder muscle activity, providing solutions for detrusor underactivity and urinary incontinence, significantly enhancing patients' quality of life (Sim et al. 2020)
Musculoskeletal system	Bone, muscle, tendons	Recording	Mechanical, electrical	Bone strain, EMG, tendon strain	Devices track mechanical and electrical signals to monitor bone strain and tendon activity. These technologies support the management of chronic musculoskeletal pain, rehabilitation from orthopedic surgeries, and development of neuroprosthetics. Such innovations enable the restoration of limb functionality and improve outcomes for patients with fractures and tendon injuries (Lu et al. 2021)

12.2.6 Industrial Perspectives on MIDDS

Since the 2008 QSP concept, experts have recognised QSP as a significant milestone since it might help in the development of the pharmaceutical industry. Since then, the modeling and simulations community's QSP critical mass has grown. Only in the last 3 years has its presence in peer-reviewed publications approached a "steady state," possibly due to the COVID-19 pandemic affecting collaboration. QSP has become a common phrase in international conferences and seminars, even taking center stage. The Netherlands has held the "International Symposium on Measurement and Kinetics of In Vivo Drug Effects" every 4 years since 1990, now dubbed the "Quantitative Systems Pharmacology Conference." QSP is also more prevalent in professional associations (QSP SIG in ISoP, Systems Pharmacology Community in ASCPT's Quantitative Pharmacology WG (Peterson and Riggs 2015), QSP Working Groups in the IQ Consortium, UK QSP Network, etc.) and the rise of successful CROs or company divisions providing QSP consultation to the pharma industry. The significant impact of QSP on industry decision-making is reflected in the growing number of QSP-based submissions to the US FDA, which double every ~1.4 years, as per published data. No matter the outcome of such submissions, the most innovative industries have used QSP as a decision-making tool (Cucurull-Sanchez 2024).

12.2.6.1 QSP and the MIDD/MID3 Paradigm

The MIDD paradigm is an industrial and regulatory framework that integrates experimental and simulation pharmacology, statistics, and biology data into decision-making. This methodology includes Model-Informed Drug Discovery and Development (MID3), not simply clinical development (Madabushi et al. 2022). QSP is slowly becoming part of the MIDD/MID3 suite of pharmacometric methods, which includes PKPD, PBPK, PopPK, disease progression models, and dose/exposure-response analyses. The 2013 Natpara submission was the first public example of QSP's impact on medication development (Manolis et al. 2017). Since 2008, QSP publications had been reported annually, but now they were growing. Both the scientific (academic or industrial) and regulatory (agency or industry) realms began to recognize the significance of this new modeling and simulation tool named "QSP." QSP's spread into regulation indicated that it was affecting industry decision-making. Given the greater complexity of regulatory milestones compared to peer-reviewed journal publishing, the literature progressed quicker than regulatory use. But 6 years later, the annual rise of QSP literature articles stopped. This stalemate may have been due to the industry's protective attitude toward QSP models' competitive intelligence or saturation of internal resources (including those needed to publish after projects ended). The overview of QSP is shown in Table 12.3 (Venkatakrishnan and van der Graaf 2022).

Table 12.3 Overview of quantitative systems pharmacology (QSP): definition, applications, and future trends

Aspect	Details
Definition of QSP	A multidisciplinary approach that integrates biology, pharmacology, and clinical science through mathematical modeling to predict drug behavior and outcomes at various levels (cellular, tissue, organ, and system) (Crisp et al. 2018)
Evolution	QSP emerged in 2008 as a method to translate preclinical discoveries into clinical progress. Its first significant regulatory impact was in 2013 during the FDA approval of Natpara, leading to its recognition in regulatory and industrial settings (Kirouac et al. 2019)
Applications	– Discovery phase: evaluates therapeutic hypotheses, prioritizes targets, and predicts human efficacy early in drug development – Clinical development: assists in dose selection, patient stratification, clinical trial design, and regulatory submissions – Post-marketing: optimizes therapeutic strategies and supports lifecycle management (Androulakis 2022)
Integration in industry	QSP has become integral in the model-informed drug development (MIDD) framework, often combined with empirical pharmacokinetics/pharmacodynamics (PKPD) modeling to optimize decision-making and resource allocation
Advantages	– Supports personalized medicine through virtual patient modeling – Enhances regulatory decision-making by providing robust simulations for dose justification, patient stratification, and drug combination therapies – Bridges preclinical and clinical phases, improving the predictability of therapeutic outcomes (Scheuher et al. 2023)
Challenges	– Lack of standard regulatory guidelines and validation requirements – Resistance within the industry due to reliance on empirical methods – High complexity in building and maintaining QSP models – Limited awareness among stakeholders about return on investment (ROI)
Regulatory milestones	– FDA'S adoption of QSP in decision-making frameworks, with increasing regulatory submissions involving QSP-based evidence – Establishment of FDA'S model-informed drug development paired meeting program – Development of ICH guidelines (M15) to standardize QSP applications
Future trends	– Increased integration with machine learning and artificial intelligence to enhance predictive accuracy – Expansion of platform QSP models for broader disease areas – Focus on regulatory harmonization to improve global adoption (Rieger et al. 2018)

12.2.7 Personalized and Patient-Centric Healthcare Solutions

Healthcare resources are stretched as the global healthcare system addresses medical, lifestyle, and personal health demands. Allowing communities to self-manage their health through health innovations improves well-being and reduces healthcare resource strain. Healthcare technology has changed dramatically in the past decade. Innovative technology use increased during the COVID-19 pandemic. If infrastructure and training are available, digital transformation offers health industry prospects (Dey et al. 2021). Regulation (EU) 2021/694 defines digital transformation as using digital technology to transform enterprises and services. Digital technology has helped combat pandemics by maintaining connectivity across borders. Digital health solutions offer remote health professional-patient engagement, enabling patient self-management and bridging the gap between patients and providers (Baashar et al. 2020). Diabetes is complicated by genetic, environmental, and lifestyle eating factors. Effective interventions also require knowledge of the patient's genetics, demographics, health history, family history, lifestyle patterns, dietary habits, real-time health data, and self-awareness of their condition. Personalization, problems, patient awareness, and adherence are sometimes difficult with current methods. A complete, patient-centered diabetes management system that adjusts to individual needs, goals, and preferences and provides timely and accurate guidance is needed. Emerging technology like digital twins could change healthcare. A digital twin can model and simulate a physical entity's behavior and environment. Digital twins, successfully used in aerospace and industry, enable tailored medicine and precision healthcare. Wearable devices, mHealth, telehealth, health IT, and telemedicine are all part of digital health. Digital health systems have shown promise for early AF identification and control. However, the widespread adoption of digital integrated care requires more high-quality data to support it (Kamel Boulos and Zhang 2021).

12.2.7.1 Digital Twins in Healthcare

Healthcare is being transformed by digital twins, virtual models that mimic physical phenomena in real time. A digital twin models a patient, organ, disease process, or healthcare system to provide an accurate and dynamic representation for simulations and forecasts. Digital twins enable extraordinarily tailored treatment. These virtual replicas integrate genomic, medical history, lifestyle, and real-time sensor data to create personalized care regimens (Ortiz-Martínez et al. 2022). More than individual treatment, digital twins can inform public health measures. Researchers are studying methods to model disease spread, predict outbreaks, and simulate public health policy effects in specific populations. Digital twins speed drug discovery and clinical trials. They allow virtual drug efficacy and safety testing, minimizing time and risk and potentially personalizing medicines to specific groups. Digital twins enhance education and training. They let doctors rehearse complex procedures, see anatomical components, and try out alternative treatment options in a safe, simulated environment. Healthcare digital twin technology is intriguing but still emerging. Digital twin production requires seamless data flow between systems and devices. Data standardization and interoperability efforts continue. As with any patient data usage, digital twin frameworks must consider privacy, security, and biases (Feldman et al. 2017).

Bioresponsive nanomedicines that respond to biological, pathological, or environmental signals are attracting researchers for precise drug delivery. These solutions have improved medicine efficacy and lessened toxicity by targeting delivery or nanoparticle payload release (Hong et al. 2023). Due to development in this sector and the clinical translation potential of many breakthrough nanomedicines, basic research must be strengthened. Bioresponsive nanomaterials for drug delivery or controlled release researches nanomaterial concepts, designs, synthesis, processing, and assessment. As an ecosystem, the oral cavity has many microenvironments with complex physicochemical features such as pH, oxygen, temperature, and redox potential (Jurasic et al. 2022). In pediatric dentistry, bioresponsive DDSs must be designed and used carefully. These devices are carefully designed to meet pediatric oral healthcare needs. The focus is on designing DDSs that are bioresponsive to oral cavity physiological changes and take into account pediatric patients' particular anatomical and physiological characteristics. Specificity optimizes therapeutic outcomes while limiting risks and side effects. This study will explore bioresponsive DDSs in a pediatric context and their tailored design and considerations for pediatric applications, shedding light on how these systems can improve therapeutic interventions for our younger patients (Broomhead et al. 2022).

12.2.8 Integration with Digital Health

Integrating vital health data demands care and protection to prevent misuse. AI is now integrating and interpreting complicated health data for tailored medicine. Open-source code is crucial since developing nations may not have access to sophisticated AI applications. Prediction models employed by artificial intelligence systems to extract trends from data for medical choices can also misuse health information. As artificial intelligence applications in healthcare increase, ethical considerations must be considered (Peralta et al. 2021). Artificial intelligence may improve diagnostic/prognostic tools, big data-based treatment customization, and healthcare workflow optimization. AI integration with healthcare is necessary because of the abundance of patient health data. Artificial intelligence systems can swiftly identify patterns and insights from many data sources, improving healthcare personalization decisions. A major cardiovascular disease epidemic in the twenty-first century is atrial fibrillation (AF), the most frequent clinically significant heart rhythm disturbance. AF causes considerable morbidity and death, burdening the healthcare system. Early detection and treatment can reduce AF problems (More et al. 2022).

12.2.8.1 Integrated Digital Atrial Fibrillation Management

Whites are more likely than Blacks, Asians, or Hispanics to get AF. Based on statistics from America and Europe, the lifetime risk (LTR) of AF was one in four for White adults aged ≥40 in the 1990s. After a decade, one in three White people over 45 is at risk (Roth et al. 2020). Lower LTRs are recorded among African Americans and Chinese individuals, with 1 in 518 and 1 in 10 at ≥40 years, respectively. AF increases with age and is more common in men. They may be due to AF risk factor differences by gender. In two North American and European groups, LTR for AF development appeared to be nearly equal in both sexes, probably reflecting women's longer life expectancy, which reaches men's cumulative incidence in later decades. Besides race/ethnicity and sex, the LTR is roughly one in five with an optimum risk factor profile and about one in three with at least one increased risk factor (Linz et al. 2021). Modifiable risk factors such as hypertension, diabetes, sleep apnea, hyperlipidemia, and lifestyle risk factors like alcohol abuse, smoking, and inactivity must be addressed to prevent new-onset and recurrent AF. In the Liraglutide Effect in Atrial Fibrillation (LEAF) study, pre-ablation weight loss (≥3–10% vs. 10 drinks/week) and an almost eightfold reduction in alcohol consumption over 6 months led to a lower AF burden (0.5% vs. 1.2%) compared to controls who continued their usual consumption (Hindricks et al. 2021).

12.2.8.2 Pericardioversion

Optimal rate management in AF patients on the waiting list for elective cardioversion or utilizing a wait-and-see technique at the ED might be difficult. Regular rate control evaluation and a straightforward preprocedural medication adjustment approach optimize pericardioversion rate control. TeleWAS-AF uses digital devices to remotely modify rate control medication and identify spontaneous sinus rhythm conversion in AF pericardioversion patients (Svennberg et al. 2021). This strategy is suitable for stable patients who report to the ED with recent-onset symptomatic AF scheduled for a wait-and-see approach and can use digital heart rate and rhythm monitoring devices. Current research is investigating if digital gadgets can help the ED manage AF and reduce ED system load (World Health Organization 2020).

12.2.8.3 Long-Term Atrial Fibrillation Management

During the COVID-19 pandemic, 40 European centers employed teleconsultation to remotely manage AF (Mitrani et al. 2022). Teleconsultation-managed patients receive remote PPG rate and rhythm monitoring with TeleCheck-AF. Patients should utilize the PPG app three times a day and 1 week before teleconsultation for symptoms. This information is utilized in teleconsultation. Data show good center and patient experience. This intervention will be tested for clinical outcomes in an RCT. Patients getting pharmacological treatment simply or left atrial appendage occlusion can use updated structured follow-up packages for home rehabilitation (mAFA III) (Weng et al. 2021).

12.2.8.4 Remote Atrial Fibrillation Management Platforms

A mHealth management system with a physician-oriented platform and patient-oriented smartphone app, myAlgos, was tested in the emPOWERED-AF RCT. Physicians got all patient-reported data (vital signs, blood pressure, heart rate, body weight, blood glucose, oxygen saturation, and medication adherence) via the platform. Five physicians and 80 patients gave the platform a mean 75% satisfaction and ease of use score. Over 6 months, control and full app version groups had similar AF-related hospitalizations. Compared to the control, the full version group had a considerable improvement in quality of life (Lazaridis et al. 2022). A prospective cohort research examined virtual post-ablation clinic appointments in 39 participants. Patients uploaded heart rate, blood pressure, blood glucose, weight, and temperature to the Ortus-iHealth app, which a clinical nurse specialist could access. Patients saved 83% on travel costs and 93% on time. Twenty AF patients undergoing antiarrhythmic therapy, cardioversion, or catheter ablation in the Atrial Fibrillation Helping Address Care with Remote Technology (AF-HEART) prospective cohort study received heart rhythm tracking (AliveCor, Kardia), weight loss, blood pressure monitoring, alcohol reduction, and sleep reduction through referrals. This enhanced the quality of life and reduced symptom severity over 6 months. The Telemedical Interventional Management in Heart Failure II (TIM-HF2) randomized heart failure patients to noninvasive remote patient management (daily transmission of body weight, blood pressure, heart rate/rhythm, peripheral capillary oxygen saturation, and self-rated health status) or usual care. AF status at randomization was examined in 571 participants in a post hoc analysis. Intervention arm AF patients had fewer days lost to unexpected cardiovascular hospital admissions or all-cause death. A 96-person prospective cohort trial examined an AF-dedicated online educational and treatment platform, while a 71-person randomized substudy examined an ambulatory single-lead ECG cardiac monitoring (AliveCor, Kardia) (Cucurull-Sanchez 2024). Patients were advised to regularly enter weight, activity, nutrition, and AF symptoms on the site. During a 6-month observation, AF symptom severity (CCS-SAF) improved, although AFSS and EQ-5D scores did not. Remote cardiac monitoring earned great satisfaction and increased quality of life in the substudy. Patients said it prevented AF-related ED visits (Lambert et al. 2021).

12.2.8.5 Mobile Health-Based Heart Rate/Rhythm Monitoring

ARCT evaluated mobile cardiac rhythm monitoring (AliveCor, Kardia) in 100 AF ablation patients during a 6-month follow-up and self-management. Hospitalizations, ED visits, and anxiety levels were similar between intervention and control groups. Control group patients needed more ambulatory heart rhythm monitoring than intervention arm patients (27% vs. 5.9%). The iPhone Helping Evaluate Atrial fibrillation Rhythm through Technology (iHEART) case-control study compared AF or atrial flutter recurrence detection in 23 patients undergoing daily smartphone ECG monitoring (AliveCor, Kardia) and 23 age- and gender-matched controls post-ablation (Trymbulak et al. 2020). During the 6-month follow-up, intervention group participants were twice as likely to develop recurrent AF/atrial flutter (HR 2.55, 95% CI 1.06–6.11). In addition, 92% of respondents said the device was helpful and 58% stated the study made them more health-concerned. Physical functioning, role physical, energetic, and mental health improved significantly. A subanalysis of the iHEART research found no significant differences in quality-adjusted life years or AF symptom severity between intervention ($n = 115$) and control ($n = 123$) groups (Guhl et al. 2020) (Table 12.4).

Table 12.4 Innovative patents in minimally invasive drug delivery and therapeutic technologies

S. no	Inventor	Work done	References
1	Steve T. Cho	To reduce breakage, the present invention provides a microneedle with a broad base relative to its height. Microneedles have fluid channels and beveled non-coring tips. Arrays of microneedles are usually made using semiconductor-derived microscale fabrication methods	Cho (2005)
2	Gillis Edward M	The catheter of the invention can be implanted and fastened without a trocar or cannula, minimizing tissue injury during implantation and over time	Gillis Edward (2001)
3	*Christopher Shaari*	The invention focuses on a nasal delivery device containing microneedles and techniques for administering compositions through them	Shaari (2011)
4	*Robert R. Gonnelli* *Devin V. Mcallister*	Microneedle devices carry chemicals, such as medicines and biological substances, across tissue. Microneedle devices achieve clinically relevant drug delivery or fluid evacuation across skin or tissue barriers without causing tissue damage, discomfort, or irritation	Gonnelli and McAllister (2011)
5	*Ronan Thornton*	Microneedles can be delivered via a needle or syringe apparatus that interfaces with existing medical devices or can be used alone. The myocardium receives at least one bioactive agent from the systems	Thornton (2011)
6	*Robert Farra*	Drug delivery devices and procedures are offered. The device had two or more housing components joined end-to-end in a fixed, linear configuration, forming a sidewall	Bright (2022)
7	*Corinne Bright*	To cure asthma, methods, equipment, and systems are disclosed for lowering sympathetic nerve innervation to and from the lungs and their arteries	Bright (2022)
8	*Elazer Edelman* *Mark Lovich*	In this invention, inotropic agents are delivered locally to specific heart regions, such as ventricles, to support myocardial contractile function during cardiothoracic surgery	Edelman and Lovich (2014)
9	*Allan Mishra*	Peripheral vascular disease can be treated with platelet-rich plasma mixtures. Platelet-rich plasma is injected into the hypovascular muscle or soft tissue to treat peripheral vascular disease	Mishra (2012)
10	*Mir Imran*	A gadget a patient can swallow into their gastrointestinal tract lumen	Imran (2018)
11	*Adel M. Malek* *Carl B. Heilman* *David A. Rezac* *Jack B. Sattell* *Alexander Bonin*	Endovascular medication delivery devices and methods for intracranial subarachnoid space delivery are described in this patent. The methods distribute the medicine using a device with a distal component in the subarachnoid space and a dural venous sinus	Malek et al. (2023)

(continued)

Table 12.4 (continued)

S. no	Inventor	Work done	References
12	*Samuel K. Sia* *Nalin Tejavibulya*	These gadgets and methods extract and transport fluids to humans and animals. Microneedles can provide fluids painlessly and without infection. This invention addresses affordability, portability, and remote use, even by inexperienced people and others	Sia and Tejavibulya (2020)
13	*Joseph Wang* *Joshua Ray Windmiller* *Roger Narayan* *Phillip Miller* *Ronen Polsky* *Thayne L. Edwards*	Methods, architectures, and systems for biosensing and medication delivery are described. A device for detecting analytes and releasing biochemicals into biological fluids may use hollowed needles	Wang et al. (2022)
14	*David J. Sogard* *Scott Smith* *Mark L. Jenson*	An extended flexible medical device is inserted through a patient's natural orifice and progresses to innervated tissue that affects renal sympathetic nerve activity	Sogard et al. (2017)
15	*Daniel R. Burnett* *Evan S. Luxon* *Marcie Hamilton* *Brian M. Neil* *Brett D. Mensh*	A housing and a flexible sensing catheter with a lumen with a plurality of apertures and an exterior surface with an analyte sensor affixed thereto can be implanted in a human subject to determine an analyte concentration in their peritoneal fluid	Burnett et al. (2017)
16	*David Zarrouk* *Moshe Shoham* *Menashe Zaaroor* *Hadas Ziso*	A robotic device for intracranial treatments with a skull-mounted baseplate and a moveable base element	de Juan et al. (2021)
17	*Eugene De Juan* *Signe Erickson* *Charles Limb* *Hugo Peris* *Nikhil Talreja* *Andrew Ayoob*	Therapies can be injected into the inner ear to address hearing loss. The tympanic membrane can temporarily receive one or more stabilizing devices that define functioning channels	de Juan et al. (2021)
18	*Leonard V. Covello*	The present invention offers minimally invasive devices and methods for sinuses and adjacent structures access for surgery and treatments. Under minimum sedation, the anterior ethmoid and maxillary sinuses are accessible and treated without postoperative complications. The sinuses and natural ostia are visually verified	Covello (2016)
19	*W. Andrew Grubbs*	A telerobotic surgery system starts with a robotic surgery station with two robot arms, each with a laparoscopic tool and motor. Each robot arm drive has a first controller	Grubbs (2020)
20	*David Thompson* *Patrick Wu* *David Lehrberg* *Mark L. Mathis* *Michael Boutillette*	A lung volume reduction system is disclosed comprising an implantable device adapted to be delivered to a lung airway of a patient in a delivery configuration and to change to a deployed configuration to bend the lung airway	Thompson et al. (2012)

(continued)

Table 12.4 (continued)

S. no	Inventor	Work done	References
21	*Paul J. Weber*	Electrosurgical lysing supplies and procedures. In some embodiments, the lysing device may contain a lysing tip with at least one bead with an electrically nonconductive surface and at least one lysing member defining at least one lysing segment extending within a recess defined, at least in part, by the bead, which protrudes both distally and proximally	Weber (2024)
22	*John Zentgraf* *David Joseph Parins* *Arun Saini*	Exchangeable heart valve repair systems are used in improved methods and apparatus for patient heart valve repair. A heart valve repair system can have a port placed in the heart wall and a catheter slidable within it. The catheter can be selectively locked to the port for heart insertion and freed with the port seated to travel distally toward the target tissue. To deploy repair devices, system pieces can be selectively removed and replaced within the port while maintaining a seal	Zentgraf et al. (2018)
23	*David Thompson* *Patrick Wu* *David Lehrberg* *Mark L. Mathis* *Michael Boutillette* *Nathan Aronson* *Jaime Vasquez*	The disclosed lung volume reduction system includes an implantable device that can be transported to a patient's lung airway and deployed to bend it	Thompson et al. (2019)
24	*Richard T. Chen* *Ming-Ting Wu* *Arun Veeramani* *Vacit Arat* *Gregory P. Schmitz*	Several tissue-cutting devices have an elongate tube with proximal and distal ends and a central axis	Chen et al. (2019)
25	*Dewan Fazlul Hoque Chowdhury*	A drug delivery device uses microneedles on a belt rotated around rollers to deliver pharmacologically active substances transdermally. The reservoir is compressed when the needles and belt contact the skin	Chowdhury (2021)
26	*Mark G. Allen* *Mark R. Prausnitz* *Devin V. Mcallister* *Florent Paul Marcel Cros*	Microneedle devices transfer substances across tissue barriers and hold microflames. Biodegradable polymer microneedles are preferred for tissue delivery	Allen et al. (2014)

(continued)

Table 12.4 (continued)

S. no	Inventor	Work done	References
27	*Jason L. Harris Taylor W. Aronhalt Dwight Henninger James W. Voegele Lee M. Kaplan Nicholas Stylopoulos*	Techniques and methods are offered to activate brown adipose tissue by targeted chemical administration. These approaches and gadgets activate BAT to boost thermogenesis, or heat generation, which can lead to weight loss and/or metabolic function improvements	Harris et al. (2018)
28	*Susanne Holm Faarbaek Karsten Hoppe Peter Boman Samuelsen Jens Branebjerg*	Accordingly, the present invention relates to a three-dimensional adhesive device to be attached to the body surface of a mammal comprising a microelectronic sensing system	Faarbaek et al. (2012)
29	*Rodney Boleyn Ezra M. Dreisbach Chuck Dulken Gust H. Bardy*	A system and method for machine learning-based atrial fibrillation detection are provided. A database is maintained that is operable to maintain a plurality of ECG features and annotated patterns of the features	Boleyn et al. (2019)
30	*Thomas Jared Bunch John David Day*	Methods for managing cardiac complex rhythm disorders in a patient may involve acquiring multiple electrical signals from a sensor system, with each signal corresponding to a distinct location on the cardiac wall of the patient's heart and each signal containing an electrogram waveform; subsequently, the electrical signals are ranked concerning one another based on at least the uniformity and frequency of the electrogram waveform of each signal	Bunch and Day (2020)

References

Aboudzadeh MA, Iturrospe A, Arbe A, Grzelczak M, Barroso-Bujans F (2020) Cyclic polyethylene glycol as nanoparticle surface ligand. ACS Macro Lett 9(11):1604–1610

Ahlschwede KM, Curran GL, Rosenberg JT, Grant SC, Sarkar G, Jenkins RB, Ramakrishnan S, Poduslo JF, Kandimalla KK (2019) Cationic carrier peptide enhances cerebrovascular targeting of nanoparticles in Alzheimer's disease brain. Nanomedicine 16:258–266

Aich K, Singh T, Dang S (2022) Advances in microneedle-based transdermal delivery for drugs and peptides. Drug Deliv Transl Res 12(7):1556–1568

Akers JC, Gonda D, Kim R, Carter BS, Chen CC (2013) Biogenesis of extracellular vesicles (EV): exosomes, microvesicles, retrovirus-like vesicles, and apoptotic bodies. J Neurooncol 113:1–11

Allen MG, Prausnitz MR, Mcallister MV, Cros FPM (2014) Microneedle devices and methods of manufacture and use thereof. US8708966B2

Androulakis IP (2022) Towards a comprehensive assessment of QSP models: what would it take? J Pharmacokinet Pharmacodyn 51(5):521–531

Apolinário AC, Naser YA, Volpe-Zanutto F, Vora LK, Sabri AH, Li M, Hutton AR, McCarthy HO, Lopes LB, Donnelly RF (2024) Novel lipid nanovesicle-loaded dissolving microarray patches for fenretinide in breast cancer chemoprevention. J Control Release 374:76–88

Baashar Y, Alhussian H, Patel A, Alkawsi G, Alzahrani AI, Alfarraj O, Hayder G (2020) Customer relationship management systems (CRMS) in the healthcare environment: a systematic literature review. Comput Stand Interfaces 71:103442

Banatvala N, Bovet P (2023) Noncommunicable diseases: a compendium. Taylor & Francis, London

Beňová E, Hornebecq V, Zeleňák V, Huntošová V, Almáši M, Máčajová M, Bergé-Lefranc D (2021) pH-responsive mesoporous silica drug delivery system, its biocompatibility and co-adsorption/co-release of 5-fluorouracil and naproxen. Appl Surf Sci 561:150011

Boehm RD, Miller PR, Hayes SL, Monteiro-Riviere NA, Narayan RJ (2011) Modification of microneedles using inkjet printing. AIP Adv 1(2):22139

Boleyn R, Dreisbach EM, Dulken C, Bardy GH (2019) System and method for machine-learning-based atrial fibrillation detection. US10463269B2

Bright C (2022) Systems and methods for cardiac plexus neuromodulation. US11446359B2

Broomhead T, Gibson B, Parkinson CR, Vettore MV, Baker SR (2022) Gum health and quality of life—subjective experiences from across the gum health-disease continuum in adults. BMC Oral Health 22(1):512

Bunch TJ, Day JD (2020) Atrial fibrillation treatment systems and methods. US10702181B2

Burnett DR, Luxon ES, Hamilton M, Neil BM, Mensh BD (2017) Methods and devices for the diagnosis and treatment of diabetes. US9603558B2

Chen ZL, Huang M, Wang XR, Fu J, Han M, Shen YQ, Xia Z, Gao JQ (2016) Transferrin-modified liposome promotes α-mangostin to penetrate the blood–brain barrier. Nanomedicine 12(2):421–430

Chen RT, Wu MT, Veeramani A, Arat V, Schmitz GP (2019) Concentric cutting devices for use in minimally invasive medical procedures. US10492822B2

Chen JC, Kan P, Yu Z, Alrashdan F, Garcia R, Singer A, Lai CE, Avants B, Crosby S, Li Z, Wang B (2022) A wireless millimetric magnetoelectric implant for the endovascular stimulation of peripheral nerves. Nat Biomed Eng 6(6):706–716

Chitkara D, Mittal A (2020) Surface functionalization of exosomes for target-specific delivery and in vivo imaging & tracking: strategies and significance. J Control Release 326:599–614

Cho ST (2005) Microneedles for minimally invasive drug delivery. US6980855B2

Chowdhury DFH (2021) Microneedle transdermal delivery device. US10953210B2

Chowdhury S, Chowdhury NS (2023) Novel anti-amyloid-beta (aβ) monoclonal antibody lecanemab for Alzheimer's disease: a systematic review. Int J Immunopathol Pharmacol 37:03946320231209839

Covello LV (2016) Devices and methods for minimally invasive access to sinuses and treatment of sinusitis. US9247950B2

Crisp A, Miller S, Thompson D, Best N (2018) Practical experiences of adopting assurance as a quantitative framework to support decision making in drug development. Pharm Stat 17(4):317–328

Cucurull-Sanchez L (2024) An industry perspective on current QSP trends in drug development. J Pharmacokinet Pharmacodyn 51(5):511–520

Cutrim ES, Vale AA, Manzani D, Barud HS, Rodriguez-Castellon E, Santos AP, Alcântara AC (2021) Preparation, characterization and in vitro anticancer performance of nanoconjugate based on carbon quantum dots and 5-fluorouracil. Mater Sci Eng C Mater Biol Appl 120:111781

de Juan E, Erickson S, Limb C, Peris H, Talreja N, Ayoob A (2021) Minimally-invasive tools and methods for accessing the middle and inner ear through the tympanic membrane. US20210228849A1

Dey N, Das N, Chaki J (2021) Digital future of healthcare. CRC, Boca Raton

Donnelly RF, Larrañeta E (2019) Slowly dissolving intradermal microneedles. Nat Biomed Eng 3(3):169–170

Economidou SN, Pere CP, Reid A, Uddin MJ, Windmill JF, Lamprou DA, Douroumis D (2019) 3D printed microneedle patches using stereolithography (SLA) for intradermal insulin delivery. Mater Sci Eng C Mater Biol Appl 102:743–755

Edelman E, Lovich M (2014) Localized delivery of cardiac inotropic agents. CA2609456C
El-Domyati M, Abdel-Wahab H, Hossam A (2018) Combining microneedling with other minimally invasive procedures for facial rejuvenation: a split-face comparative study. Int J Dermatol 57(11):1324–1334
Faarbaek SH, Hoppe K, Samuelsen PB, Branebjerg J (2012) A three-dimensional adhesive device having a microelectronic system embedded therein. EP1871218B1
Feldman AL, Long GH, Johansson I, Weinehall L, Fhärm E, Wennberg P, Norberg M, Griffin SJ, Rolandsson O (2017) Change in lifestyle behaviors and diabetes risk: evidence from a population-based cohort study with 10 year follow-up. Int J Behav Nutr Phys Act 14(1):39
Ghasemiyeh P, Mohammadi-Samani S (2019) Hydrogels as drug delivery systems; pros and cons. Trends Pharmacol Sci 5(1):7–24
Gillis Edward M (2001) Minimally invasive drug delivery catheter. WO-03002170-A2
Gonnelli RR, McAllister DV (2011) Microneedle array patch. US8162901B2
Gothwal A, Lamptey RN, Singh J (2023) Multifunctionalized cationic chitosan polymeric micelles polyplexed with pVGF for noninvasive delivery to the mouse brain through the intranasal route for developing therapeutics for Alzheimer's disease. Mol Pharm 20(6):3009–3019
Grubbs WA (2020) Telerobotic surgery system using minimally invasive surgical tool with variable force scaling and feedback and relayed communications between remote surgeon and surgery station. US10813710B2
Gu Y, Han J, Jiang C, Zhang Y (2020) Biomarkers, oxidative stress and autophagy in skin aging. Ageing Res Rev 59:101036
Guhl E, Althouse AD, Pusateri AM, Kimani E, Paasche-Orlow MK, Bickmore TW, Magnani JW (2020) The atrial fibrillation health literacy information technology trial: pilot trial of a mobile health app for atrial fibrillation. JMIR Cardio 4(1):e17162
Hamzehlou S, Aboudzadeh MA (2021) Special issue on "multifunctional hybrid materials based on polymers: design and performance". PRO 9(8):1448
Harland B, Aqrawe Z, Vomero M, Boehler C, Cheah E, Raos B, Asplund M, O'Carroll SJ, Svirskis D (2022) A subdural bioelectronic implant to record electrical activity from the spinal cord in freely moving rats. Adv Sci 9(20):2105913
Harris JL, Aronhalt TW, Henninger D, Voegele JW, Kaplan LM, Stylopoulos N (2018) Methods and devices for activating brown adipose tissue with targeted substance delivery. US10111770B2
Hindricks G, Potpara T, Dagres N, Arbelo E, Bax JJ, Blomström-Lundqvist C, Boriani G, Castella M, Dan GA, Dilaveris PE, Fauchier L (2021) 2020 ESC guidelines for the diagnosis and management of atrial fibrillation developed in collaboration with the European Association for Cardio-Thoracic Surgery (EACTS) the task force for the diagnosis and management of atrial fibrillation of the European Society of Cardiology (ESC) developed with the special contribution of the European heart rhythm association (EHRA) of the ESC. Eur Heart J 42(5):373–498
Hong T, Shen X, Syeda MZ, Zhang Y, Sheng H, Zhou Y, Xu J, Zhu C, Li H, Gu Z, Tang L (2023) Recent advances of bioresponsive polymeric nanomedicine for cancer therapy. Nano Res 16(2):2660–2671
Huang J, Li Q, Sun D, Lu Y, Su Y, Yang X, Wang H, Wang Y, Shao W, He N, Hong J (2007) Biosynthesis of silver and gold nanoparticles by novel sundried Cinnamomum camphora leaf. Nanotechnology 18(10):105104
Huang J, Yap N, Walter M, Green A, Smith C, Johnson J, Saigal R (2022) 3D-printed polypyrrole microneedle arrays for electronically controlled transdural drug release. ACS Biomater Sci Eng 8(4):1544–1553
Ilyas NRA, Putri APD, Pratama FA, Abdullah DAP, Azzahra KS, Permana AD (2024) Implantable trilayer microneedle transdermal delivery system to enhance bioavailability and brain delivery of rivastigmine for Alzheimer treatment: a proof-of-concept study. Eur J Pharm Biopharm 201:114382. https://doi.org/10.1016/j.ejpb.2024.114382
Imran M (2018) Swallowable drug delivery device. EP2515992B2
Ita K (2015) Transdermal delivery of drugs with microneedles—potential and challenges. Pharmaceutics 7(3):90–105

Jeong WY, Kwon M, Choi HE, Kim KS (2021) Recent advances in transdermal drug delivery systems: a review. Biomater Res 25(1):24

Jin L, Zeng X, Liu M, Deng Y, He N (2014) Current progress in gene delivery technology based on chemical methods and nano-carriers. Theranostics 4(3):240

Jin X, Zhang X, Li Y, Xu M, Yao Y, Wu Z, He Y, Gao J, Li B (2022) Long-acting microneedle patch loaded with adipose collagen fragment for preventing the skin photoaging in mice. Biomater Adv 135:212744

Jurasic MM, Gibson G, Orner MB, Wehler CJ, Jones JA, Cabral HJ (2022) Topical fluoride effectiveness in high caries risk adults. J Dent Res 101(8):898–904

Kalluri R, LeBleu VS (2020) The biology, function, and biomedical applications of exosomes. Science 367(6478):eaau6977

Kamel Boulos MN, Zhang P (2021) Digital twins: from personalised medicine to precision public health. J Pers Med 11(8):745

Kathe C, Michoud F, Schönle P, Rowald A, Brun N, Ravier J, Furfaro I, Paggi V, Kim K, Soloukey S, Asboth L (2022) Wireless closed-loop optogenetics across the entire dorsoventral spinal cord in mice. Nat Biotechnol 40(2):198–208

Kim G, Kim M, Lee Y, Byun JW, Lee M (2020) Systemic delivery of microRNA-21 antisense oligonucleotides to the brain using T7-peptide decorated exosomes. J Control Release 317:273–281

Kirouac DC, Cicali B, Schmidt S (2019) Reproducibility of quantitative systems pharmacology models: current challenges and future opportunities. CPT Pharmacometrics Syst Pharmacol 8(4):205–210

Kulkarni D, Damiri F, Rojekar S, Zehravi M, Ramproshad S, Dhoke D, Musale S, Mulani AA, Modak P, Paradhi R, Vitore J (2022) Recent advancements in microneedle technology for multifaceted biomedical applications. Pharmaceutics 14(5):1097

Lambert CT, Patel D, Bumgarner JM, Kanj M, Cantillon D, Saliba W, Hussein A, Baranowski B, Dresing T, Chung MK, Rickard J (2021) Atrial fibrillation future clinic. Novel platform to integrate smart device electrocardiogram into clinical practice. Cardiovasc Digit Health J 2(2):92–100

Lazaridis C, Bakogiannis C, Mouselimis D, Tsarouchas A, Antoniadis AP, Triantafyllou K, Fragakis N, Vassilikos VP (2022) The usability and effect of an mHealth disease management platform on the quality of life of patients with paroxysmal atrial fibrillation—the emPOWERD-AF study. Health Informatics J 28(4):14604582221139053

Li H, Cao Y, Ye J, Yang Z, Chen Q, Liu X, Zhang B, Qiao J, Tang Q, Yang H, Li J (2023) Engineering brain-derived neurotrophic factor mRNA delivery for the treatment of Alzheimer's disease. Chem Eng J 466:143152

Liang JX, Liao X, Li SH, Jiang X, Li ZH, Wu YD, Xiao LL, Xie GH, Song JX, Liu HW (2020) Antiaging properties of exosomes from adipose-derived mesenchymal stem cells in photoaged rat skin. Biomed Res Int 2020(1):6406395

Ligon SC, Liska R, Stampfl J, Gurr M, Mülhaupt R (2017) Polymers for 3D printing and customized additive manufacturing. Chem Rev 117(15):10212–10290

Linz D, Hermans A, Tieleman RG (2021) Early atrial fibrillation detection and the transition to comprehensive management. Europace 23(Supplement_2):ii46–ii51

Lombardo SM, Schneider M, Türeli AE, Türeli NG (2020) Key for crossing the BBB with nanoparticles: the rational design. Beilstein J Nanotechnol 11(1):866–883

Lu W, Bai W, Zhang H, Xu C, Chiarelli AM, Vázquez-Guardado A, Xie Z, Shen H, Nandoliya K, Zhao H, Lee K (2021) Wireless, implantable catheter-type oximeter designed for cardiac oxygen saturation. Sci Adv 7(7):eabe0579

Lyons A, Roy J, Herrmann J, Chipps L (2018) Treatment of décolletage Photoaging with fractional microneedling radiofrequency. J Drugs Dermatol 17(1):74–76

Madabushi R, Seo P, Zhao L, Tegenge M, Zhu H (2022) Role of model-informed drug development approaches in the lifecycle of drug development and regulatory decision-making. Pharm Res 39(8):1669–1680

Malek AM, Heilman CB, Rezac DA, Sattell JB, Bonin A (2023) Catheter systems and methods for medical procedures using catheters, US Patent Application 2 0200 069 927, assigned to CereVasc LLC

Manolis E, Brogren J, Cole S, Hay JL, Nordmark A, Karlsson KE, Lentz F, Benda N, Wangorsch G, Pons G, Zhao W (2017) Commentary on the MID3 good practices paper. CPT Pharmacometrics Syst Pharmacol 6(7):416–417

Matusiak M, Rurarz BP, Kadłubowski S, Wolszczak M, Karczmarczyk U, Maurin M, Kolesińska B, Ulański P (2021) Synthesis and properties of targeted radioisotope carriers based on poly(acrylic acid) nanogels. Pharmaceutics 13(8):1240

Mishra A (2012) Compositions and minimally invasive methods for treating peripheral vascular disease. US8088371B2

Mitrani LR, Goldenthal I, Leskowitz J, Wan EY, Dizon J, Saluja D, Creber RM, Turchioe MR, Sciacca RR, Garan H, Hickey KT (2022) Risk factor management of atrial fibrillation using mHealth: the atrial fibrillation–helping address care with remote technology (AF-HEART) pilot study. Cardiovasc Digit Health J 3(1):14–20

More VP, Hugar SM, Sogi S, Bhambar RS, Suganya M, Hugar S (2022) Comparative evaluation of the efficacy of chlorhexidine, fluoride and the combined use of chlorhexidine and fluoride varnishes on salivary Streptococcus mutans count in children with mixed dentition: an in vivo study. Int J Clin Pediatr Dent 15(3):267

Ortiz-Martínez M, González-González M, Martagón AJ, Hlavinka V, Willson RC, Rito-Palomares M (2022) Recent developments in biomarkers for diagnosis and screening of type 2 diabetes mellitus. Curr Diab Rep 22(3):95–115

Ouyang W, Lu W, Zhang Y, Liu Y, Kim JU, Shen H, Wu Y, Luan H, Kilner K, Lee SP, Lu Y (2023) A wireless and battery-less implant for multimodal closed-loop neuromodulation in small animals. Nat Biomed Eng 7(10):1252–1269

Peralta M, Jannin P, Baxter JS (2021) Machine learning in deep brain stimulation: a systematic review. Artif Intell Med 122:102198

Peterson MC, Riggs MM (2015) FDA advisory meeting clinical pharmacology review utilizes a quantitative systems pharmacology (QSP) model: a watershed moment? CPT Pharmacometrics Syst Pharmacol 4(3):189–192

Pomatto M, Gai C, Negro F, Cedrino M, Grange C, Ceccotti E, Togliatto G, Collino F, Tapparo M, Figliolini F, Lopatina T (2021) Differential therapeutic effect of extracellular vesicles derived by bone marrow and adipose mesenchymal stem cells on wound healing of diabetic ulcers and correlation to their cargoes. Int J Mol Sci 22(8):3851

Prosheva M, Aboudzadeh MA, Leal GP, Gilev JB, Tomovska R (2019) High-performance UV protective waterborne polymer coatings based on hybrid graphene/carbon nanotube radicals scavenging filler. Part Part Syst Charact 36(7):1800555

Pu J, Han J, Yang J, Yu L, Wan H (2024) Anaerobic glycolysis and ischemic stroke: from mechanisms and signaling pathways to natural product therapy. ACS Chem Nerosci 15(17):3090–3105

Ramadon D, Permana AD, Courtenay AJ, McCrudden MT, Tekko IA, McAlister E, Anjani QK, Utomo E, McCarthy HO, Donnelly RF (2020) Development, evaluation, and pharmacokinetic assessment of polymeric microarray patches for transdermal delivery of vancomycin hydrochloride. Mol Pharm 17(9):3353–3368

Rieger TR, Allen RJ, Bystricky L, Chen Y, Colopy GW, Cui Y, Gonzalez A, Liu Y, White RD, Everett RA, Banks HT (2018) Improving the generation and selection of virtual populations in quantitative systems pharmacology models. Prog Biophys Mol Biol 139:15–22

Rizvi SA, Saleh AM (2018) Applications of nanoparticle systems in drug delivery technology. Saudi Pharm J 26(1):64–70

Roth GA, Mensah GA, Johnson CO, Addolorato G, Ammirati E, Baddour LM, Barengo NC, Beaton AZ, Benjamin EJ, Benziger CP, Bonny A (2020) Global burden of cardiovascular diseases and risk factors, 1990–2019: update from the GBD 2019 study. J Am Coll Cardiol 76(25):2982–3021

Sabir F, Qindeel M, Zeeshan M, Ul Ain Q, Rahdar A, Barani M, González E, Aboudzadeh MA (2021) Onco-receptors targeting in lung cancer via application of surface-modified and hybrid nanoparticles: a cross-disciplinary review. PRO 9(4):621

Salunkhe S, Basak M, Chitkara D, Mittal A (2020) Surface functionalization of exosomes for target-specific delivery and in vivo imaging & tracking: strategies and significance. J Control Release 326:599–614

Scheuher B, Ghusinga KR, McGirr K, Nowak M, Panday S, Apgar J, Subramanian K, Betts A (2023) Towards a platform quantitative systems pharmacology (QSP) model for preclinical to clinical translation of antibody drug conjugates (ADCs). J Pharmacokinet Pharmacodyn 51(5):429–447

Shaari C (2011) Microneedle nasal delivery device. US20130085472A1

Shakouri S, Pourmadadi M, Abdouss M, Rahdar A, Pandey S (2023) pH-responsive double emulsion system for targeted anticancer therapy based on polyacrylic acid-polyvinyl pyrrolidone containing carbon nanotubes for 5-fluorouracil as an anticancer medication. Inorg Chem Commun 158:111494

Shin J, Yan Y, Bai W, Xue Y, Gamble P, Tian L, Kandela I, Haney CR, Spees W, Lee Y, Choi M (2019) Bioresorbable pressure sensors protected with thermally grown silicon dioxide for the monitoring of chronic diseases and healing processes. Nat Biomed Eng 3(1):37–46

Sia SK, Tejavibulya N (2020) Fluid extraction and drug delivery system and methods using microneedles. US10542922B2

Sim K, Ershad F, Zhang Y, Yang P, Shim H, Rao Z, Lu Y, Thukral A, Elgalad A, Xi Y, Tian B (2020) An epicardial bioelectronic patch made from soft rubbery materials and capable of spatiotemporal mapping of electrophysiological activity. Nat Electron 3(12):775–784

Sogard DJ, Smith S, Jenson ML (2017) Minimally invasive access for renal nerve ablation. US9668811B2

Sun Z, Hou X, Zhang J, Li J, Wu P, Yan L, Qian H (2022) Diagnostic and therapeutic roles of extracellular vesicles in aging-related diseases. Oxid Med Cell Longev 2022(1):6742792

Svennberg E, Friberg L, Frykman V, Al-Khalili F, Engdahl J, Rosenqvist M (2021) Clinical outcomes in systematic screening for atrial fibrillation (STROKESTOP): a multicentre, parallel group, unmasked, randomised controlled trial. Lancet 398(10310):1498–1506

Thompson D, Wu P, Lehrberg D, Mathis ML, Boutillette M (2012) Delivery of minimally invasive lung volume reduction devices. US8142455B2

Thompson D, Wu P, Lehrberg D, Mathis ML, Boutillette M, Aronson N, Vasquez J (2019) Improved and/or longer minimally invasive lung volume reduction devices and their delivery. CA2737186C1

Thornton R (2011) Microneedle Delivery Device and Methods of Using Same. US20110014181A1

Tian T, Zhang HX, He CP, Fan S, Zhu YL, Qi C, Huang NP, Xiao ZD, Lu ZH, Tannous BA, Gao J (2018) Surface functionalized exosomes as targeted drug delivery vehicles for cerebral ischemia therapy. Biomaterials 150:137–149

Trymbulak K, Ding E, Marino F, Wang Z, Saczynski JS (2020) Mobile health assessments of geriatric elements in older patients with atrial fibrillation: the Mobile SAGE-AF study (M-SAGE). Cardiovasc Digit Health J 1(3):123–129

Valencia-Lazcano AA, Hassan D, Pourmadadi M, Behzadmehr R, Rahdar A, Medina DI, Díez-Pascual AM (2023) 5-fluorouracil nano-delivery systems as a cutting-edge for cancer therapy. Eur J Med Chem 246:114995

Venkatakrishnan K, van der Graaf PH (2022) Toward project Optimus for oncology precision medicine: multi-dimensional dose optimization enabled by quantitative clinical pharmacology. Clin Pharmacol Ther 112(5):927–932

Wang KC, Jeng JY, Shen PS, Chang YC, Diau EW, Tsai CH, Chao TY, Hsu HC, Lin PY, Chen P, Guo TF (2014) P-type mesoscopic nickel oxide/organometallic perovskite heterojunction solar cells. Sci Rep 4(1):4756

Wang J, Windmiller JR, Narayan R, Miller P, Polsky R, Edwards TL (2022) Microneedle arrays for biosensing and drug delivery. US20220151516A1

Weber PJ (2024) Apparatus and systems for minimally invasive dissection of tissues. US12053228B2

Weng W, Blanchard C, Reed JL, Matheson K, McIntyre C, Gray C, Sapp JL, Gardner M, AbdelWahab A, Yung J, Parkash R (2021) A virtual platform to deliver ambulatory care for patients with atrial fibrillation. Cardiovasc Digit Health J 2(1):63–70

Wong E, Liao GP, Chang JC, Xu P, Li YM, Greengard P (2019) GSAP modulates γ-secretase specificity by inducing conformational change in PS1. Proc Natl Acad Sci U S A 116(13):6385–6390

World Health Organization (2020) Screening programmes: a short guide. Increase effectiveness, maximize benefits and minimize harm. World Health Organization, Geneva

Xu X, Robles-Martinez P, Madla CM, Joubert F, Goyanes A, Basit AW, Gaisford S (2020) Stereolithography (SLA) 3D printing of an antihypertensive polyprintlet: case study of an unexpected photopolymer-drug reaction. Addit Manuf 33:101071

Yang Y, Fan L, Jiang J, Sun J, Xue L, Ma X, Kuai L, Li B, Li Y (2024) M2 macrophage-polarized anti-inflammatory microneedle patch for accelerating biofilm-infected diabetic wound healing via modulating the insulin pathway. J Nanobiotechnology 22(1):489

Ye Z, Gastfriend BD, Umlauf BJ, Lynn DM, Shusta EV (2022) Antibody-targeted liposomes for enhanced targeting of the blood-brain barrier. Pharm Res 39(7):1523–1534

Yoon Y, Shin H, Byun D, Woo J, Cho Y, Choi N, Cho IJ (2022) Neural probe system for behavioral neuropharmacology by bi-directional wireless drug delivery and electrophysiology in socially interacting mice. Nat Commun 13(1):5521

Zafar A, Alruwaili NK, Imam SS, Alsaidan OA, Alharbi KS, Yasir M, Elmowafy M, Mohammed EF, Al-Oanzi ZH (2021) Formulation of chitosan-coated piperine NLCs: optimization, in vitro characterization, and in vivo preclinical assessment. AAPS PharmSciTech 22:1–6

Zauška Ľ, Beňová E, Urbanová M, Brus J, Zeleňák V, Hornebecq V, Almáši M (2022) Adsorption and release properties of drug delivery system naproxen-SBA-15: effect of surface polarity, sodium/acid drug form and pH. J Funct Biomater 13(4):275

Zentgraf J, Parins DJ, Saini A (2018) Exchangeable system for minimally invasive beating heart repair of heart valve leaflets. US10130474B2

Zhang J, Bahar D, Ong HL, Arnold P, Zhang M, Jiang Y, Tao R, Haworth L, Yang X, Brain C, Rahmati M (2024a) Flexible surface acoustic wave technology for enhancing transdermal drug delivery. Drug Deliv Transl Res 15:1363–1375

Zhang J, Wu T, Wang Z, Xu S, Jing X, Zhang Z, Lin J, Zhang H, Liu D, Zhou R, Guo L (2024b) Plasma-generated RONS in liquid transferred into cryo-microneedles patch for skin treatment of melanoma. Redox Biol 75:103284

Zhao S, Tang X, Tian W, Partarrieu S, Liu R, Shen H, Lee J, Guo S, Lin Z, Liu J (2023) Tracking neural activity from the same cells during the entire adult life of mice. Nat Neurosci 26(4):696–710

Zhou A, Santacruz SR, Johnson BC, Alexandrov G, Moin A, Burghardt FL, Rabaey JM, Carmena JM, Muller R (2019) A wireless and artefact-free 128-channel neuromodulation device for closed-loop stimulation and recording in non-human primates. Nat Biomed Eng 3(1):15–26

13 Intellectual Property Rights in Minimally Invasive Drug Delivery System

Prajwal Shetty and Shaila Angela Lewis

Abstract

Minimally invasive drug delivery systems (MIDDS) are becoming a game changer in the healthcare sector due to their offering of targeted, efficacious, and patient-friendly treatment options. Such systems utilize advanced technologies such as microneedles, transferosomes, sonophoresis, and other implantable devices to deliver medications directly to the desired site, which lessens systemic side effects and improves therapeutic outcomes.

The story of innovation in this domain stands on the strong foundation of IP strategies that cover novel designs, compositions, and methods of delivery. Patents encourage research by fostering proprietary technologies while enabling partnerships and licensing opportunities, and trademarks protect distinctive signs, symbols, words, or designs that identify and distinguish the goods or services of one entity from those of others. Nonetheless, in view of the rapid evolution of MIDDS, various pertinent and crucial challenges arise: navigating the patent landscape, infringement risks, and others entailing ethical issues regarding access and affordability.

This article summarizes the patent and trademark in the sector of the transdermal drug delivery system and the recent advancements in the microneedle and other technologies registered in USPTO (US Patent and Trademark Office).

P. Shetty · S. A. Lewis (✉)
Manipal College of Pharmaceutical Sciences, Manipal Academy of Higher Education, Manipal, Karnataka, India
e-mail: s.lewis@manipal.edu

M. R. Babu et al. (eds.), *Transdermal Applications of Minimally Invasive Drug Delivery Systems*, https://doi.org/10.1007/978-981-96-6275-3_13

Keywords

Minimally invasive drug delivery system (MIDDS) · Patent · Trademark · Microneedle · USPTO (US Patent and Trademark Office)

13.1 Introduction

13.1.1 Minimally Invasive Drug Delivery

First getting into minimally invasive delivery, it is necessary to know about the invasive delivery. The conventional needle syringe or catheters are examples for invasive delivery. This device-based delivery crosses the skin barrier by piercing the cutaneous layer. Invasive delivery has its own advantages and disadvantages—when it comes to disadvantage, pain in administering, possibility of transmission of blood-borne bacteria, safety concerns, and inability for patients' self-medication. To overcome these hurdles, the noninvasive drug delivery is opted, which does not involve any piercing through the skin (Amarnani and Shende 2022).

The minimally invasive drug delivery system involves delivering the drug across the subcutaneous layer, but it does not involve any metal needles; when it comes to technicality, this system virtually breaches the subcutaneous layer. Unlike invasive delivery, it does not cause any pain or any other drawbacks. Minimally invasive drug delivery systems (MIDDS) are advanced technologies designed to administer medications with minimal disruption to the body's tissues. These systems aim to enhance therapeutic efficacy, improve patient comfort, and reduce recovery times associated with traditional delivery methods (Guy and Hadgraft 2003).

The evolution of transdermal drug delivery systems (TDDS) has seen significant advancements over the years, driven by the need for more effective and patient-friendly methods of drug administration. Here's a detailed overview of the enhanced TDDS technologies that have emerged:

1. *Chemical enhancers*: Early TDDS primarily relied on chemical enhancers to increase skin permeability. These substances modify the stratum corneum (the outermost skin layer) to facilitate drug absorption. However, their effectiveness has been limited, especially for macromolecules, and they often require careful formulation to avoid skin irritation.
2. *Microneedles*: This technology has gained considerable attention as a minimally invasive method painlessly pierces the skin's stratum corneum, creating temporary hydrophilic pathways for drug infusion to enhance transdermal delivery. Microneedles create micro-sized pores in the skin, allowing for the passage of larger molecules and vaccines that would otherwise not penetrate the skin barrier. They can be made from various materials, including metals, polymers, and ceramics, and can be coated with drugs or used to deliver drugs in a dissolving format. They can deliver small molecules, nanoparticles, macromolecules, vaccines, and even insulin with rapid onset and sustained delivery.

Fabricated as hollow, coated, or dissolvable microneedles, they offer methods like "poke and flow," "coat and poke," and "poke and release" for drug administration. While microneedles reduce risks of needle reuse and enhance vaccination efficiency, challenges include limited drug dosage, mechanical strength, and specific material requirements. The development of microneedles has been supported by advances in manufacturing techniques, such as 3D bioprinting, which allows for precise control over microneedle design and drug loading.

3. *Iontophoresis*: This method uses a small electric current to drive charged drug molecules through the skin. Iontophoresis can enhance the delivery of both small and large molecules and is particularly useful for delivering peptides and proteins. The technology has been integrated into wearable devices, making it more convenient for patients. Iontophoresis is a noninvasive drug delivery method that uses a low-intensity electric current (0.1–1.0 mA/cm^2) to transport ionized drug molecules across the skin via electrodes. It is effective for localized and systemic drug delivery, with applications in conditions like acne, inflammation, and cystic fibrosis. For example, pilocarpine is iontophoretically delivered to induce sweat for cystic fibrosis diagnosis, and lidocaine patches (e.g., LidoSite®) provide rapid anesthesia. While traditionally limited to low molecular weight drugs (under 7 kDa), innovations such as chemical enhancers and microneedle formulations now enable the delivery of larger biomolecules like insulin and cytochrome *c*. Factors like drug pH, current strength, and application time influence delivery efficiency, with FDA-approved devices demonstrating sustained and targeted drug release for conditions like migraines.
4. *Electrophoresis*: Electroporation is a noninvasive drug delivery technique that uses short, high-voltage electrical pulses (5–500 V) to temporarily create pores in the skin's stratum corneum, enhancing drug permeability. It enables the delivery of both low molecular weight (e.g., doxorubicin, timolol) and high molecular weight drugs (e.g., peptides, oligonucleotides, heparin). The process, typically safe and painless, is also useful for gene transfer and can be combined with microneedles for improved macromolecule delivery. However, challenges include limited drug amounts, potential cell damage, and risks to heat-sensitive drugs like proteins.
5. *Electroosmosis*: Electroosmosis is a process where an electric field creates an osmotic gradient across a charged membrane, causing fluid to flow from the drug reservoir to the skin. Often combined with iontophoresis, this mechanism enhances drug delivery by increasing skin permeability and moving both charged and neutral molecules through the skin. The bulk fluid flow typically occurs from the anode to the cathode, enabling the delivery of negatively charged entities like proteins. While effective, skin damage is a potential concern with this technique.
6. *Sonophoresis*: Similar to iontophoresis, sonophoresis employs ultrasound waves to enhance skin permeability. The mechanical vibrations disrupt the stratum corneum, allowing for increased drug penetration. Ultrasound-assisted drug delivery uses low-frequency ultrasonic waves (20 kHz to 16 MHz) to enhance transdermal drug absorption by creating water-filled pathways in the

skin through cavitation. Drugs mixed with coupling agents like gels or creams transfer the ultrasonic energy to the skin, increasing permeability and facilitating drug diffusion. The process also generates heat, further aiding drug penetration. Approved in 2004, the FDA-cleared SonoPrep® device successfully delivers drugs like mannitol and insulin, regardless of their solubility or ionization properties. While effective, challenges include device availability, exposure duration, and potential side effects like burns. The precise mechanisms remain under study. This method is particularly effective for hydrophilic drugs and has been explored for various therapeutic applications.

7. *Photomechanical waves*: Photomechanical waves enhance skin permeability by creating transient channels in the stratum corneum, allowing drugs to pass through. Generated by low radiant exposure (5–7 J/cm^2) or single laser pulses, these waves enable controlled ablation to target specific skin depths (50–400 μm). This technique has successfully delivered macromolecules like 40 kDa dextran and 20 nm latex particles within minutes, offering a precise and efficient method for transdermal drug delivery.
8. *Jet injectors*: Needle-free injection devices deliver drugs or vaccines using high-pressure jet streams generated by compressed air, springs, or gas. These devices, like the Biojector 2000, Intraject®, and Crossjet®, offer painless and efficient drug delivery without needles, targeting subcutaneous or intramuscular tissues. Applications include insulin, vaccines, pain relief, and protein therapies. While fast and user-friendly, challenges include potential nozzle contamination, drug backsplash in reusable devices, and unpleasant odors from certain models. Despite some limitations, they are valuable for mass vaccination, personal use, and specific therapeutic needs.
9. *Magnetophoresis*: Magnetophoresis enhances drug delivery by using a magnetic field to improve drug diffusion and penetration. Studies have shown increased drug absorption, such as lidocaine delivery via magnetophoretic patches at field strengths of 30–300 mT, driven by the magnetokinesis phenomenon. This technique improves drug solubility and bioavailability compared to non-magnetic patches. Similarly, thermally assisted drug delivery (TADD) systems use controlled heat to boost blood flow and drug absorption. Devices like CHADD apply heat generated through oxidation reactions, significantly enhancing the effectiveness of transdermal patches, as seen with nicotine and nitroglycerin, achieving up to 13-fold increases in drug delivery.
10. *Microporation*: This technique involves creating micropores in the skin using various methods, including laser or mechanical devices. Microporation can enhance the delivery of larger molecules and is being researched for applications in vaccine delivery and other therapeutic areas.
11. *Laser ablation*: Laser ablation enhances transdermal drug delivery by selectively removing the stratum corneum (SC) without harming underlying tissues. This process creates micro-channels and facilitates the absorption of both lipophilic and hydrophilic drugs. Factors such as wavelength, pulse energy, exposure duration, and tissue properties determine its effectiveness. The method has been successfully used for high molecular weight drugs and shown to provide

rapid drug delivery, as demonstrated in the fast onset of lidocaine effects after laser pretreatment.

12. *Transferosomes*: Transferosomes are ultra-deformable lipid vesicles designed to enhance drug delivery through the skin and other biological membranes. They consist of phospholipids, edge activators (such as surfactants), and an aqueous core that can encapsulate both hydrophilic and hydrophobic drugs. The unique flexibility of transferosomes allows them to squeeze through narrow intercellular spaces, overcoming the barrier properties of the stratum corneum in the skin. This property makes them highly effective for transdermal drug delivery, offering improved bioavailability, targeted drug release, and reduced systemic side effects. Applications of transferosomes include the delivery of small-molecule drugs, peptides, proteins, and vaccines, making them a promising technology in noninvasive therapeutics.
13. *Ethosomes*: Ethosomes are novel lipid-based vesicular carriers designed for the enhanced delivery of drugs through the skin or other barriers. Composed of phospholipids, ethanol in high concentrations, and water, ethosomes exhibit remarkable flexibility and deformability, allowing them to penetrate deep into the skin layers. The high ethanol content increases the fluidity of the lipid bilayers and enhances the permeability of the stratum corneum, making ethosomes highly efficient for transdermal and topical drug delivery. They are capable of encapsulating a wide range of drugs, including lipophilic and hydrophilic molecules, and have applications in delivering small molecules, peptides, proteins, and even DNA. Ethosomes hold great promise for improving drug bioavailability while offering a noninvasive alternative for therapeutic delivery.
14. *Invasomes*: Invasomes are advanced vesicular carriers designed for transdermal drug delivery, characterized by their enhanced ability to penetrate biological membranes. They consist of phospholipids, ethanol, and penetration enhancers such as terpenes, which improve their deformability and membrane interaction properties. Invasomes have a flexible structure that enables them to navigate through the stratum corneum, enhancing drug transport efficiency. They can encapsulate both hydrophilic and lipophilic drugs, providing improved bioavailability and targeted delivery. These carriers are particularly useful for the noninvasive delivery of therapeutic agents, offering potential applications in areas like dermatology, pain management, and systemic drug administration (Maibach).
15. *Nanocarriers*: The use of nanocarriers, such as liposomes, niosomes, and solid lipid nanoparticles, has emerged as a promising approach to enhance transdermal delivery. These carriers can encapsulate drugs, protect them from degradation, and facilitate their transport across the skin barrier. Nanocarriers can also improve the solubility and stability of poorly soluble drugs, enhancing their therapeutic efficacy.
16. *3D-Printed patches*: Advances in 3D printing technology have enabled the development of customized transdermal patches that can deliver multiple drugs simultaneously or release drugs in a controlled manner. This technology allows

for precise dosing and can be tailored to individual patient needs, improving adherence and treatment outcomes.

17. *Smart transdermal systems*: The integration of sensors and electronic components into transdermal systems has led to the development of smart patches that can monitor physiological parameters and adjust drug delivery accordingly. These systems can provide real-time feedback to patients and healthcare providers, enhancing treatment personalization.
18. *Hybrid systems*: Combining various delivery methods, such as microneedles with iontophoresis or sonophoresis, has shown promise in enhancing drug delivery efficiency. These hybrid systems can leverage the advantages of multiple technologies to overcome the limitations of traditional TDDS.
19. *Biodegradable polymers*: The use of biodegradable polymers in transdermal patches has improved the safety and efficacy of drug delivery systems. These materials can provide sustained release of drugs while minimizing the risk of irritation and allergic reactions associated with traditional adhesives (Akhtar et al. 2020; Jeong et al. 2021).

13.1.2 Intellectual Property (IP)

Intellectual property (IP) refers to original creations of the human mind, such as artistic, literary, technical, or scientific works. Intellectual property rights (IPR) are legal protections granted to creators or inventors, allowing them exclusive use of their creations for a specific period. These rights ensure that innovators can benefit from their efforts, recover costs, and earn profits to reinvest in further research and development.

In today's economy, IP plays a critical role. The high costs of research and development (R&D) and bringing new technologies to market make protecting intellectual property essential. IPR safeguards creators' investments, time, and effort by preventing unauthorized use of their work. This protection not only promotes innovation and healthy competition but also supports industrial development and economic growth. This review highlights the importance of IPR, with a focus on its relevance in the pharmaceutical industry.

13.1.3 What Is Patent

13.1.3.1 Patentability Criteria in USPTO and EPO

A patent is a legal title that grants the holder exclusive rights to exploit an invention for a limited period, allowing them to prevent others from using it without permission. This exclusivity enables inventors to recoup their investments while publicly disclosing their inventions, thus fostering innovation after the patent expires. In the United States, patents are regulated by the US Patent and Trademark Office (USPTO n.d.), which recognizes three types oftents: design patents for new and ornamental designs, plant patents for distinct and new plant varieties, and utility patents for new and useful processes, machines, or compositions of matter (Nath Saha and Bhattacharya 2011).

Not all inventions are patentable; exclusions include writings, computer programs, methods for human and animal treatments, unique symbols, mathematical algorithms, and abstract ideas. In Europe, the European Patent Office (EPO) administers patent grants through the European Patent Convention (EPC), ensuring protection across multiple member states. This system encourages commerce and innovation, as the public gains access to pioneering inventions after a few years of exclusivity (Nath Saha and Bhattacharya 2011).

Usually, the European patent is similar to that of US patent conferring to the patent holder's exclusivity over his innovation that is 20 years, but sometimes, this condition changes in case of medical and plant patents because the approval procedure takes longer time for the approval.

Evolving transdermal drug delivery technologies are revolutionizing therapeutic practices by offering noninvasive, controlled, and sustained drug administration methods, enhancing patient compliance and bioavailability. Innovations like microneedles, ethosomes, transferosomes, and smart patches are overcoming the challenges of delivering large molecules, vaccines, and biologics through the skin barrier. Intellectual property rights (IPR) play a crucial role in fostering advancements in this field by protecting novel formulations, designs, and processes. Patents enable innovators to secure their technologies, such as the composition of delivery systems, fabrication methods, and drug release mechanisms, driving investment and collaboration in research. However, navigating IPR complexities, including overlapping patents and global filing requirements, is critical for the successful commercialization of transdermal innovations. The landscape continues to grow, with leading pharmaceutical companies and academic institutions filing patents to advance the development of next-generation transdermal technologies.

13.1.4 Microneedle

13.1.4.1 Regulatory Guidelines

To address the challenges posed by the novel nature of MN systems, the US FDA issued two key documents in 2017:

1. **Regulatory Considerations for Microneedling Devices**.
2. **Scientific Considerations for Microneedling Products**.

These guidelines define MN systems as medical devices under Section 201(h) of the Federal Food, Drug, and Cosmetic Act. They classify MN systems intended to cure, treat, or monitor a specific condition as **Class II medical devices** under 21 CFR 878.4430. This classification requires premarket notification and regulatory compliance with specific quality and safety standards.

13.1.5 Regulatory Oversight

MN systems fall under the regulatory purview of:

- **CDRH (Center for Devices and Radiological Health):** For devices classified solely as medical devices.
- **CDER (Center for Drug Evaluation and Research):** For combination products where the primary mode of action is pharmaceutical.
- **CBER (Center for Biologics Evaluation and Research):** For MN systems delivering biological products like vaccines.

13.1.5.1 Required Specifications

FDA guidelines stipulate that MN system manufacturers provide detailed information in product labeling and submissions, including:

- **Device specifications:** Needle geometry, insertion depth, and design details.
- **Intended use:** Specific diseases or conditions the device targets.
- **Treatment course:** Recommended duration and frequency of use.
- **Sterility:** MN devices must be sterile since they come into contact with viable skin cells.
- **Disposal instructions:** Procedures for single-use and reusable MN devices to ensure safety.
- **Expiration details:** Information on shelf life and product stability.

13.1.5.2 Challenges in Approval and Commercialization

The complexity of MN systems introduces several challenges for regulatory approval:

- **Sterility and Safety:** Since MN systems penetrate the skin, sterility is a critical requirement to prevent infections.
- **Metabolism and Elimination Studies:** Comprehensive studies are needed to understand the pharmacokinetics of drugs delivered via MN systems.
- **Delayed Commercialization:** The extensive testing and regulatory approvals required can slow down market entry.
- **Reusable Systems:** Reusable MN devices require detailed processing instructions to avoid cross-contamination.

13.1.5.3 Industry and Research Implications

The inclusion of MN systems as Class II devices emphasizes the importance of robust research and development. Studies on materials, manufacturing processes, and drug delivery efficacy are essential to meet regulatory standards. Additionally, global regulatory bodies may require harmonization of guidelines to streamline international commercialization (Guy and Hadgraft 2003).

Transdermal Drug Delivery Systems

Transdermal delivery enables systemic effects by allowing drugs to permeate the skin and enter the bloodstream, while topical applications provide localized therapeutic action. Despite challenges like the skin's natural barrier properties, recent innovations in formulation technologies, penetration enhancers, and delivery devices have significantly improved drug efficacy and broadened the scope of applications. The review categorizes patents from 1997 to 1999 into two main areas: methods to enhance transdermal absorption and new pharmacological formulations.

Enhancers for Skin Permeability

- **Macrochem (WO9909954, 1999).**
  - Enhanced NSAID delivery using **1,3-dioxolanes or 1,3-dioxanes**.
  - Key components: Enhancers (5–10%), propylene glycol (1–15%), ethanol (55–70%), and water.
  - Outcome: Dose-dependent penetration improvements were shown, measured using HPLC.
- **Alza Corporation (WO9640259, 1997).**
  - **Dual enhancer system**: Glyceryl monolaurate and lauryl acetate for delivering drugs like testosterone and alprazolam.
  - Applications: Treatments for migraines, fungal infections, and systemic conditions.
- **Minnesota Mining & Manufacturing (WO9837871, 1998).**
  - Terpene and fatty acid derivatives for **testosterone delivery**.
  - Target conditions: Male hypogonadism, AIDS-related wasting.

Transdermal Patches

- **Rotta Research BV (WO9723227, 1997).**
  - Multilayered HRT patch delivering estradiol and norethisterone acetate.
  - No enhancers were used, relying on diffusion mechanisms for delivery.
- **Rutgers University (WO9739743, 1997).**
  - Contraceptive transdermal system delivering estradiol and progestin with permeation enhancers (dimethyl sulfoxide, lauryl lactate).
- **Minnesota Mining & Manufacturing (WO9723205, 1997).**
  - **Flurbiprofen patch** for chronic pain relief.
  - Components: Isopropyl myristate and copolymers for effective adhesion and delivery.
- **SmithKline Beecham (WO9639136, 1996).**
  - Parkinson's disease patch delivering **ropinirole**.
  - Free-base form of the drug resulted in 40-fold higher penetration compared to its salt forms.

Electrically Assisted Delivery

- **General Advances.**
  - **Iontophoresis**: Low-current application for controlled molecule delivery.

- **Electroporation**: High-voltage application for macromolecule delivery (e.g., insulin and peptides).

Topical Formulations

Skin Disorders

- **Avon Products (US5621008, 1997).**
  - Neutral pH exfoliators (N-acyl derivatives) for treating **xerosis, dandruff, and ichthyosis**.
  - Benefits: Reduced skin irritation compared to glycolic acid-based formulations.
- **Schering AG (WO9925332, 1999).**
  - Gel containing azelaic acid, EDTA, and benzoic acid with improved penetration properties.
  - Results: Higher concentrations in the stratum corneum compared to lipophilic creams.
- **Fujisawa Pharmaceutical (WO9836747, 1998).**
  - Tacrolimus (FK506) topical emulsions for atopic dermatitis.
  - Key components: Hydrophilic substances (carboxyvinyl polymer) and oily bases (isopropyl myristate).

Hormonal and Steroidal Treatments

- **Estée Lauder (WO9710255, 1997).**
  - Steroid esters (e.g., dehydroepiandrosterone) for aging, psoriasis, acne, and hyperpigmentation.
- **Schwarz Pharma AG (WO9901116, 1999).**
  - Long-lasting transdermal therapeutic system (TTS) for **estrogens and androgens**, ensuring bioavailability for up to 7 days.

Antimicrobials

- **Bristol Myers Squibb (WO9816104, 1998).**
  - Lipid-based formulations for reducing skin microbial adherence (e.g., *Candida albicans*, *Staphylococcus* spp.)
  - Key ingredient: **Palmitoleic acid**.
- **Ganeden Biotech (WO9847374, 1998).**
  - Probiotic-based formulations combining fructooligosaccharides (FOS) and *Bacillus* spp. spores for infection prevention.

Neurological Applications

- **Alza Corporation (WO9640139, 1996).**
  - Transdermal delivery of **pergolide** for Parkinson's disease to maintain consistent plasma levels.
- **Britannia Pharmaceuticals (WO9749396, 1997).**
  - ADHD treatment through controlled release of **lofexidine** and **clonidine**, delivered via skin patches or oral solutions.

Immunosuppressants

- Fujisawa Pharmaceutical (WO9836747, 1998).
  - Tacrolimus formulations for **autoimmune skin disorders** and inflammation.
  - Delivery: Aqueous emulsified forms tailored for better absorption.

Anti-Inflammatory and Analgesic Agents

- **Minnesota Mining & Manufacturing (WO9723205, 1997).**
  - Flurbiprofen delivery via adhesive patches for pain management.
  - Enhanced bioavailability through polyvinylpyrrolidone-based adhesive.
- **Alza Corporation (WO9932096, 1999).**
  - κ-opioid agonist **asimadoline** for chronic pain via multilayered patches (Chen and Fang 2000).

13.1.6 Patent Innovations in Transdermal Drug Delivery (TDD): A Comprehensive Analysis

Transdermal drug delivery (TDD) systems have evolved significantly, becoming a cornerstone of pharmaceutical innovation due to their ability to provide controlled and efficient drug administration through the skin. The patent landscape in this domain reflects a surge in technological advancements, with numerous innovations addressing the challenges of drug permeation, sustained release, and patient compliance. This section explores the trends, key players, technologies, and applications derived from a thorough bibliometric and patent analysis spanning the past two decades.

13.1.6.1 Global Trends in TDD Patents

Between 2000 and 2022, a total of 7275 patents were recorded, grouped into 2997 patent families, with 314 granted patents. The filing trends illustrate a robust increase in activity during the early 2000s, characterized by a focus on enhancing skin permeation techniques and developing novel delivery systems. However, the growth rate plateaued in recent years, reflecting a maturation of core technologies and the need for incremental innovation.

Jurisdictional data reveal that the United States leads patent filings, followed by China and Europe. This trend is driven by the presence of leading pharmaceutical and biotechnology companies in these regions, alongside robust regulatory and intellectual property frameworks. Key markets such as Canada, South Korea, and Japan also feature prominently, indicating global interest in TDD technologies. Table 13.1 displays recently approved patents for transdermal drug delivery systems.

13.1.6.2 Technological Classifications and Key Innovations

The Cooperative Patent Classifications (CPCs) of TDD patents highlight the focus on pharmaceutical formulations, devices, and enhancement techniques. Common CPCs include **A61K** (pharmaceutical product administration and sterilization methods) and **A61M** (medical and drug delivery devices).

Table 13.1 Recently patented transdermal drug delivery systems (PPUBS, Trademark Search, U.S Food & Drug Administration)

	Title	Inventor(s)	Patent no.	Date approved
Transdermal	Combinations of Copanlisib with anti-PD-1 antibody	Liu; Ningshu	US-20240382495-A1	2024/11/21
Microneedle	Immunotherapeutic treatment of cancer	MARTYN; GLEN	US-20230050258-A1	2023/02/16
Sonophoresis	Modulators of STING (stimulator of interferon genes)	GAJIWALA; Ketan Satish et al.	US-20240383903-A1	2024/11/21
Magnetophoretic	Magneto-patterned-cell-laden hydrogel materials and methods of making and using same	Zlotnick; Hannah M. et al.	US-12138366-B2	2024/11/12
Ethosomes	TRPV1 modulator compounds	Devesa Giner; Isabel et al.	US-11274073-B2	2022/03/15
Transferosomes	Complement component Irna compositions and methods of use thereof	Borodovsky; Anna et al.	US-20240376477-A1	2024/11/14

Chemical and Physical Enhancers

- **Chemical Penetration Enhancers**: Patents describe the use of surfactants, solvents, and fatty acids to temporarily disrupt the skin's stratum corneum and improve drug permeation. For example, formulations incorporating liposomes and transfersomes are protected for their ability to encapsulate hydrophilic and hydrophobic drugs, enhancing skin absorption.
- **Electrically Assisted Methods**: Patented techniques such as iontophoresis, electroporation, and sonophoresis enable noninvasive delivery of macromolecules like insulin and peptides. These methods utilize controlled electrical or ultrasound stimuli to increase skin permeability.

Electrically Assisted Methods

- **Patent number**: US-8617071-B2 (filed 2007/06/21, granted 2013/12/31).
- **Owner**: Abbott Diabetes Care Inc.
- **Description**: An electrochemical sensor-based device integrating analyte monitoring with transdermal delivery, enabling optimized drug dosing.
- **Applications**: Management of diabetes and dynamic drug adjustment.

Advanced Delivery Systems

- **Microneedle Technologies**: Microneedles represent a transformative innovation in TDD. Numerous patents detail minimally invasive devices that combine arrays of micro-protrusions with drug reservoirs, allowing for efficient delivery of vaccines, hormones, and pain management drugs. Microneedles have been extensively patented by both academic institutions and companies, reflecting their versatility and growing clinical applications.

Microneedles

- Patent Number.
- US-8372040-B2 (filed 2006/05/24, granted 2013/02/12).
- US 8,523,791 B2 (filed 2009/08/11, granted 2013/09/03).
- **Owners**: Chrono Therapeutics Inc. and CareWear Corp.
- **Description**: Microneedle-based systems for minimally invasive transdermal drug delivery.
- **Applications**: Delivery of vaccines, hormones (e.g., estradiol, testosterone), and pain management drugs.
- **Controlled-Release Systems**: Patents emphasize systems that ensure sustained drug release over extended periods, enhancing therapeutic efficacy and patient compliance. Examples include multilayered patches for hormone replacement therapy and biodegradable polymers for chronic disease management.

Controlled-Release Patches

- **Patent number**: US 9,375,529 B2 (filed 2009/09/02, granted 2016/06/28).
- **Owner**: Becton Dickinson and Company.
- **Description**: Extended-use medical device with a controlled-release mechanism.
- **Applications**: Hormone replacement therapy (HRT), chronic disease treatment.

Integrated Systems

Emerging patents integrate biosensors and controllers with TDD devices, enabling real-time monitoring and automated adjustments of drug administration. For instance, Abbott Diabetes Care's highly cited patent (US 8,617,071 B2) describes an analyte monitoring device paired with a transdermal delivery module to optimize therapy dynamically.

Sensor-Enhanced Devices

- **Patent number**: US 8,617,071 B2 (filed 2007/06/21, granted 2013/12/31).
- **Owner**: Abbott Diabetes Care Inc.
- **Description**: Combines drug reservoirs, sensors, and controllers to enable real-time monitoring and adaptive transdermal delivery.
- **Applications**: Diabetes management, personalized medicine.

13.1.6.3 Leading Innovators and Applications

Key players in TDD patenting include pharmaceutical giants and specialized technology firms:

- **Noven Pharmaceuticals**: Known for transdermal patches for ADHD (Daytrana) and menopause symptom relief (Minivelle), Noven leads in the development of advanced patch technologies.
- **Alza Corporation**: Pioneers of microneedle patches and controlled-release systems, Alza has significantly influenced the field, with its portfolio now part of Johnson & Johnson.
- **3M Innovative Properties**: A leader in patch-based drug delivery, 3M has patented adhesives, enhancers, and novel reservoir systems for improved skin permeation.

These patents target a wide range of medical conditions, from chronic pain and neurological disorders to vaccination and hormonal therapies. The adaptability of TDD technologies to both local and systemic applications makes them highly versatile and clinically valuable.

13.1.6.4 Emerging Trends in TDD Patents

- **Biodegradable Materials**: Recent patents describe the use of biocompatible polymers in microneedles and drug reservoirs, ensuring safe degradation after drug delivery.
- **Nanotechnology**: Encapsulation of drugs within nanoparticles or nanocarriers has become a focus, with patents highlighting applications in cancer therapy and anti-inflammatory treatments.
- **Wearable Devices**: Integration of wearable TDD systems with digital health platforms is an emerging trend, offering opportunities for personalized medicine (Cammarano et al. 2023).

13.2 Trademarks

A trademark is a distinctive sign that identifies the goods or services of one enterprise and sets them apart from those of others. Protected under intellectual property rights, trademarks can be secured at both national and international levels. Nationally or regionally, protection is obtained by registering the trademark with the appropriate office and paying the required fees. Internationally, businesses can either apply for protection in each target country individually or streamline the process through WIPO's Madrid System, which facilitates trademark registration across multiple jurisdictions.

Registering a trademark grants its owner exclusive rights to its use, providing legal certainty and stronger protection in cases of infringement or litigation. Owners can also license their trademarks to others for a fee, creating additional commercial opportunities. Trademark registrations typically last for 10 years but can be renewed indefinitely upon payment of renewal fees. A wide range of elements can qualify as trademarks, including words, letters, numerals, symbols, drawings, shapes, packaging, colors, sounds, and even fragrances, offering virtually limitless options for brand differentiation (WIPO n.d.).

Pharmaceutical trademarks are crucial as they directly impact public health by helping users identify desired products, thereby enhancing the reputation of pharmaceutical companies and motivating them to maintain high quality. By reducing name-related prescription errors, trademarks safeguard patient health and safety, as any confusion in a drug's name, appearance, packaging, or trade style could lead to the purchase of the wrong medication, potentially resulting in fatal consequences. Courts take these issues seriously, prioritizing public welfare when addressing pharmaceutical trademark disputes. The registration process for pharmaceutical trademarks is time-intensive due to stringent business and regulatory factors. Key steps include listing potential trademarks, conducting thorough clearance searches to avoid confusion with existing names, monitoring for infringements, and filing for registration with the assistance of IP professionals. Beyond protection, trademarks incentivize manufacturers to invest in new treatments, assist health professionals in minimizing medication errors, empower consumers to select appropriate medications, and enable manufacturers to monitor the safety of their drugs.

13.2.1 Trademarking Transdermal Products

Pharmaceutical trademarks encompass brand names, logos, and sometimes even the appearance or packaging of a product. In the context of transdermal drug delivery, trademarks are critical for distinguishing innovative products in a competitive market while ensuring patient safety by reducing confusion between similar treatments. Figs. 13.1, 13.2 and 13.3 are examples of trademarks obtained for various transdermal drug delivery devices.

Key examples of trademarked transdermal products include:

- **Nicoderm®**: A nicotine patch designed to aid smoking cessation. The brand name, a combination of "nicotine" and "dermal," emphasizes the transdermal delivery mechanism and the product's purpose.

Fig. 13.1 Trademark registration by companies for microneedle (USPTO n.d.)

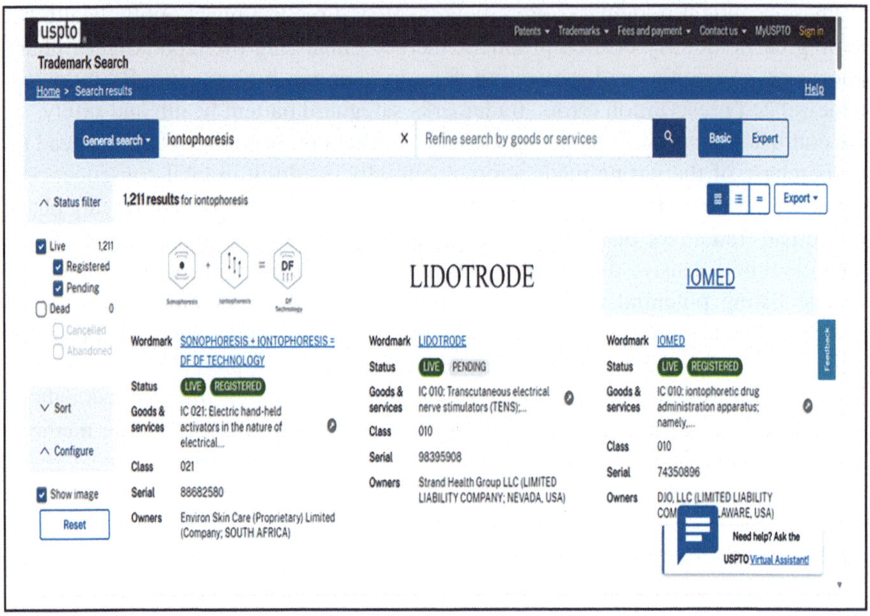

Fig. 13.2 Trademark registration by companies for iontophoresis (USPTO n.d.)

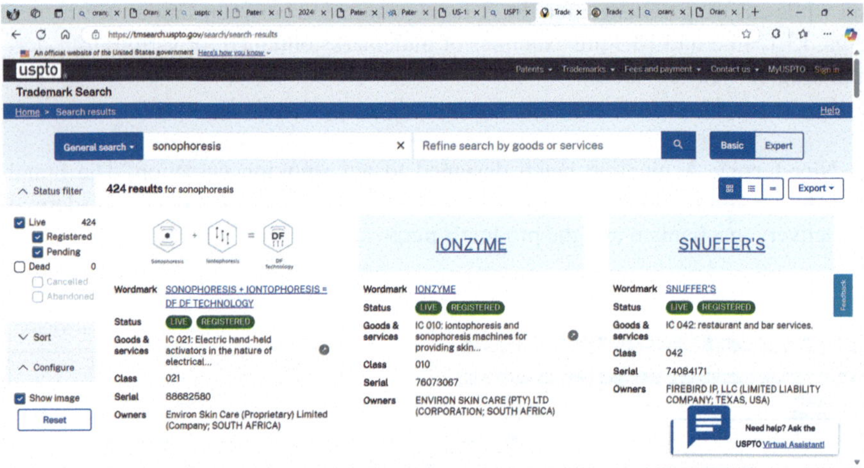

Fig. 13.3 Trademark registration by companies for sonophoresis (USPTO n.d.)

- **Lidoderm®**: A transdermal patch delivering lidocaine for localized pain relief. The trademark reflects the active ingredient and its dermal application, ensuring clear identification.

13.2.2 The Role of Trademarks in Patient Safety

Trademarks are more than commercial tools; they play a pivotal role in ensuring patient safety in the pharmaceutical sector. Unique and clear branding reduces the risk of dispensing errors, particularly in transdermal products, where the dosage form may look similar across brands. Regulatory bodies often scrutinize trademarks for their potential to cause confusion, highlighting their importance in safeguarding public health (Nimje n.d.).

13.3 Conclusion

The advancements in minimally invasive drug delivery systems (MIDDS) have revolutionized therapeutic practices by addressing the limitations associated with conventional invasive methods. These innovative technologies, such as microneedles, iontophoresis, sonophoresis, and nanocarriers, have demonstrated significant potential in enhancing patient compliance, ensuring controlled and sustained drug release, and overcoming the challenges of delivering complex molecules such as vaccines and biologics through the skin.

Intellectual property rights (IPR), including patents and trademarks, are pivotal in fostering these innovations. Patents protect the underlying technologies, formulations, and devices, incentivizing research and investment in this field. Trademarks, on the other hand, ensure product differentiation, enhance brand recognition, and safeguard patient safety by minimizing prescription errors. The integration of smart technologies, hybrid systems, and 3D printing further underscores the dynamic evolution of MIDDS and TDDS, paving the way for personalized and efficient drug delivery solutions.

This confluence of technological advancement and robust IPR frameworks not only enhances therapeutic efficacy but also drives commercial viability, ensuring that these groundbreaking solutions reach the market effectively. As the landscape continues to evolve, collaboration between pharmaceutical companies, regulatory bodies, and researchers will be critical in overcoming remaining challenges and shaping the future of noninvasive drug delivery.

References

Akhtar N, Singh V, Yusuf M, Khan RA (2020) Non-invasive drug delivery technology: development and current status of transdermal drug delivery devices, techniques and biomedical applications. Biomed Tech 65(3):243–272

Amarnani R, Shende P (2022) Microneedles in diagnostic, treatment and theranostics: an advancement in minimally-invasive delivery system. Biomed Microdevices 24(1):4

Cammarano A, Dello Iacono S, Meglio C, Nicolais L (2023) Advances in transdermal drug delivery systems: a bibliometric and patent analysis. Pharmaceutics 15(12):2762

Chen HY, Fang JY (2000) Therapeutic patents for topical and transdermal drug delivery systems [internet]. Exp Opin Ther Patents 10(7):1035–1043. http://www.ashley-pub.com

Guy RH, Hadgraft J (eds) (2003) Transdermal drug delivery. A textbook drug and the pharmaceutical sciences, vol 123, 2nd edn. Marcel Dekker, New York

Jeong WY, Kwon M, Choi HE, Kim KS (2021) Recent advances in transdermal drug delivery systems: a review. Biomater Res 25(1):24

Nath Saha C, Bhattacharya S (2011) Intellectual property rights: an overview and implications in pharmaceutical industry. J Adv Pharm Technol Res 2:88–93

Nimje P (n.d.) The need for trademark in pharmaceutical industry. https://sagaciousresearch.com/blog/the-need-for-trademark-in-pharmaceutical-industry/

Patent Public Search Basic (PPUBS Basic) (n.d.). https://ppubs.uspto.gov/pubwebapp/static/pages/ppubsbasic.html

Trademark Search (n.d.). https://tmsearch.uspto.gov/search/search-results

U.S. Food & Drug Administration (n.d.) Orange Book: Approved drug products with therapeutic equivalence evaluations. https://www.accessdata.fda.gov/scripts/cder/ob/index.cfm

USPTO (n.d.) United State Patent and Trademark Office. https://www.uspto.gov/

WIPO (n.d.). https://www.wipo.int/en/web/trademarks

Clinical Approaches of Minimally Invasive Drug Delivery Systems

14

Amit Raj Singh, Paramvir Singh Tomar, Malakapogu Ravindra Babu, Biplab Pal, Sweta Kumari, Shaik Abdul Rahaman, and Ananda Kumar Chettupalli

Abstract

This chapter focuses on microneedles, hydrogels, thermal ablation, sonophoresis, and iontophoresis, highlighting current developments and future directions in MIDDS. Problems that have arisen during the clinical development of MIDDS range from issues of medication absorption to formulation stability, regulatory hurdles, and variations in patient compliance in this article. This throws light on the various therapeutic applications for which each of these MIDDS technologies is undergoing studies and gives an overview of the ongoing clinical trials involving those technologies. Advances in the areas of bio-responsive microneedles, smart integration of sonophoresis and iontophoresis, targeted delivery through novel designs in nanoparticles, responsive hydrogels, and thermal ablation techniques are expected to brighten the future prospects of MIDDS. The introduction of wireless electronics, artificial intelligence, and machine learning is going to make drug distribution, treatment personalization, and targeting efficiency take giant strides. Overall goals include obtaining better treatment results and better patient compliance by removing biological obstacles and attaining preprogrammed pulsatile release. Paper-based conclusions show that MIDDS improve patient compliance, improve patients' adherence to treatment protocols, and also reduce overall healthcare costs since the said delivery procedures are noninvasive, rapid, and painless.

A. R. Singh · P. S. Tomar · M. R. Babu (✉) · S. A. Rahaman · A. K. Chettupalli
Department of Pharmacy, School of Medical and Allied Sciences, Galgotias University, Greater Noida, Uttar Pradesh, India

B. Pal · S. Kumari
School of Pharmaceutical Sciences, Lovely Professional University, Phagwara, Punjab, India

M. R. Babu et al. (eds.), *Transdermal Applications of Minimally Invasive Drug Delivery Systems*, https://doi.org/10.1007/978-981-96-6275-3_14

Keywords
Minimally invasive drug delivery systems (MIDDS) · Microneedles · Sonophoresis · Iontophoresis · Hydrogel · Thermal ablation · Clinical trials

14.1 Introduction

Painless medication administration techniques that involve the delivery of drugs across biological barriers, such as the skin or mucosal lining, are sometimes referred to as noninvasive drug delivery. The benefits of noninvasive drug delivery methods over injection are clear; patients prefer them, and they have the potential to increase medicine accessibility, therapy adherence, and drug safety. Moreover, healthcare providers can avoid the high expenses and risks connected with needle-related issues by administering drugs in a noninvasive manner [1]. The safety, ease, and high rate of patient compliance with noninvasive medication delivery devices make them a good choice [2]. An overview of the current state of the art in the varied, active, and continuously developing area of noninvasive drug delivery research is the goal of this special issue. Various methods for drug delivery through the skin [3–7], nasal passages [8], respiratory system [9, 10], and oral cavity [11–14] have been discussed. In addition, a wide range of drug delivery strategies and therapies were discussed and reviewed, including nano-systems such as silica nanoparticles, which allow for noninvasive administration of biologics and tiny pharmacological molecules [15].

14.1.1 Challenges in Clinical Development of MIDDS

The efficacy and application of noninvasive drug administration in medical practice are threatened by many obstacles encountered during the clinical development of minimally invasive drug delivery systems. Issues with medication absorption, formulation instability, regulatory hurdles, and inconsistent patient adherence are all examples of such obstacles. If noninvasive therapies are to improve clinical outcomes, these must be taken into account. While early diagnosis and multicenter collaboration are crucial for effective condition treatment and the execution of trials for more robust therapies, they can restrict the growth of MIDDS in broader clinical contexts [16].

The principal problems in the clinical development of MIDDS during clinical trials are as follows:

14.1.1.1 Challenges in Drug Absorption

Minimally Invasive Drug Delivery Systems (MIDDS) face a significant challenge in the clinical advancement of drug absorption. There are a number of drug absorption-related issues that can affect how well these trials work. Systemic absorption and bioavailability may be impeded by the drug's physicochemical properties and the physiological processes controlling its absorption. This necessitates extensive research into better modes of pharmaceutical delivery, especially for substances that are not well absorbed through oral and noninvasive routes [17]. Overcoming biological obstacles to medication uptake is a major challenge in the advancement of noninvasive drug delivery methods. The skin's and the gut's barrier properties significantly impact the drug's bioavailability upon administration; this is especially true for oral and transdermal routes [18]. Due to the complexity of human physiology, absorption rates can and do vary; thus, we must develop cutting-edge technology to improve permeability and guarantee efficient medication delivery.

14.1.1.2 Concerns Regarding Formulation Stability

Minimally invasive devices and delivery systems (MIDDS) stability is an important consideration throughout clinical development. Clinical trial stability has a major impact on the safety and effectiveness of a pharmaceutical formulation. It is crucial to closely monitor several factors that cause instability, including changes in mesh sizes and parameter fluctuations, in order to maintain the drug's therapeutic advantages [19]. Formulation stability ranks high among the many issues linked with minimally invasive medication delivery. Most APIs are sensitive to environmental conditions, such as temperature and humidity, and this will degrade their potency with time [20]. In addition, it is also essential that a formulation retains stability and potency from its preparation to clinical use because instability may lead to degradation of the drug and, therefore, loss of therapeutic benefit.

14.1.1.3 Regulatory Obstacles

Minimally invasive devices and delivery systems (MIDDS) face significant challenges in their clinical advancement due to regulatory hurdles. These hurdles sometimes involve complex and different approval procedures in different countries, which can greatly slow down research and make it harder for patients to get new treatments. When clinical research organizations (CROs) are involved, the already complex process of conducting a clinical trial is further complicated by inconsistent laws and their interpretations [21]. For as long as there have been noninvasive formulations, the regulatory landscape has presented the greatest obstacle. While introducing the new formulation, all sorts of detailed documentation is needed for the proof of the safety and efficacy of the formulation by the regulators [22]. The series of clinical trials become cumbersome, time-consuming, and expensive, especially while introducing a new technology or delivery mechanism.

14.1.1.4 Variability in Patient Compliance

The inconsistency in patient adherence poses a significant obstacle in the clinical advancement of minimally invasive devices and delivery systems (MIDDS). Patient compliance with trial protocols, encompassing accurate dosing and regimen implementation, is essential for obtaining precise and significant outcomes from clinical trials. However, fluctuations in adherence can substantially impact the dependability of these results, underscoring the necessity for enhanced measurement and reporting standards [23]. Minimally invasive drug delivery systems strictly depend on patient adherence. The factors involved are ease of administration, perceived efficacy, and possible side effects that may cause patient adherence to prescribed treatment [24]. For an alternative to being elevated in noninvasive, it must be intuitive and offer some sort of distinction from traditional techniques.

14.1.1.5 Innovations and Solutions

Innovations and solutions have been developed and applied to tackle the obstacles in the clinical development of minimally invasive devices and delivery systems (MIDDS). Improvements in clinical trial designs are crucial for addressing challenges associated with recruitment rates and standardization, particularly in acute ischemic stroke therapies. Incorporating lead-in stages, meticulous patient selection, and transparent data access can markedly enhance the efficiency and results of these trials [25]. Researchers are studying new methods and technologies that may overcome these hurdles. Nanocarriers and advanced polymers used in drug formulations could enhance the solubility and permeability of drugs [26]. Clinical trials that are targeted to the specific needs of the patient populations being treated may speed up approval and improve the chances of effective treatment.

14.2 Ongoing Clinical Trial on MIDDS

Current clinical trials for minimally invasive drug delivery devices present significant breakthroughs in therapeutics by increasing patient comfort and enhancing treatment efficacy. Numerous tactics and advances, especially in biologics and specialized devices, are being investigated to optimize delivery pathways and improve treatment results.

14.2.1 Microneedles

In recent years, advancements in molecular pharmaceuticals, materials chemistry, and biomedical devices have positioned the MNDDS as a promising innovative technique for subcutaneous medication administration [27] (Fig. 14.1) (Table 14.1).

14.2.2 Sonophoresis

Sonophoresis, employing ultrasonography as a way to improve the physical systemic medication delivery, has demonstrated efficacy in delivering diverse drug types irrespective of their electrical properties and can be seamlessly integrated with other transdermal drug delivery technologies to augment drug delivery rates [38–40] (Table 14.2).

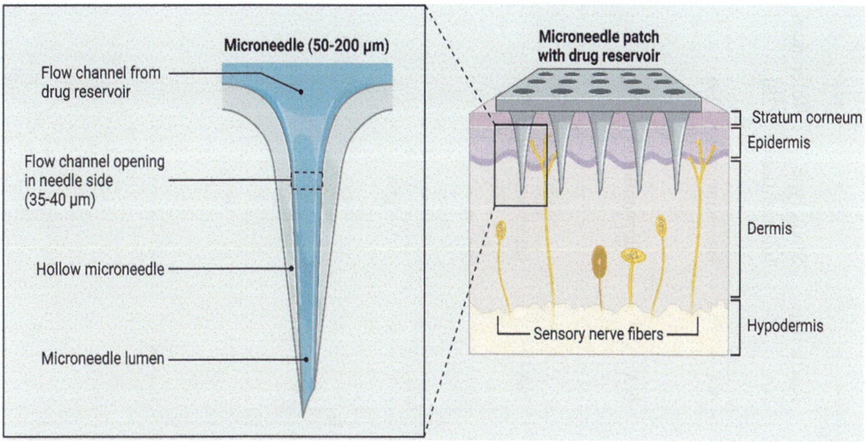

Fig. 14.1 Microneedle drug delivery system

14.2.3 Iontophoresis

Iontophoresis is a technique used to apply a small amount of electric current to the skin to enhance the transdermal transfer of drugs.

Iontophoresis improves the penetration of both charged and neutral chemicals by means of electromigration and electro-osmosis, offering a way to distribute medications in a planned manner [43] (Fig. 14.2) (Table 14.3).

14.2.4 Thermal Ablation

Heat energy is applied to tissue throughout all thermal ablation therapies, which causes tumor death and tissue necrosis. The treatments fall under the category of heat-based modalities, which include laser ablation, high-intensity focused ultrasound (HIFU), microwave (MW), radiofrequency (RF), and the tissue-freezing technique known as cryoablation [53] (Table 14.4).

14.2.5 Hydrogel

Three-dimensional networks of cross-linked hydrophilic polymers, known as hydrogels, have the ability to reversibly swell or de-swell in water while holding large amounts of liquid in their swollen state. Hydrogels can be designed to react to changes in the external environment by exhibiting controlled responses, including expansion or contraction [64] (Fig. 14.3) (Table 14.5).

Table 14.1 An overview of recent clinical trials of microneedles

S. no	Title of the study	State of disease/ condition	Intervention	Phase	Recruiting	Study type	References
1	Phase 1 evaluation of the H1 influenza vaccine MIMIX MAP has delivered	Influenza	Biological: H1 influenza antigen	Phase 1 completed	NA	Interventional	[28]
2	Radiofrequency microneedle treatment for photoaging skin	Pigmentation Pigmentation disorder	Device: RF microneedle Drug: Combination cream	NA	NA	Interventional	[29]
3	Treatment of skin aging in adult women of Indonesia using stem cell Secretome: A comparison of fractional CO_2 laser and microneedles	Skin aging Transepidermal water loss	Device: Fractional CO_2 laser Device: Microneedle	NA	NA	Interventional	[30]
4	Improving topical anesthesia with a transdermal microneedle patch in children with thalassemia	Thalassemia in children	Device: Microneedle Drug: 1 fingertip unit (FTU) EMLA cream (30-min application time) Drug: 1 fingertip unit (FTU) EMLA (15-min application time) Drug: 0.5 fingertip unit (FTU) EMLA (30-min application time) Device: Sham patch	Phase 2 completed	NA	Interventional	[31]

5	The bioequivalence of two Abaloparatide-sMTS treatments: A randomized control trial in healthy women	Osteoporosis	Combination product: Abaloparatide-sMTS (Patheon) Combination product: Abaloparatide-sMTS (Kindeva) Combination product: Abaloparatide-sMTS (Patheon) Combination product: Abaloparatide-sMTS (Patheon)	Phase 1 completed	NA	Interventional	[32]
6	Pilocarpine microneedles for sweat induction (PMN-SI)	Cystic fibrosis	Device: Pilocarpine microneedle patch Device: Pilocarpine iontophoresis	NA	NA	Interventional	[33]
7	Phases 1 and 2 of the microneedle patch vaccine for measles and rubella	Measles Rubella Vaccination Healthy	Biological: Measles-rubella vaccine (MRV-SC) Biological: MRV-MNP Others: PLA-MNP	Phase 2 completed	NA	Interventional	[34]
8	For adults suffering from nodular basal cell carcinoma, this study aims to assess the efficacy and safety of D-MNA	Nodular basal cell carcinoma	Drug: D-MNA 200 mcg Drug: D-MNA 100 mcg Others: Placebo	Phase 2 completed	Recruiting	Interventional	[35]
9	Investigating the effects of microneedle radiofrequency and Oral isotretinoin on moderate to severe acne in a clinical trial	Moderate to severe acne vulgaris	Drug: Oral isotretinoin Others: Oral isotretinoin combined with microneedle radiofrequency therapy	NA	Recruiting	Interventional	[36]
10	Botulinum toxin microneedle patches for the Management of Enlarged Facial Pores	Enlarged pores	Device: Microneedle patch Device: Intradermal injection	Phase 3 completed	Active, not recruiting	Interventional	[37]

Table 14.2 An overview of recent clinical trials of sonophoresis

S. no	Title of the study	State of disease/ condition	Intervention	Phase	Recruiting	Study type	References
1	Osteoarthritis-related knee pain symptomatic treatment with sustained acoustic medicine (SAM)	Osteoarthritis Knee arthritis	Device: Sustained acoustic device with 2.5% diclofenac patch Device: Sustained acoustic device with 0% diclofenac patch Drug: 2.5% diclofenac patches	Phase 2 completed	NA	Interventional	[41]
2	Sustained acoustic medicine for symptomatic treatment of knee pain related to osteoarthritis	Osteoarthritis Knee arthritis	Device: Sustained acoustic device with 2.5% diclofenac patch Device: Sustained acoustic device with 0% diclofenac patch Drug: 2.5% diclofenac patches Others: SAM patch	Phase 2 completed	Not yet recruiting	Interventional	[41]
3	Utilizing sustained acoustic medicine (SAM) to alleviate bone fracture pain symptoms	Bone fracture Fresh fracture	Device: Sustained acoustic device with 2.5% diclofenac patch Drug: 2.5% diclofenac patches	Phase 1 completed	Recruiting	Interventional	42
4	Enhancing jaw mobility and quality of life for survivors of head and neck cancer via Photobiomodulation and Sonophoresis	Head and neck cancer Trismus	Combination product: Photobiomodulation therapy with mandibular exercises Combination product: Sonophoresis with mandibular exercises	Unknown status	NA	Interventional	[42]

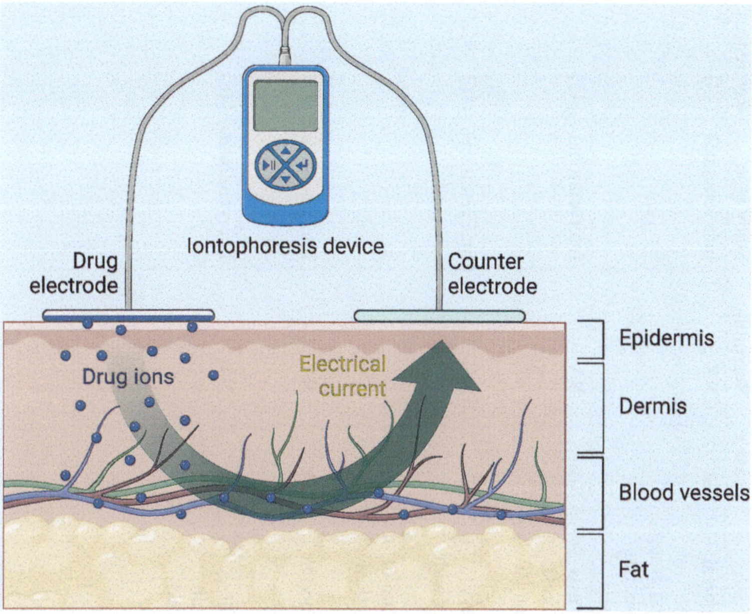

Fig. 14.2 Iontophoresis as a pharmaceutical delivery method

14.3 Future Perspectives

We should expect to see major developments in minimally invasive drug delivery systems in the near future. The development of bio-responsive and dissolvable microneedles is one example of how technology is progressing toward better targeting and more comfortable needles for patients. The use of smart technologies will enhance the selectivity and precision of sonophoresis and iontophoresis by optimizing drug delivery rates and creating personalized treatments. Improvements in nanoparticle design to avoid immune capture and attain improved targeting, especially for intravascular and extravascular disorders, will lead to the continued pivotal role of systems based on nanomaterials. Lipid nanoparticles and PEGylation are two examples of how these systems can improve the delivery of macromolecules and biologics. Such systems will allow use of hydrogels sensitive to physiological conditions and ensure the medications are released there accurately at the point in time needed. Such a combination will elevate their efficacy through conjunction with another delivery mode, including usage of microneedles. Better performance by nanoparticle-based thermal ablation using latest imaging technologies would allow safer, minimally invasive ways of these ablation treatments. New technologies such as wireless electronics, machine learning, and artificial intelligence could be introduced to improve pharmacokinetics of drugs, individualize therapies, and

Table 14.3 An overview of recent clinical trials of iontophoresis

S. no	Title of the study	State of disease/ condition	Intervention	Phase	Recruiting	Study type	References
1	Extracorporeal shock wave therapy and ultrasound iontophoresis for lateral epicondylitis	Lateral epicondylitis	Others: Extracorporeal shock wave group Others: Ultrasound group Others: Dexamethasone iontophoresis group	NA	NA	Interventional	[44]
2	Comparison of hydrocortisone iontophoresis and Phonophoresis for post-surgery scar treatment	Scar Hydrocortisone Iontophoresis Phonophoresis	Others: Iontophoresis Others: Phonophoresis Others: Traditional PT program	NA	NA	Interventional	[45]
3	Effects of iontophoresis on plantar fasciitis: A comparison with ultrasound and iontophoresis	Iontophoresis Ultrasound therapy, complications Fasciitis, plantar Pain, acute	Drug: Application of lidocaine together with dexamethasone Device: Application of ultrasound as physical therapy	NA	NA	Interventional	[46]
4	Evaluation of laser iontophoresis vs. tap water iontophoresis for hyperhidrosis	Hyperhidrosis	Device: Laser acupoint Device: Tap water iontophoresis	NA	NA	Interventional	[47]
5	Alternative methods for treating chronic mechanical neck pain (DN): Dry needling vs. magnesium sulfate iontophoresis	Chronic pain	Others: Dry needling Others: Magnesium sulfate iontophoresis Others: Stretching exercise	NA	NA	Interventional	[48]

6	Comparison of shock wave therapy with iontophoresis for the treatment of knee osteoarthritis (shockwave)	Knee osteoarthritis	Device: Radial extracorporeal shock wave Device: Iontophoresis Others: Exercise	NA	NA	Interventional	[49]
7	Magnesium sulfate electromotive drug administration for hypertrophic scars	Hypertrophic scars	Device: Magnesium sulfate iontophoresis Others: Traditional physical therapy program	NA	Not yet recruiting	Interventional	[50]
8	Treatment of chronic unilateral knee osteoarthritis: A comparison of iontophoresis and electrical dry needling	Osteoarthritis of the knee	Others: Conventional physical therapy Others: Electrical dry needling Others: Glucosamine sulfate iontophoresis	NA	Recruiting	Interventional	[51]
9	Evaluation of acetylcholine iontophoresis against low extra-corporeal shock waves in the treatment of type 2 diabetes neuropathy	Diabetic nephropathy	Device: Low-intensity extracorporeal shock wave	NA	Not yet recruiting	Interventional	[51]
10	A New wireless iontophoresis device for transdermal administration	Spinal cord injuries Constipation Fecal incontinence Neurogenic bowel	Drug: Combination of neostigmine and glycopyrrolate Device: iBox by Dynatronics	Early phase 1	Recruiting	Interventional	[52]

Table 14.4 An overview of recent clinical trials of thermal ablation

S. no	Title of the study	State of disease/ condition	Intervention	Phase	Recruiting	Study type	References
1	Modifications in cytokine expression following thermal ablation for hepatocellular carcinoma	Hepatocellular carcinoma	Device: Radiofrequency ablation or microwave ablation	NA	NA	Interventional	[54]
2	Endobronchial thermal liquid ablation (ETLA) as a treatment for emphysema: The REDUCE EU study	Emphysema or COPD Emphysema Emphysema, pulmonary COPD COPD (chronic obstructive pulmonary disease)	Device: Endobronchial thermal liquid ablation (ETLA)	NA	Not yet recruiting	Interventional	[55]
3	Transbronchial Bronchoscopic thermal vapor ablation (BTVA) for heterogeneous emphysema: A safe and effective method	Heterogeneous emphysema	Procedure: Transbronchial endoscopic thermal vapor ablation	NA	Not yet recruiting	Interventional	[56]
4	Donafenib as a post-ablation adjuvant treatment for residual HCC	Hepatocellular carcinoma Microwave ablation Donafenib Recurrent tumor	Procedure: Thermal ablation Drug: Donafenib	NA	Recruiting	Interventional	[57]
5	A prospective investigation on the EPIBONE study's feasibility, safety, and accuracy	Bone metastases Fractures, bone Bone tumor	Procedure: CT-guided bone percutaneous procedure	NA	Not yet recruiting	Interventional	[58]

(continued)

6	Oligometastatic diseases: A longitudinal study on local ablative therapy (OLIGO-DK)	Oligometastatic disease Metastases Ablation techniques Radiotherapy Stereotactic radiation Surgery	Procedure: Local ablative therapy (LAT)	NA	Recruiting	Interventional	[59]
7	Results from hyperthermia procedures for noncancerous thyroid nodules guided by ultrasound	Thyroid nodules	Procedure: Ultrasound-guided thermal ablation	NA	Not yet recruiting	Interventional	[60]
8	A comparison of TA and LEEP in Mozambique for the prevention of cervical cancer in women living with HIV/AIDS	HIV infections HPV infection CIN 2/3	Device: Thermal ablation Procedure: *loop electrosurgical excision procedure*	NA	Recruiting	Interventional	[61]
9	Endoscopic mucosal resection of large non-pedunculated colorectal polyps followed by a prophylactic double thermal ABLATION procedure (ABLATION)	Colorectal cancer Polyp of the colon	Procedure: Hybrid argon plasma coagulation (h-APC) Procedure: Snare tip soft coagulation (STSC)	NA	Recruiting	Interventional	[62]
10	Examining the functional results of benign prostatic hyperplasia-related obstruction patients following bilateral, 3D ultrasound-guided focal thermal ablation of the prostate transition zone (BETTANY)	Benign prostatic hyperplasia	Procedure: Targeted microwave ablation	Phase 2 completed	Not yet recruiting	Interventional	[63]

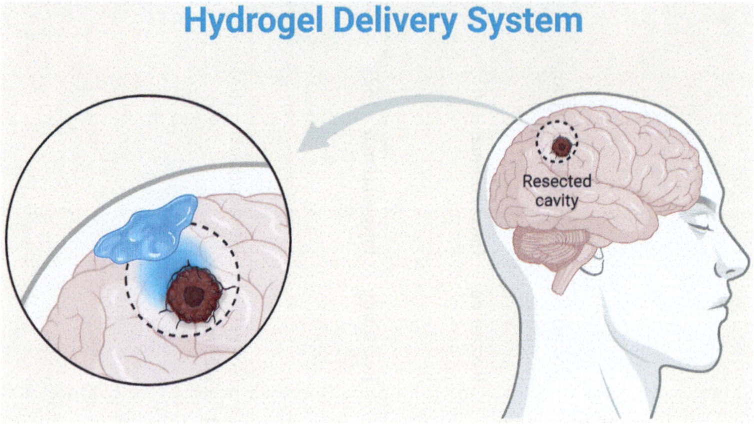

Fig. 14.3 Hydrogel drug delivery system

enhance targeting efficiency. All these developments, which seek to bypass the biological barrier of the blood-brain barrier and to achieve pulsatile release that is programmed, aim for bettering of treatment outcomes and improvement in patient compliance.

14.4 Conclusion

The MIDDS revolutionizes the medical world, making the administration of drugs rapid, painless, and noninvasive. Patient compliance, adherence to treatment, and healthcare costs are positively influenced when compared with traditional injection practices. This chapter explores various non-invasive drug delivery technologies, including microneedles, hydrogels, thermal ablation, iontophoresis, and sonophoresis. Some of the challenges, which MIDDS clinical development has to face, are drug absorption problems, concerns over formulation stability, barriers from regulatory agencies, and variability in patient compliance. Researchers are meeting these head-on with creative solutions ranging from tailored clinical trials and smart technologies to nanocarriers. Infectious disorders, cancer, and pain management are just a few of the therapeutic areas where MIDDS is showing promise in ongoing clinical trials. More advancements in MIDDS, resulting in better treatment outcomes and more patient comfort, are likely to occur as technology progresses.

Table 14.5 An overview of recent clinical trials of hydrogel

S. no	Title of the study	State of disease/ condition	Intervention	Phase	Recruiting	Study type	References
1	Clinical performance of two frequent replacement silicone hydrogel multifocal contact lenses	Presbyopia	Device: Serafilcon A multifocal contact lenses Device: Senofilcon A multifocal contact lenses Device: CLEAR CARE® cleaning & disinfecting solution	NA	NA	Interventional	[65]
2	Intra-articular polyacrylamide hydrogel in gonarthrosis	Osteoarthritis	Device: Hydrous biopolymer with silver ions “Argiform”	NA	NA	Interventional	[66]
3	Evaluation of the clinical performance of daily disposable silicone hydrogel multifocal Toric contact lenses	Visual acuity	Device: JJVC investigational multifocal Toric contact lens manufactured in senofilcon A (C3) material with UV blocker/ HEV filter	NA	NA	Interventional	[67]
4	Chitosan Nano-hydroxyapatite hydrogel in non-surgical treatment of Intrabony defect	Chitosan Nano-hydroxyapatite hydrogel Intrabony defect	Others: Nano-HAP/chitosan hydrogel	NA	NA	Interventional	[68]
5	Assessment of wound healing, cooling efficacy and local tolerability of a wound Care hydrogel	Wound	Device: Zoralan wound	NA	NA	Interventional	[69]

(continued)

Table 14.5 (continued)

S. no	Title of the study	State of disease/ condition	Intervention	Phase	Recruiting	Study type	References
6	A prospective, randomized, double-blind, parallel, active-controlled clinical study of silicone hydrogel monthly soft contact lens for the safety and effectiveness in visual correction	Myopia	Device: Soft contact lens	NA	NA	Interventional	[70]
7	Clinical assessment of a daily disposable soft silicone hydrogel contact lens	Refractive errors	Device: Delefilcon A contact lenses	NA	NA	Interventional	[71]
8	Clinical performance evaluation of two frequent replacement silicone hydrogel Toric contact lenses	Ametropia Myopia Hyperopia Astigmatism	Device: Serafilcon A toric contact lenses Device: Senofilcon A toric contact lenses Device: CLEAR CARE	NA	NA	Interventional	[72]
9	Performance evaluation of two silicone hydrogel Toric lens designs in habitual soft contact lens wearers	Astigmatism	Device: Lens A (comfilcon A toric lens) Device: Lens B (lehfilcon A toric lens)	NA	NA	Interventional	[73]
10	Clinical efficiency of xanthan hydrogels containing local anesthetics encapsulated in nanostructured lipid carries	Dental trauma Local anesthesia Oral cavity disease	Drug: Topical anesthetic	Phase 1 completed	NA	Interventional	[74]

Bibliography

1. Vllasaliu D (2021) Non-invasive drug delivery systems. Pharmaceutics 13(5):611
2. Akhtar N, Singh V, Yusuf M, Khan RA (2020) Non-invasive drug delivery technology: development and current status of transdermal drug delivery devices, techniques and biomedical applications. Biomed Tech 65:243–272
3. Jeong SY, Park JH, Lee YS, Kim YS, Park JY, Kim SY (2020) The current status of clinical research involving microneedles: a systematic review. Pharmaceutics 12(11):1113
4. Molinaro R, Gagliardi A, Mancuso A, Cosco D, Soliman ME, Casettari L et al (2019) Development and in vivo evaluation of multidrug ultradeformable vesicles for the treatment of skin inflammation. Pharmaceutics 11(12):644
5. Kitaoka M, Oka A, Goto M (2020) Monoolein assisted oil-based transdermal delivery of powder vaccine. Pharmaceutics 12(9):814
6. Yang H, Kang G, Jang M, Um DJ, Shin J, Kim H et al (2020) Development of lidocaine-loaded dissolving microneedle for rapid and efficient local anesthesia. Pharmaceutics 12(11):1067
7. Esposito L, Barbosa AI, Moniz T, Costa Lima S, Costa P, Celia C et al (2020) Design and characterization of sodium alginate and poly (vinyl) alcohol hydrogels for enhanced skin delivery of quercetin. Pharmaceutics 12(12):1149
8. Froelich A, Osmałek T, Jadach B, Puri V, Michniak-Kohn B (2021) Microemulsion-based media in nose-to-brain drug delivery. Pharmaceutics 13(2):201
9. Liang W, Pan HW, Vllasaliu D, Lam JKW (2020) Pulmonary delivery of biological drugs. Pharmaceutics 12(11):1025
10. Alabsi W, Al-Obeidi FA, Polt R, Mansour HM (2020) Organic solution advanced spray-dried microparticulate/nanoparticulate dry powders of lactomorphin for respiratory delivery: physicochemical characterization, in vitro aerosol dispersion, and cellular studies. Pharmaceutics 13(1):26
11. New R (2020) Oral delivery of biologics via the intestine. Pharmaceutics 13(1):18
12. Bahman F, Taurin S, Altayeb D, Taha S, Bakhiet M, Greish K (2020) Oral insulin delivery using poly (styrene co-maleic acid) micelles in a diabetic mouse model. Pharmaceutics 12(11):1026
13. Carobolante G, Mantaj J, Ferrari E, Vllasaliu D (2020) Cow milk and intestinal epithelial cell-derived extracellular vesicles as systems for enhancing oral drug delivery. Pharmaceutics 12(3):226
14. McCartney F, Rosa M, Brayden DJ (2019) Evaluation of sucrose laurate as an intestinal permeation enhancer for macromolecules: ex vivo and in vivo studies. Pharmaceutics 11(11):565
15. Ways TM, Ng KW, Lau WM, Khutoryanskiy VV (2020) Silica nanoparticles in transmucosal drug delivery. Pharmaceutics 12(8):751
16. Sweney MT, Silver K, Gerard-Blanluet M, Pedespan JM, Renault F, Arzimanoglou A et al (2009) Alternating hemiplegia of childhood: early characteristics and evolution of a neurodevelopmental syndrome. Pediatrics 123(3):e534–e541
17. Majumdar S, Mitra AK (2009) Approaches towards enhanced transepithelial drug delivery. Discov Med 6(36):229–233
18. Fatima A, Apte SS (2022) Formulation and development of transdermal patches. GSC Biol Pharm Sci 19(1):346–352 https://api.semanticscholar.org/CorpusID:273114206
19. Geart CW, Tu KW (1974) The effect of variable mesh size on the stability of multistep methods. SIAM J Numer Anal 11(5):1025–1043. https://doi.org/10.1137/0711079
20. Loke YH, Jayakrishnan A, Razif MRFM, Yee KM, Kee PE, Goh BH et al (2024) A comprehensive review of challenges in oral drug delivery systems and recent advancements in innovative design strategies. Curr Pharm Des 31(5):360–376 https://api.semanticscholar.org/CorpusID:273239310
21. Cagnazzo C (2022) New trends in clinical trials—between complexity and the need for renewal. AboutOpen 9(1):42–44
22. Asanova B, Davidova E, Vukikjevikj N (2022) Challenges in the development and registration of generic topical products. Maced Pharm Bull 68:105–106 https://api.semanticscholar.org/CorpusID:255640856

23. Pullar T, Kumar S, Feely M (1989) Compliance in clinical trials. Ann Rheum Dis 48(10):871–875
24. Bava B, Sharma K, Yadav V (2024) Intranasal drug delivery system: a review. Res J Sci Technol 16(1):51–58 https://api.semanticscholar.org/CorpusID:269284954
25. Zaidat OO, Liebeskind DS, Edgell RC, Amlie-Lefond CM, Kalia JS, Alexandrov AV (2012) Clinical trial design for endovascular ischemic stroke intervention. Neurology 79(13_supplement_1):S221–S233
26. Babaie S, Taghvimi A, Charkhpour M, Zarebkohan A, Keyhanvar P, Hamishehkar H (2020) Optimization of influential variables in the development of buprenorphine and bupivacaine loaded Invasome for dermal delivery. Adv Pharm Bull 11:522–529 https://api.semanticscholar.org/CorpusID:222115295
27. Kim YC, Park JH, Prausnitz MR (2012) Microneedles for drug and vaccine delivery. Adv Drug Deliv Rev 64(14):1547–1568 https://www.sciencedirect.com/science/article/pii/S0169409X12001251
28. Study Details. Phase 1 evaluation of H1 influenza vaccine delivered by MIMIX MAP. ClinicalTrials.gov
29. Study Details. Effects of RF microneedle on photoaging skin. ClinicalTrials.gov
30. Study Details. Microneedle versus fractional CO_2 laser for skin aging treatment with stem cell secretome in Indonesian adult women. ClinicalTrials.gov
31. Study Details. Efficacy of transdermal microneedle patch for topical anesthesia enhancement in Paediatric thalassemia patients. ClinicalTrials.gov
32. Study Details. Study to evaluate and compare the bioequivalence of two Abaloparatide-sMTS treatments in healthy women. ClinicalTrials.gov
33. Study Details. Pilocarpine microneedles for sweat induction (PMN-SI). ClinicalTrials.gov
34. Study Details. Measles and rubella vaccine microneedle patch phase 1-2 age De-escalation trial. ClinicalTrials.gov
35. Study Details. The purpose of this study is to evaluate the safety and efficacy of D-MNA in adult patients with nodular basal cell carcinoma. ClinicalTrials.gov
36. Study Details. Clinical trial of microneedle radiofrequency combined with Oral isotretinoin in moderate to severe acne. ClinicalTrials.gov
37. Study Details. Microneedle patch with botulinum toxin in Management of Enlarged Facial Pores. ClinicalTrials.gov
38. Merino G, Kalia YN, Guy RH (2003) Ultrasound-enhanced transdermal transport. J Pharm Sci 92(6):1125–1137
39. Mitragotri S (2000) Synergistic effect of enhancers for transdermal drug delivery. Pharm Res 17:1354–1359
40. Langer R. New methods of drug delivery. Science (1979). 1990;249(4976):1527–1533
41. Study Details. Sustained acoustic medicine (SAM) for symptomatic treatment of knee pain related to osteoarthritis. ClinicalTrials.gov
42. Study Details. Photobiomodulation and Sonophoresis improving jaw mobility and quality of life among head and neck cancer survivors. ClinicalTrials.gov
43. Priya B, Rashmi T, Bozena M (2006) Transdermal iontophoresis. Expert Opin Drug Deliv 3(1):127–138. https://doi.org/10.1517/17425247.3.1.127
44. Study Details. Lateral epicondylitis extracorporeal shock wave therapy, ultrasound iontophoresis. ClinicalTrials.gov
45. Study Details. Effect of hydrocortisone iontophoresis versus hydrocortisone phonophoresis on post surgical scar. ClinicalTrials.gov
46. Study Details. Effect of iontophoresis vs. ultrasound and iontophoresis in plantar fasciitis. Plantar fasciitis. ClinicalTrials.gov
47. Study Details. Laser acupuncture versus tap water iontophoresis on hyperhidrosis. ClinicalTrials.gov
48. Study Details. Dry needling versus magnesium Sulphate iontophoresis in subjects with chronic mechanical neck pain. ClinicalTrials.gov
49. Study Details. Shock wave versus iontophoresis in treatment of subjects with knee osteoarthritis. ClinicalTrials.gov

50. Study Details. Electromotive drug administration of magnesium sulphate on hypertrophic scar. ClinicalTrials.gov
51. Study Details. Electrical dry needling versus iontophoresis in treating chronic unilateral knee osteoarthritis. ClinicalTrials.gov
52. Study Details. Transdermal administration by a novel wireless iontophoresis device. ClinicalTrials.gov
53. Webb H, Lubner MG, Hinshaw JL (2011) Thermal ablation. Semin Roentgenol 46(2):133–141 https://www.sciencedirect.com/science/article/pii/S0037198X10000568
54. Study Details. Cytokine change in hepatocellular carcinoma treated by thermal ablation. ClinicalTrials.gov
55. Study Details. The REDUCE EU study—endobronchial thermal liquid ablation (ETLA) for the treatment of emphysema. ClinicalTrials.gov
56. Study Details. Efficacy and safety of Transbronchial Bronchoscopic thermal vapor ablation (BTVA) in heterogeneous emphysema. ClinicalTrials.gov
57. Study Details. Donafenib as adjuvant therapy following ablation for recurrent HCC. ClinicalTrials.gov
58. Study Details. EPIBONE study: a prospective study on feasibility, safety and accuracy. ClinicalTrials.gov
59. Study Details. Longitudinal study of local ablative therapy in Oligometastatic disease. ClinicalTrials.gov
60. Study Details. The outcomes of ultrasound-guided thermal ablation for benign thyroid nodules. ClinicalTrials.gov
61. Study Details. Assessment of the effectiveness of TA versus LEEP for cervical cancer risk reduction in WLHIV in Mozambique. ClinicalTrials.gov
62. Study Details. Prophylactic double thermal ablation after endoscopic mucosal resection of large non-pedunculated colorectal polyps. ClinicalTrials.gov
63. Study Details. Evaluation of the functional outcomes after bilateral, 3D ultrasound-guided focal thermal ablation of the prostate transition zone, in patients with benign prostatic hyperplasia-related obstruction. ClinicalTrials.gov
64. Ahmed EM (2015) Hydrogel: preparation, characterization, and applications: a review. J Adv Res 6(2):105–121 https://www.sciencedirect.com/science/article/pii/S2090123213000969
65. Study Details. Clinical performance of two frequent replacement silicone hydrogel multifocal contact lenses. ClinicalTrials.gov
66. Study Details. Intra-articular polyacrylamide hydrogel in Gonarthrosis. ClinicalTrials.gov
67. Study Details. Evaluation of the clinical performance of daily disposable silicone hydrogel multifocal Toric contact lenses. ClinicalTrials.gov
68. Study Details. Chitosan nano-hydroxyapatite hydrogel in non-surgical treatment of Intrabony defect. ClinicalTrials.gov
69. Study Details. Assessment of wound healing, cooling efficacy and local tolerability of a wound care hydrogel. ClinicalTrials.gov
70. Study Details. A prospective, randomized, double-blind, parallel, active-controlled clinical study of silicone hydrogel monthly soft contact lens for the safety and effectiveness in visual correction. ClinicalTrials.gov
71. Study Details. Clinical assessment of a daily disposable soft silicone hydrogel contact lens. ClinicalTrials.gov
72. Study Details. Clinical performance evaluation of two frequent replacement silicone hydrogel Toric contact lenses. ClinicalTrials.gov
73. Study Details. Performance evaluation of two silicone hydrogel Toric lens designs in habitual soft contact lens wearers. ClinicalTrials.gov
74. Study Details. Clinical efficiency of xanthan hydrogels containing local anesthetics encapsulated in nanostructured lipid carries. ClinicalTrials.gov

The manufacturer's authorised representative in the EU is Springer Nature Customer Service Centre GmbH, Europaplatz 3, 69115 Heidelberg, Germany. If you have any concerns regarding our products, please contact ProductSafety@springernature.com

Printed and bound by CPI Group (UK) Ltd, Croydon, CR0 4YY
01/10/2026
02225921-0008